Niessner, Schäffer
Organic Trace Analysis
De Gruyter Graduate

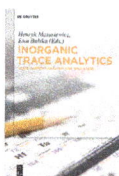

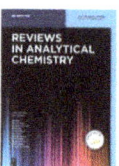

Niessner, Schäffer

Organic Trace Analysis

Reinhard Niessner, Technical University of Munich
Andreas Schäffer, RWTH Aachen University

DE GRUYTER

DOI 10.1515/9783110441154-201

Authors
Prof. Dr. Reinhard Niessner
Institute of Hydrochemistry, Chair of Analytical Chemistry
Technical University of München
Marchioninistr. 17
D-81377 München
reinhard.niessner@mytum.de

Prof. Dr. Andreas Schäffer
RWTH Aachen University
Institute for Environmental Research
Worringerweg 1
D-52074 Aachen
andreas.schaeffer@bio5.rwth-aachen.de

ISBN 978-3-11-044114-7
e-ISBN (PDF) 978-3-11-044115-4
e-ISBN (EPUB) 978-3-11-043293-0

Library of Congress Cataloging-in-Publication Data
A CIP catalog record for this book has been applied for at the Library of Congress.

Bibliographic information published by the Deutsche Nationalbibliothek
The Deutsche Nationalbibliothek lists this publication in the Deutsche Nationalbibliografie; detailed bibliographic data are available on the Internet at http://dnb.dnb.de.

© 2017 Walter de Gruyter GmbH, Berlin/Boston
Typesetting: Integra Software Services Pvt. Ltd.
Printing and binding: CPI books GmbH, Leck
Cover image: Andrew Brookes, National Physical Laboratory/Science Photo Library
♾ Printed on acid-free paper
Printed in Germany

www.degruyter.com

Preface

Our book *Organic Trace Analysis* is dedicated to advanced students of analytical chemistry as core discipline first. Beyond them, also those learning and practising food chemistry, pharmaceutical chemistry, life sciences, or environmental sciences in general, will benefit from this textbook.

Of course, many excellent monographs on chromatography, mass spectrometry etc. are available, but we had the strong feeling, that they do not cover *teaching* the specialties and peculiarities comprising trace analysis of organic compounds: sampling from the very diverse matrices, sample pre-treatment, analyte enrichment, critical assessment of chromatography or separation techniques, up to the application of bioanalytical tools.

Organic trace analysis is full of pitfalls and surprise! The enormous complexity and molecular diversity of organic molecules asks for sometimes strange methodologies. Compared to inorganic compounds the analytes are often labile, but some of them biologically extremely effective. Even more diverse are the matrices from where they have to be isolated covering soil, water, and atmosphere, but also organisms living in these compartments. Hence, organic trace analysis is characterized as a refined craftsmanship, dealing with a multitude of techniques.

Knowledge on spectroscopy, physical chemistry, molecular biotechnology, chemical engineering, and statistics are the basics for it.

We had the intention to provide a balanced mixture of analytical basics, news from the forefront of (bio)analytical chemistry and separation science, completed by a bunch of approved methods applied to everlasting trace analytical problems.

The authors like to thank the extremely helpful team of de Gruyter! They made our manuscript finally readable. Many thanks also to Prof. Totaro Imasaka, Kyushu University at Fukuoka, who hosted RN for longer periods. There, a considerable part of this textbook became written. AS is grateful for the patience of his wife and family tolerating many weekends and vacations when the other half of the manuscript was developed.

Reinhard Niessner
Munich, April 2017

Andreas Schaeffer
Aachen, April 2017

DOI 10.1515/9783110441154-202

Contents

DOI 10.1515/9783110441154-203

1 Overview

1.1 General Remarks

This textbook has the intention to give an introduction and information to those scientists facing analytical problems connected with organic molecules in very tiny amounts, and this in different matrices. We call it *organic trace analysis*. Since dealing with this is very different from what is practiced in inorganic trace analysis, it seems justified to present a comprehensive view of how to handle this.

1.2 What Does "Organic Traces" Mean?

Trace analysis is defined as the qualitative and quantitative determination of a very small amount of substances within any matrix. Organic trace compounds stem from organic or bioorganic chemistry. We will not consider inorganic carbon-containing substances like carbonates or different forms of elemental carbon, despite their importance. Of course, a clear separation is impossible, for example, organometallic compounds of transition elements are often rated as a part of inorganic chemistry. The same goes for microbial matter, which may partly be seen as biopolymers. In the following sections, we will show some examples for further understanding.

Trace analysis is relative in terms of concentration. First of all, we should recognize the difference between "trace analysis methods" and those belonging to "microanalysis" techniques, or nowadays "nanoanalytics." Trace analysis methods deal with concentration determination within a given sample, whereas a microanalysis technique identifies a compound localized at a selected spot. The latter does not analyze the total volume.

Frequently, trace concentrations are presented as fractions in a dimensionless notation. Foremost, one has to decide whether a mass-to-mass fraction ($m_{analyte}/m_{sample}$) or a mass-to-volume fraction ($m_{analyte}/V_{sample}$) is meant. Fractions given as $m_{analyte}/V_{sample}$ need additional parameters, as the volume depends on pressure and temperature. For aqueous samples, the difference between m/v and m/m is negligible.

These ratios are shown in Table 1.1.

Let's get an impression of these amounts, 1 ppq is equivalent to 1 drop (50 μL) of water diluted into a cube of water measuring approximately 3,680 m on one side (50 billion cubic meters). This is roughly the volume of Lake Constance, the second largest fresh water lake in Northern Europe. This concentration ratio is currently the lower limit in organic trace analysis. The US-Environmental Protection Agency (US-EPA) has set its threshold for safe dioxin exposure at a toxicity equivalence (TEQ) of 0.7 pg/kg of body weight per day. This corresponds to 0.7 ppq. The upcoming threshold value for endocrine disruptors, substances blocking the function of natural endogenic hormones, is discussed to be in the lower ppt–ppq range, too, like the example

DOI 10.1515/9783110441154-001

Table 1.1: Abbreviations for relative quantity measures as fractions in parts per notation.

ppm = parts per million	10^{-6} = mg/kg = µg/g = g/t = 10^{-4}%
ppb = parts per billion	10^{-9} = µg/kg = ng/g = mg/t = 10^{-7}%
ppt = parts per trillion	10^{-12} = ng/kg = pg/g = µg/t = 10^{-10}%
ppq = parts per quadrillion	10^{-15} = pg/kg = fg/g = ng/t = 10^{-13}%

mg, µg, ng, pg and fg are abbreviations for milli-, micro-, nano-, pico- and femtogram, respectively.

of ethinylestradiol (5–6 ng/L) in a Canadian lake leading to the impairment of fish populations after chronic exposure [1].

1.3 Importance of Organic Trace Analysis

An easy way to reveal the importance of trace analysis of organic compounds is a look into the databank SciFinder (see Table 1.2). The frequency of citations reflects the importance of xenobiotic organic trace compounds quite well.

This is a direct consequence of the physiological or toxic impact of many organic compounds even in smallest concentrations. Additionally, the ongoing search for bio-markers in clinical chemistry or continuing clarification of metabolic pathways will enlarge the need for analyzing minute amounts of chemicals considerably. We also observe a trend to a first so-called nontarget screening for (almost) everything present in an environmental matrix. This first became established in water and air analysis some years ago, when multidimensional chromatography came into operation, and by the hyphenation of strong separation schemes with high-resolution mass spectro-metry. By such nontarget analysis, one likes to detect any possible contamination, for example, within water effluent from a sewage treatment plant which often still con-tains trace amounts of synthetic chemicals. It is clear that with increasing resolution and/or detection power of the analytical technique, the number of so far unknown contaminants will also rise in parallel.

Table 1.2: Citation frequency of analyses of important analytes in SciFinder databank (year 2015).

Trace analysis of	References found	Analytical conc. range
Pesticides	50,786	ppt
Polycyclic aromatic hydrocarbons	25,884	ppt–ppb
Dioxins	5,939	ppq
Explosives	20,315	ppt–ppb–ppm
Mycotoxins	4,638	ppt
Endocrine disruptors	2,320	ppq–ppt
Benzene, toluene, xylene	14,627	ppt–ppb–ppm

Table 1.3: List of selected environmental regulations and monitoring programs.

Regulation	Location	Substances
Decision 2455/2001/EC on a first list of priority substances	EU	33 substances or group of substances shown to be of major concern for European Waters
Regulation (EC) No 1107/2009	EU	Regulation of the European Parliament and of the Council concerning the placing of plant protection products on the market
National primary/secondary drinking water regulations	US-EPA	Disinfectant, disinfectant by-product, organic chemicals, foaming agents
Soil Remediation Circular 2009	The Netherlands	Aromatic compounds, PAHs, chlorinated hydrocarbons, pesticides
Atlantic RBCA for Petroleum-Impacted Sites in Atlantic Canada Version 3	Canada	BTEX, aromatic fractions, heavy ($>C_{32}$) hydrocarbon fraction
Directive 2013/39/EU watch list for water	EU	Ten substances, which include three pharmaceutical substances (Diclofenac, 17-beta-estradiol (E2) and 17-alpha-ethinylestradiol (EE2)
MAK values (maximum workplace concentration) and of BAT values (biological workplace tolerance values)	Germany	Collection for Occupational Health and Safety; numerous accepted and suspected carcinogenic substances
CADAMP	California, USA	Dioxins, furans, PCB congeners, brominated diphenyl ethers

PAH, polycyclic aromatic hydrocarbon; RBCA, risk-based corrective action; CADAMP, California Ambient Dioxin Air Monitoring Program; EU, European Union; US-EPA, US-Environmental Protection Agency; PCBs, polychlorinated biphenyls.

The importance of organic trace compounds and their analysis is seen best in the numerous regulations published by national or supranational organizations. A set of the most important environmental regulations and monitoring programs is depicted in Table 1.3.

But not only for monitoring the environmental trace analysis of organic substances is essential. Rigorous quality regulations also exist for the pharmaceutical drug production or food production. The direct impact of unwanted contaminations has led to strict monitoring of pesticides, toxins and so on in raw products.

Nowadays, possible bioterroristic attacks ask for rapid development and application of extremely sensitive and rapid detection methods for microbial pollution and derived toxins such as ricin, *Botulinus* toxins or aflatoxins. The rapid distribution of freshly produced dairy products (milk, yoghurt, etc.) serves as a template for such attack. The response time for any countermeasure is only hours.

1.4 Peculiarities with Organic Trace Analysis

One has to recognize the many difficulties faced in organic trace analysis:
- Per year, more than 10^6 new substances become synthesized by organic chemists; so, there is simply no analytical capacity to follow up. The characterization techniques used by the organic chemists in most cases do not allow a trace analysis.
- There is no opportunity to apply a simple separation scheme as it was possible with inorganic substances, for example, separation by precipitation of insoluble sulfides. In the last century, food analysts have started to create separation schemes for organic food ingredients, but the performance was insufficient due to higher solubility of organic compounds and this was stopped.
- Development of trace analytical determination protocols never happened for reasons of precaution but instead always afterward, i.e. event driven. Once a new intoxication caused by a harmful organic substance becomes known, the development of a trace analytical technique starts. An example for this is the recent acrylamide contamination of fried food. The substance and its chemical characterization have been published decades ago, but no trace analytical technique was at hand when needed.
- A special challenge is the chemical analysis of trace contaminants in "difficult" matrices, such as pesticide residues in plants forming strongly bound or entrapped residues or in insects; an example for the latter is the assumed intoxication of bees with neonicotinoids: sub-ng/bee quantities have to be analyzed [2, 3].
- Many compounds possess nearly identical physicochemical properties (vapor pressure, solubility, mass, spectral features, chirality, etc.), but nevertheless exhibit completely different physiological effects.
- The matrices containing organic trace compounds may change a lot and this is often unexpected. Polycyclic aromatic hydrocarbons (PAHs) can be dissolved in a liquid, but in the presence of dispersed particles they may become attached to their surfaces, exhibiting now very different partition behavior in separation. PAH sorbed to resuspended sediment particles may still lead to toxicity symptoms in exposed fish [4].
- In organic trace analyses, the unambiguous identification of individual molecules is requested, whereas inorganic trace analysis is often only restricted to determining the elemental composition. In some cases, the individual isomers, for example enantiomers, have to be measured. A prominent example for this was the Contergan scandal: the leprosy drug thalidomide exists as two optical isomeric, (R) and (S), forms. Thalidomide is racemic; the individual enantiomers can racemize. So administering the pure optical isomer does not help, since racemization happens in the vein already. It is assumed later, that it is responsible for terrible teratogenic malformations of the fetus after treatment of pregnant women. At this time, the possibility of enantiomeric analysis of small quantities was not as developed today.

– The instability of many organic substances is a tremendous challenge. Not only does this hamper sample treatment and storage but also asks for special caution in usage of standard dilutions for calibration. Known examples are the oxidation-sensitive vitamins or certain PAH molecules (e.g., pyrene). Aside from thermal instability, illumination by sunlight can trigger a change of isomeric forms or may start degradation by photooxidation. Photo-dimerization of unsaturated compounds has been observed too.

1.5 Essential Matrices in Organic Trace Analysis

Organic substances may occur in all matrices and derived compartments as well. As an example, we can take the path of environmental compounds within the water cycle (see Figure 1.1).

At the beginning of the water cycle, water vapor becomes condensed to ultrafine dispersed matter in ambient air (=aerosol) forming a cloud. Without such aerosol particles, no cloud or fog formation can happen. Depending on the origin of such particles, the freshly condensed water is already contaminated at its initial state of existence. During the ongoing water condensation, the water droplets start to rain out. When falling down, there are plenty of opportunities to pick up other air constituents (gases, small particles, etc.). Also photooxidation starts, hence changing the composition of an individual droplet. The pollutant may undergo transformation and translocation processes.

With the advent of rain droplets at ground surface, the dissolved and dispersed micropollutants follow the way of seeping water. Already the first contact with solid matter (plant leaves, soil, rock material) offers again many opportunities for the pollutant. Either it is simply attached by adsorption, or it is fixed by ion exchange, depending on the pH within the aqueous phase, polarity and presence of exchangeable functional groups. In contact with humic substances or clay particles, the pollutant may intercalate into their inner spacings, again fixed by ion exchange or with time by covalent bonding [5]. Since soil microorganisms need organic compounds to gain energy through assimilation, microbial organisms are happy to incorporate pollutants for further enzymatic digestion. In the course of this, the organic compound becomes at last metabolized. The often more polar metabolites may behave completely differently in their transport potentials. When fixed at a surface, air and light may also start similar decomposition reactions. The original micropollutant might become enriched within such cells. Other routes go via the roots of plants. Within the rhizosphere, the pollutant can be solubilized by natural surfactants and through this enters the water-conducting capillaries, xylem and phloem, of a growing plant. Thus, the pollutants become dumped with time in the plant material.

The partly transformed pollutants dissolved in water or attached to nanoparticles will follow the water flow through unsaturated and saturated soil layers until they reach an impermeable layer, forming groundwater. Groundwater serves in many

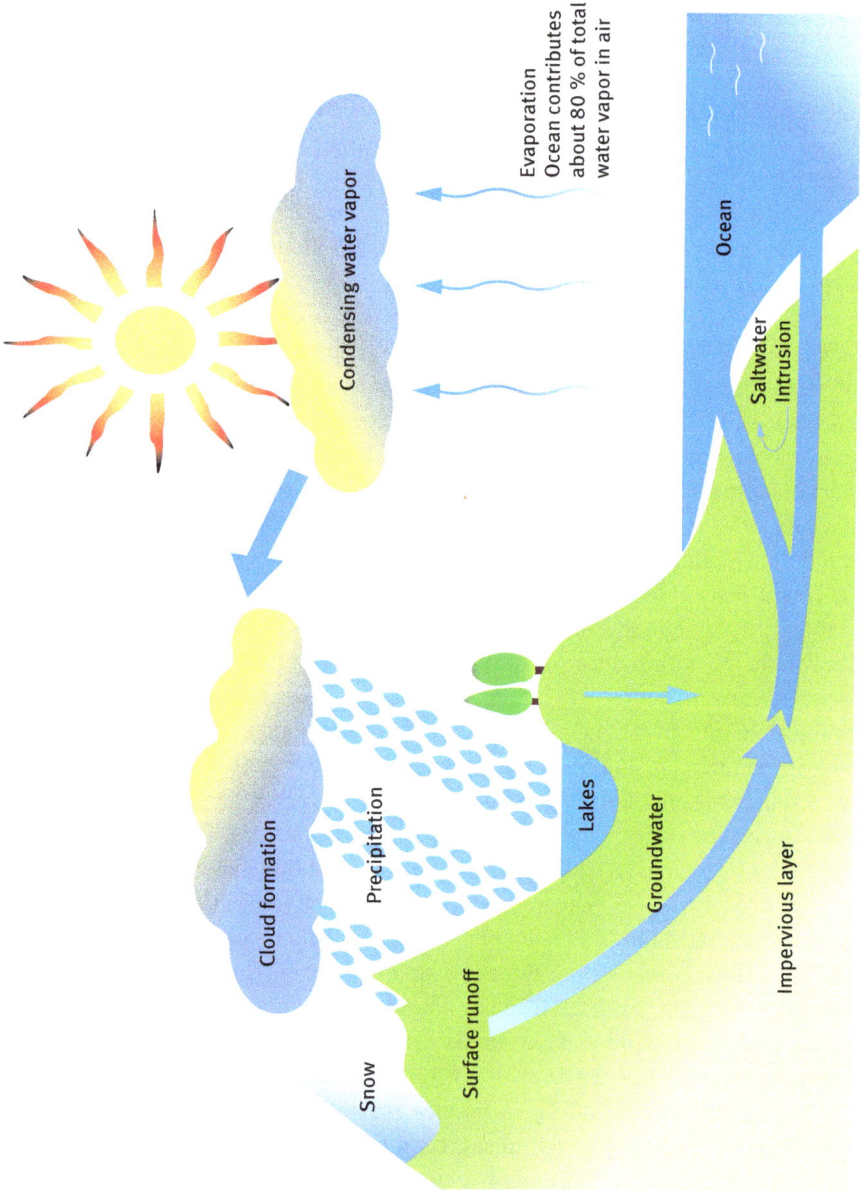

Figure 1.1: Water cycle as host and reactor for contaminants within different compartments.

countries as drinking water reservoirs. When pumped out, it is used for irrigation purposes or production of drinking water. For drinking water pollutants, there exists a
nonzero probability to become bound after ingestion to human receptors with corresponding physiological consequences, or the contaminant becomes translocated again
through excrements reaching soils [6].

In essence, a micropollutant will have many options to make its way through the
different compartments. As a consequence, each of them poses analytically very different challenges for sampling, sample storage, pre-enrichment of the wanted analyte
and selection of the appropriate and tolerant analytical instrumentation.

1.6 Aims of Organic Trace Analysis

The questions to be answered by organic trace analysis may be grouped as follows:

Which substance? How much of it?

From the presented characteristics and peculiarities of organic trace analysis, it can be
concluded that the foremost aim will be the unambiguous identification of one single
compound in its unique structure within its surrounding matrix.

This is currently unfeasible, since it also would mean availability of standard compounds in measurable quantity and of highest purity. Otherwise, calibration of the
analytical process would be impossible. Sometimes, synthesis of even tiny amounts
of such standards, especially if isotope labeling is required, becomes extremely cumbersome and expensive. Very often toxicity prohibits handling of such compounds. We
also do not always have the analytical instrumentation with the needed performance
to meet the requirements for this. Nevertheless, toxicological considerations ask for individual identification and quantification, when within a whole group of isomers only
a few are relevant. Examples are polychlorinated biphenyls (PCBs) or polyhalogenated
compounds in general (see Chapter 11).

Bioavailability of pollutants

In order to estimate the uptake of pollutants in exposed organisms, for example earthworms or plants in soil, the question arises how much of the pollutant is present in a
bioavailable form rather than to analyze its total concentration. This can be addressed
by extraction with aqueous solutions like 0.1 M $CaCl_2$ or using solid phase extraction
disks/fibers. Obviously, the resulting concentrations are often well below the total concentrations, which can be determined by thorough extraction using organic solvents
under drastic conditions (Soxhlet or accelerated pressurized extraction).

More realistic in many cases is to find an answer to:

Which substance group or pattern of individual molecules is present? What is their total amount?

This is sometimes simplifying the analytical task. As an example, as the 209 congeners of the PCB family aren't analytically feasible, and also only some of them are of toxicological relevance, one can either measure the sum of chlorine bound to the biphenyl structure as a substitute, or the presence and amount of selected PCB patterns. Alternatively, often only lead substances are analyzed, for example, benzo[a]pyrene (BaP) serves as the lead substance in water and air pollution monitoring, representing a class of PAHs (see Chapter 11). BaP is also toxicologically the most important representative of this class. The same is found for toxins, furans and dioxins.

Once we have this information, we can answer the questions

Does the found concentration exceed a maximum tolerable threshold? Is the found analytical signal differentiable in a unique manner from a background?

Both questions have severe economic and public relevance. Based on the toxicological risk management, the findings will directly influence human society. Dangerous occupational hygiene and environmental pollution will directly and indirectly impact living and working conditions. Production lines will become closed when emissions permanently exceed a standard value. Food has to be destroyed when a tolerable pesticide contamination level is exceeded. Pesticide application conditions have to be changed or its usage restricted if groundwater monitoring reveals concentrations above the (European) threshold of 0.1 μg/L. Maintenance of hazardous production facilities may be stopped. In an area where already high contamination levels are present, new industrial locations may be forbidden. Contaminated ground may prevent construction of new homes. Sewage sludge is no longer applicable as fertilizer in agriculture. These are just a few examples to show that trace-level contaminants may lead to various personal, professional and legal consequences.

Further Reading

Barcelo D. Environmental analysis: techniques, applications and quality assurance. In Techniques and instrumentation in analytical chemistry. Amsterdam, The Netherlands: Elsevier, 1993.

Haghi AK, Carvajal-Millan E. Food composition and analysis: methods and strategies. Oakville, ON: Apple Academic Press Inc., 2014.

Marsili R. Flavor, fragrance, and odor analysis, 2nd ed. Boca Raton: CRC Press, 2012.

Namiesnik J, Szefer P. Airborne measurements for environmental research: methods and instruments. Weinheim, Germany: Wiley-VCH Verlag GmbH & Co. KGaA, 2013.

Pico Y. Chemical analysis of food: techniques and applications publisher. Waltham, MA: Elsevier Inc., 2012.

Quevauviller P, Maier EA, Griepink B. Quality assurance for environmental analysis. Method evaluation within the measurements and testing program (BCR). In Techniques and instrumentation in analytical chemistry. Amsterdam, The Netherlands: Elsevier, 1995.

Seifert B, van de Wiel HJ, Dodet B, O'Neill IK. Environmental carcinogens, methods of analysis, and exposure measurement. Int Agency Res Cancer, Lyon, France 1993;12:109.

Sunahara, G. Environmental analysis of contaminated sites: tools to measure success or failure. Chichester: John Wiley & Sons Ltd., 2002.

Wang J, MacNeil JD, Kay JF. Chemical analysis of antibiotic residues in food. Hoboken, New York: John Wiley & Sons, Inc., 2012.

Wendisch M, Brenguier J-L. Airborne measurements for environmental research: methods and Instruments. Weinheim, Germany: Wiley-VCH Verlag GmbH&Co. KGaA, 2013.

Bibliography

[1] Kidd KA, Blanchfield PJ, Mills KH, Palace VP, Evans RE, Lazorchak JM, Flick RW. Collapse of a fish population after exposure to a synthetic estrogen. Proc Natl Acad Sci U.S.A. 2007;104(21):8897–901.

[2] Jin N, Klein S, Leimig F, Bischoff G, Menzel R. The neonicotinoid clothianidin interferes with navigation of the solitary bee Osmia cornuta in a laboratory test. J Exp Biol 2015;218(18):2821–5.

[3] Williams GR, Troxler A, Retschnig G, Roth K, Yanez O, Shutler D, Neumann P, Gauthier L. Neonicotinoid pesticides severely affect honey bee queens. Sci Rep 2015;5: 1–8.

[4] Brinkmann M, Hudjetz S, Kammann U, Hennig M, Kuckelkorn J, Chinoraks M, Cofalla C, Wiseman S, Giesy JP, Schaeffer A, Hecker M, Woelz J, Schuettrumpf H, Hollert H. How flood events affect rainbow trout: evidence of a biomarker cascade in rainbow trout after exposure to PAH contaminated sediment suspensions. Aquat Toxicol 2013;128: 13–24.

[5] Kaestner M, Nowak KM, Miltner A, Trapp S, Schaeffer A. Classification and modelling of nonextractable residue (NER) formation of xenobiotics in soil – a synthesis. Crit Rev Environ Sci Technol 2014;44(19):2107–71.

[6] Junge T, Classen N, Schaeffer A, Schmidt B. Fate of the veterinary antibiotic C-14-difloxacin in soil including simultaneous amendment of pig manure with the focus on non-extractable residues. J Environ Sci Health Part B Pesticides Food Contam Agric Wastes 2012;47(9):858–68.

2 Statistical Evaluation

2.1 General Remarks

As the results of analytical determinations often serve as a base for further decisions, or are used for validation of theoretical considerations in modeling, the statistical certainty of the measured quantities must be of the highest confidence level. Among the decisions to be made can be, "Does the found concentration for a toxic compound exceed the given threshold value?" or "Does the measured signal for a searched quantity of an analyte show a difference to the statistical noise of a blank sample?"

Especially for organic trace analysis, we have to be aware of stressing the maximum possible performance of the selected methods in terms of selectivity and sensitivity. For example, to determine 2,3,7,8-tetrachlorodibenzo-*p*-dioxin in the ppq range in flue gas, tremendous efforts to enrich and to separate this analyte from all other matrix constituents is necessary. Due to the law of error propagation, all necessary handling steps (sampling, enrichment, standard addition of the labeled analyte, chromatographic separation, further mass spectral separation and detection) with its individual uncertainties will unavoidably lead to a large accumulated total uncertainty of data. In pharmaceutical technology, only a narrow concentration level of a highly efficient active ingredient may be acceptable, whose validity must be guaranteed by the producer.

To make it clear to the analyst, any decision based on analytical results is always connected with the intrinsic statistical probability level selected. Hence, this has to be provided (together with the measured result) by the analyst to the decision-making officer! Never report an analytical result without its domain of uncertainty.

Subsequently, we will make you familiar with the most important statistical elements, which are essential for an analyst. This is especially of relevance when rating an analytical technique described in literature, or when a new determination technique has to be established. For this, there is a minimum of statistical qualification of analytical methods needed.

It would be beyond the scope of this book to provide a deep insight into statistical data treatment. We only will show some fundamental basics, which should make the reader sensitive.

2.2 Calibration Function

In analytical chemistry, the measured signal y depends on the analyte concentration c by the *analytical function f*:

$$c = f(y) \tag{2.1}$$

f is developed by the analyst through a *calibration*. To do so, the analyst has to apply known standard concentrations of the requested analyte within the same analytical

DOI 10.1515/9783110441154-002

procedure. In favorable cases, a linear relationship between observed signals y and used concentrations c is

$$y = g\,(c) \tag{2.2}$$

There are some fundamental requirements to obey for such calibration functions:
- continuously differentiable (no saltus function);
- differential coefficient $\neq 0$ (or inflexion point within function g);
- possibly of linear character $y = bc + d$;
- or, at least it is mathematically linearizable (by applying appropriate linearization models).

The slope b of a calibration line is important. A very steep slope b gives rise to a very narrow working range. On the other hand, a too less inclined slope b produces insensitivity in the signal y as a function of changing concentration c. Whether a calibration is applicable is decided by the analyst, according to the needed concentration range.

2.3 Sensitivity

Mathematically, the sensitivity of a method is the slope of the calibration function

$$S = g' = \frac{dy}{dc} \tag{2.3}$$

which is the differential quotient of the calibration function g. It becomes obvious why function g should possess $g' \neq 0$, since this would not represent a useful calibration function g at all. Nowadays, by using electronic data treatment, it seems easy to produce an appropriate *sensitivity*. This neglects the underlying physicochemical relationship, which remains valid despite any mathematical operation.

2.4 Types of Error

Let's consider the usual outcome of an analytical determination, represented by a typical calibration graph (see Figure 2.1). Shown is the relationship between standard calibration points ("calibrator") on the abscissa and the observed analyses' signals y on the ordinate. Again, each calibration point, if repeated several times, shows its own statistical uncertainty of reproducibility, expressed by the indicated range of standard deviation for a single standard concentration. This is caused by a *nonsystematic random error*. Usually, one *standard deviation* (abbreviated as 1s) above and below the mean $\bar{y}_s$ is depicted.

Obviously, repeated determination of the same analysis at a constant concentration does not yield exactly the same result. Random deviations will occur in both the negative and the positive direction. Assuming a Gaussian distribution of such data

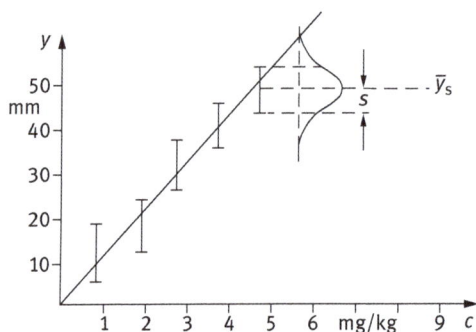

Figure 2.1: Graphical presentation of an analytical calibration function generated by multiple standards.

scattering around the *mean $\bar{y}$* (exemplarily shown in Figure 2.2) allows the calculation of the mean value $\bar{y}$ and its *standard deviation s* (=for a limited size of sample) for a scheduled probability level. *s* is used here instead of δ (=for infinite size of sample), since in practice we handle always a limited number of data (we make an estimation), whereas δ is valid only for an infinite number range.

As shown in Figure 2.2, the most probable data are represented by the *mean $\bar{y}$* but due to the spread of the data scattering around the mean value, there are also values found more or less far away from the mean (with positive and negative deviation). Depending on the selected *probability level*, for example 3*s*, such range will cover 99% of the values found (area under the distribution function). In a juridical trial, very often probability levels of 99.9% are claimed, which in turn requires an extremely high reproducibility of an analytical procedure. We also recognize that a 100% probability level will ask for indefinitely wide standard deviation ranges (limits of integration ranging from $-\infty$ to $+\infty$)!

Figure 2.3 shows a frequent type of calibration function with an intercept deviating from zero.

This onset indicates the presence of a *constant systematic error*. Often this is attributable to a mismatch in the instrumental setting or a constant loss of substance within the operations. If this remains constant throughout the working period, one can handle this by ruling out the intercept value mathematically. In photometric

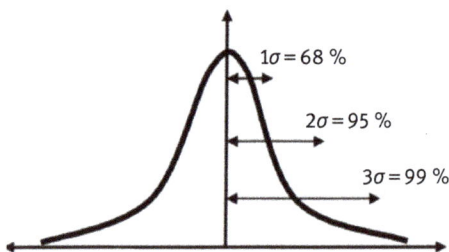

Figure 2.2: Gaussian distribution function describing error distribution around a mean value due to nonsystematic random data scattering.

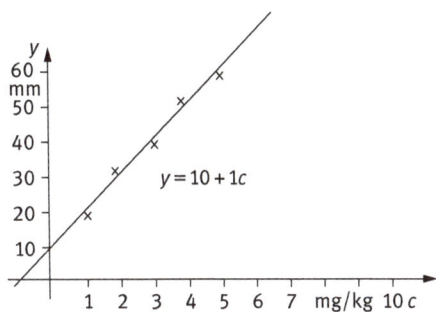

Figure 2.3: Linear calibration function with a constant systematic error.

$y = 10 + 1c$

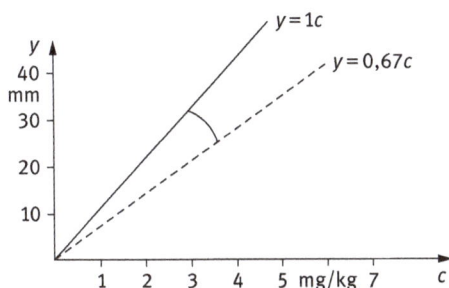

$y = 1c$

$y = 0{,}67c$

Figure 2.4: Example for a systematic proportional error in calibration.

determinations, this can be done by alternating measurements of a probe cell and a reference cell filled with blank solution.

Much more difficult is the presence of a *proportional systematic error* (see Figure 2.4). This may not be recognized as a source of error at a first glance. The linearity might be even perfect, but the slope is different from what is expected. This is revealed by interlaboratory comparison tests (round robin tests), where different analysts from different labs compare their methods subjected to the same sample. Such an error may stem from a mistake that frequently happens to the beginner. When preparing a master standard solution and storing it in a refrigerator at 6 °C, the volume contraction must be considered. Withdrawing a small aliquot from the cooled master solution by a volumetric pipette and rapid subsequent dilution with solvent at ambient temperature produces a nonnegligible error (volumetric flasks are calibrated at 20 °C).

Producing a series of standards by such a procedure generates a remarkable deviation from what would be valid. The same deviation happens when a defined alcoholic extract mixes with water (due to nonlinear volume contraction). Even more drastic in this relationship is the usage of an instable or impure compound for preparing standard solutions. This is a crucial point in organic trace analysis, where many compounds are instable when treated with oxygen or heat. Hence, knowledge of the purity of the analyte substance used for standardization is essential. Another source of error can happen when freezing an aqueous sample: if the solution is warmed up, a gradient in concentration may establish in the sample container, for example, a test tube. If the sample after thawing is not thoroughly mixed up, the upper layer may contain much lower concentrations of the analyte than lower layers.

All error types occur in lab practice and may happen in parallel. Therefore, it is advisable to establish a checking routine regularly, for example, by participating in inter-comparison programs or usage of an in-house standard sample as reference.

2.5 Limit of Detection/Limit of Quantification

One of the most important questions raised is for the detection limit of an offered analytical technique. The *limit of detection* (LOD) is by IUPAC definition the concentration x_{LOD}, which can be discriminated from the blank mean value $\bar{x}_{bl}$ (obtained by analyzing a blank sample) at a scheduled signal distance of three times the standard deviation of the blank s_{bl}:

$$x_{LOD} = \bar{x}_{bl} + k \cdot s_{bl} \qquad k = 3 \tag{2.4}$$

Graphically, this is seen for $k = 3$ from Figure 2.5. Assuming the same distribution of data scattering of the blank and the lowest detectable analyte concentration, this means that 50% of the data can be already part of the blank.

The *limit of quantification* (LOQ) uses $k = 10$, in order to obtain a higher confidence level as also seen from Figure 2.5. The α error (probability of false positive) is small.

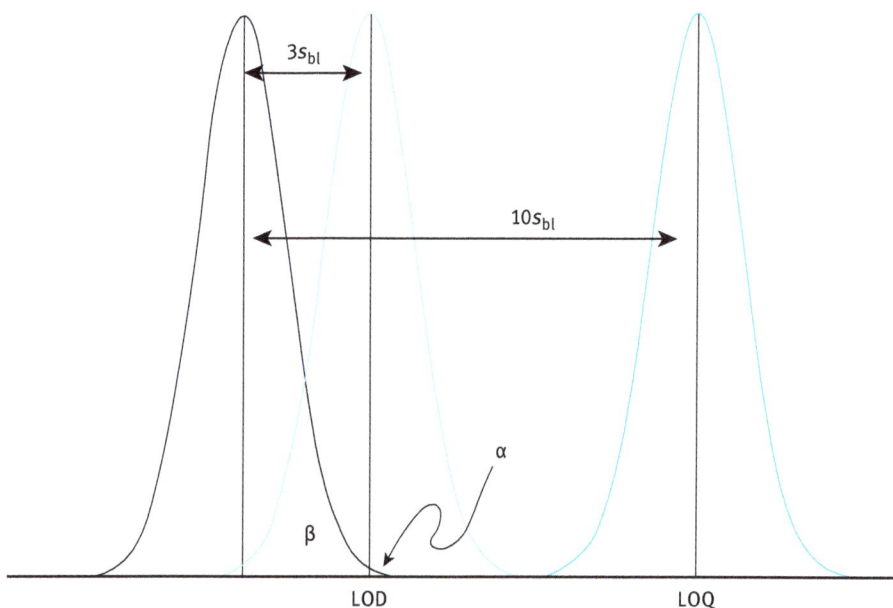

Figure 2.5: Overlap of blank signal and analyte signal near LOD and LOQ.

However, the β error (probability of a false negative) is 50% for a sample that has a concentration at the LOD.

Typically, a minimum of six blank samples have to be analyzed in order to determine the detection limit. To understand the demand to analyze as many samples as possible for such a task, one has to see the general mathematical expression to determine LOD from a calibration function.

$$x_{\text{LOD}} = \frac{s_y}{b} \cdot t \cdot \sqrt{\frac{1}{n} + \frac{1}{m} + \frac{\bar{x}^2}{\sum_{i=1}^{n} (x_i - \bar{x})^2}} \qquad (2.5)$$

where x_{LOD} is the limit of detection, $t = t(f;\ \alpha)$ is the tabulated value of t – distribution function for a given value of degree of freedom f and selected confidence level α, b is the constant of calibration function, n is the number of analyzed samples, m is the number of parallel determinations, s_y is the standard deviation of the tests, and x the measurement signal.

The higher the number n of analyzed samples is, the lower the detection limit will be.

2.6 Precision

Precision is the degree of reproducibility of repeated measurements. Figure 2.6 shows four situations from target shooting, where obviously the shooter scored in a differently reproducible manner. Figure 2.6 (b) depicts high precision in target shooting. Nevertheless, the shooter did not meet the mark of target obviously due to a systematic error. We can conclude that high precision does not necessarily mean accuracy.

2.7 Accuracy

Accuracy is the degree of accordance of the measured value with the "true" value of a sample. Since the true value of a real analytical sample is usually unknown, the

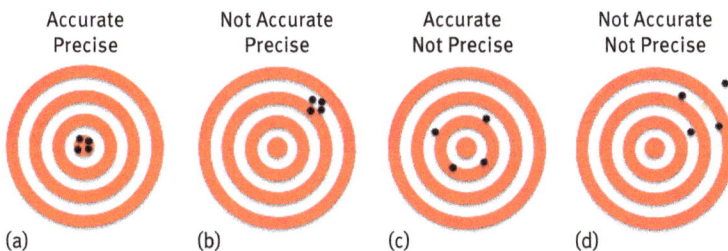

| Accurate Precise | Not Accurate Precise | Accurate Not Precise | Not Accurate Not Precise |

(a)   (b)   (c)   (d)

Figure 2.6: Graphical demonstration of precision and accuracy from target shooting.

only way is estimation of it by a round robin test through parallel analysis by several experienced labs. It is reasonable to assume that the mean value of such testing is approximating the true value. Figure 2.6 shows examples where one shooter is acting imprecisely (c), but the mean hits the center point, whereas an other shooter practices highest precision and accuracy (a). Example (b) shows the result of precise but not accurate shooting, and in (d) both precision and accuracy are missing.

2.8 Correlation and regression

Correlations are used everywhere in analytical chemistry. Already preparing a calibration line is mathematically nothing else than correlating an observation (of the analytical signal) with the given standard concentrations. As an example, we show the graphical presentation of such calibration for an electrochemical immunosensor. Accomplished is the calibration by producing a regression line (shown in Figure 2.7) between the measured sensor signal and the applied aflatoxin B_1 concentration.

Given is the squared correlation coefficient R^2, and the data pair n used for calculation. Often shown in publications is R in order to demonstrate the quality of a linear relationship. The reader should be aware of the dependency of R on data pair number n as a measure of good correlation. Best correlation value R is achieved for $n = 2$, since R is then 1.000. Obviously, one has to know the underlying data volume in correctly using correlation coefficients. Therefore, always look for data pair n too. Without n, the value R is only of qualitative character.

Regression lines offer a graphical impression, whether two variables, for example analysis data obtained from two independent techniques, are linearly connected.

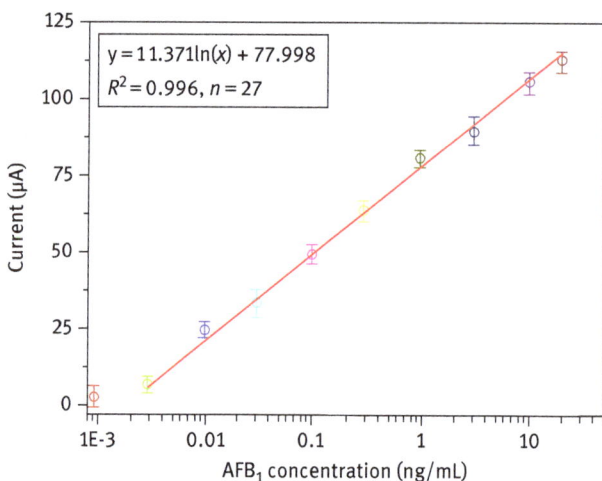

Figure 2.7: Calibration curve of the competitive-type electrochemical immunoassay toward different concentrations of aflatoxin B_1 standards.

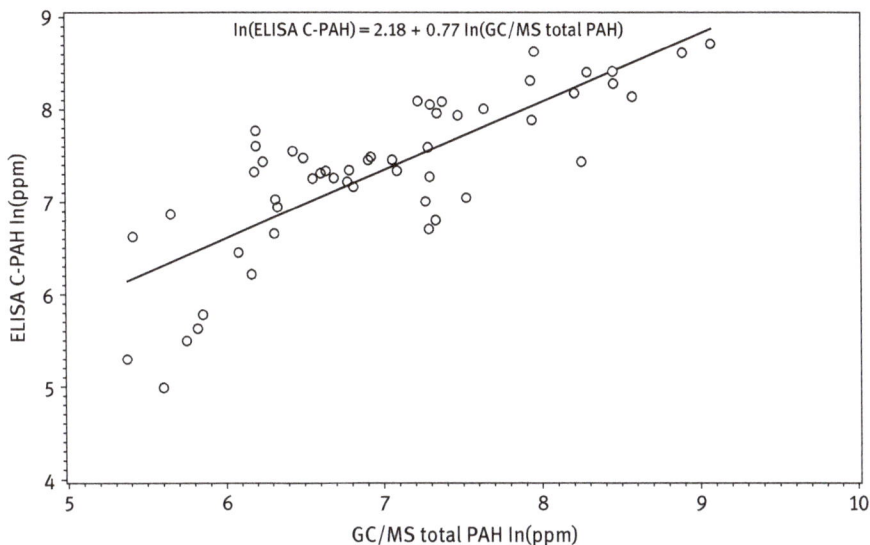

Figure 2.8: Comparison of immunoassay and GC–MS to the same soil samples toward total PAH concentrations.

One example is shown in Figure 2.8. It was the purpose of a study to compare two techniques. The ideal case would be to see a regression line on the bisecting line of an angle. This would indicate a perfect match between the two different techniques.

The presented data for 52 polycyclic aromatic hydrocarbon (PAH)-contaminated soil samples, in parallel analyzed by independent techniques, here enzyme-linked immunoassay versus gas chromatography–mass spectrometry (GC–MS), offer at a first

Table 2.1: Statistical data for regression analysis

	Regression analysis results for soil samples			
	Raw data		ln-transformed data	
	ELISA vs. sum of B2-PAH	ELISA vs. sum of all PAH	ELISA vs. sum of B2-PAH	ELISA vs. sum of all PAH
N	52	52	52	52
R^2 (%)	68	68	61	64
Intercept	609.78	926.37	2.17	2.18
Slope	2.78	0.67	0.87	0.74
P-value	<0.001	<0.001	<0.001	<0.001

Note: The form of the regression equation is: ELISA-derived concentration = intercept + slope × (GC-MS concentration).

glance a relatively weak correlation. Nevertheless, detailed statistical analysis (parameters see Table 2.1) reveals a linear correlation at a confidence level >99%. What is also evident from the nonzero intercept for the immunological technique is an influence by other compounds. It looks like the immunoassay is also sensitive to other compounds, such as humic substances or metabolites, which can't be discriminated by the applied antibodies. But for quick screening purposes, the immunoassay could be used instead of the time-consuming GC–MS procedure.

Further Reading

Anderson R. Practical statistics for analytical chemistry. New York: Van Nostrand Reinhold, 1987

Devore J. Probability and statistics for engineering and the sciences. Boston: Brooks/Cole, 2012

Doerffel K. Statistik in der analytischen Chemie Weinheim: Wiley – VCH, 1990

Mickey R, Dunn O, Clark V. Applied statistics: Analysis of Variance and Regression. New York: Wiley, 2004

Graham RC. Data analysis for the chemical sciences – a guide to statistical techniques New York: VCH, 1993

Meier P, Zünd R. Statistical methods in analytical chemistry. New York: Wiley, 2000

Miller JC, Miller JN. Statistics and chemometrics for analytical chemistry. Harlow: Pearson Education, 2010

3 Quality Control Strategies

3.1 General Remarks

Quality control (QC) in trace analysis is a must in general. The fact that organic compounds often are not stable due to degradation asks for a thorough QC ranging from the sampling step until the final determination by any analytical procedure.

Over the years, several routes became feasible for this.

3.2 Artificially Generated Matrix Surrogates

Depending on the matrix wherein the organic traces are to be analyzed, one way could be the preparation of synthetic samples (e.g., gaseous or aqueous phase, soil or any chemically or physically well-defined solid material), but *except* the analyte. The substance to be analyzed is then added to this artificial matrix in a known amount and should be found to a recovery rate of 100%.

The reason for this is the often observed susceptibility of analytical procedures to matrix constituents, for example, by a detector system in chromatography, to an extent which may seriously limit the qualitative and quantitative identification of a compound.

The production of artificial matrices is often inadequate. Only in the gas or liquid phase is homogeneous mixing easily achieved. Within solid phase matrices, the inhomogeneity, for example, caused by polydisperse particle distribution remains the main problem. Depending on the selected analytical determination, the required subsample might be of different composition compared to others.

Even more difficult are analyses of biological material such as animal tissue or plants. Here, binding to natural polymers, for example peptides and lignin in plants, mask the presence of analytes. The same happens within natural soil samples: For example, pesticides often become covalently linked to humic substances or incorporated within the spacing of layered clay minerals (e.g., montmorillonite). Aging plays an enormous role and may lead to reduced recoveries, a process which occurs after very short times for some compounds and which cannot be mimicked in the production of artificial soil samples so far.

Producing an artificial gas phase seems easier, but depending on the analytical task, this can become complicated. One example is producing an artificial exhaust gas, here needed to substitute real combustion engine exhaust gas in a procedure to measure soot reactivity [1]. This is shown in Figure 3.1.

The analytical determination of soot reactivity is accomplished by thermally programed oxidation of a few milligrams of deposited soot on a metal fiber filter within the artificial gas matrix. Important for this is the presence of water vapor, nitrogen,

DOI 10.1515/9783110441154-003

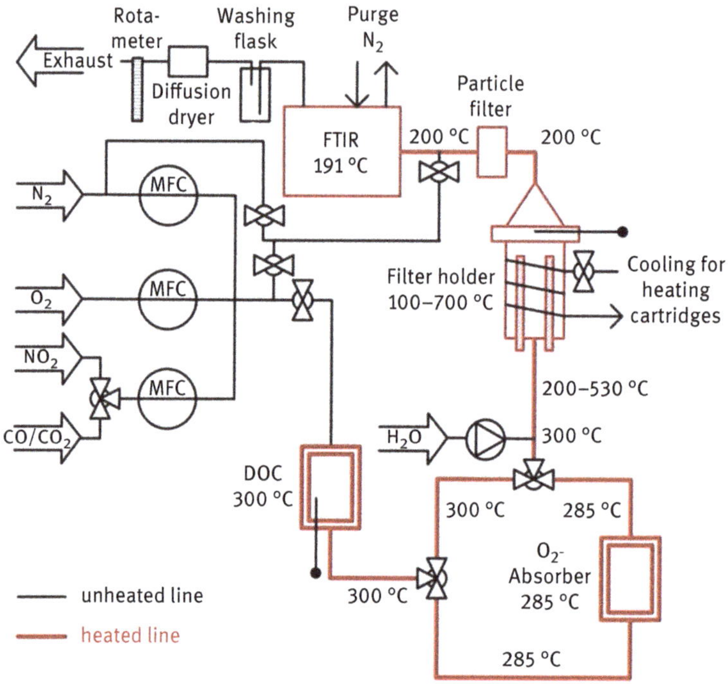

Figure 3.1: Schematic overview on the major components of the model catalytic converter system. MFC = mass flow controller; DOC = diesel oxidation catalyst; FTIR = Fourier transform infrared spectroscopy.

CO/CO_2 and nitrogen dioxide. Depending on temperature, gas phase composition and soot nanostructure, the soot combustion will onset at different temperatures, which can be followed by FTIR CO/CO_2 spectroscopy for gas phase concentration analysis. In this procedure, the gas concentrations have to be fixed, but of course in reality, these matrix constituents may change in concentration in an unpredictable way, depending on the loading conditions of an engine. This shows exemplarily how insufficient such artificial generation of a sample matrix can be.

A similar scenario is given for the aqueous matrix. Natural water systems do not contain only water. As a rough estimate, surface water also contains several hundred thousands of particles per mL, mainly in the nanometer range, which may change the dissolution equilibrium of dissolved trace organic analytes dramatically depending on its physicochemical properties. It is well known that, for example, polycyclic aromatic hydrocarbons (PAHs), whose representatives typically range from naphthalene up to coronene, become immediately adsorbed to particles to widely differing degrees, depending on the nature of such nanomaterials and, of course, on the presence of other constituents of the aqueous phase. Namely, natural (e. g., rhamnolipids) and synthetic surfactants or tensides change such adsorption equilibria. It is therefore nearly impossible to produce an artificial matrix which can approximate a natural matrix.

We have to conclude that only in rare cases, such as in a reproducible industrial production, or when consensus exists, such artificial synthetic generation of a test matrix is achieved.

Problems with the fortification of analytes in soils and sediments may occur due to the presence of significant amounts of black carbon – originating from natural and manmade combustion processes and nowadays present in many regions of the world – which provides a strongly sorbing matrix for pollutants.

3.3 Addition of an Internal Standard

A widely used practice is to check an unknown sample for matrix interferences by the so-called standard addition procedure.

Usually, the unknown sample becomes split into two equal parts. To one part, a known amount of analyte is added and subsequently thoroughly mixed until homogeneous distribution of the spiked amount is achieved. In the next step, both parts – spiked (A) and non-spiked subsample (B) – then become analyzed. The result of the spiked sample A will show the sum of added amount of analyte and its natural content of analyte (received by analysis of sample B), if there is no matrix interference of the analytical determination. In case a deviation from the expected value in A is determined, often a linear correction is applied to sample B. Whether this represents possible interferences by unknown constituents in the original sample B remains questionable. The main problem with this is the assumption of a linear additivity of effects on the analytical determination over a wide concentration range, which may not be given. Natural adsorption systems (soil, sand grains, airborne particles, etc.) always show a nonlinear adsorption dependency (Freundlich adsorption isotherm, eq. (3.1)) due to differently strong adsorption sites:

$$c_{ads} = K_f c_e^{1/n} \qquad (3.1)$$

c_{ads} is the mass of adsorbate divided by the mass of adsorbent ($= mass_{adsorbate}/mass_{adsorbent}$), c_e is the concentration of the adsorbate in solution after equlibrium is reached, $1/n$ is the exponent describing the deviation from the linear adsorption dependency and K_f is the Freundlich adsorption coefficient.

A typical example for this is the QC by standard addition when PAHs are to be measured within exhaust aerosol. Here, the partially high amount of strongly adsorbing soot particles can produce enormous difficulties in recovering additionally spiked PAHs, such as triphenylene, or other rarely occurring PAHs like p-quaterphenyl [2].

One must be aware that from a statistical point of view, standard addition is not the ideal procedure. The correction by applying a recovery rate is based on the mathematical treatment of the found values of samples A and B, with their individual statistical variations. So, since the corrected value is always the addition of the variances squared of values found for A and B, the total uncertainty becomes larger (direct consequence of the law of error propagation).

There are three ways for the usage of internal standards:

A. fortification of the pure analyte within the sample, with no change in molecular structure or isotopic constitution;

B. sample spiking by addition of a slightly altered analyte with similar physical chemical properties as the analyte, for example, synthesized by changing or addition of functional groups like OH or alkyl groups;

C. spiking by adding of the pure analyte, but substitution of one or more atoms within the molecular structure by a stable/unstable (radioactive) isotope ("isotope labeling").

In case of procedure A, sometimes the synthesis of high-purity substances needed for spiking is not achieved easily. Examples are dioxins and dibenzofurans, which are not available in larger quantities. Also, labile species (e.g., amines) may pose a problem.

A way out may be route B, where slightly altered analyte molecules are used which are well discriminated by the selected analytical procedure. This was often used in clinical chemical analyses [3]. In Table 3.1, several examples are compiled.

Such internal standards (method B) are useful in chromatographic separations *without* mass spectrometry (MS) as detector. Especially, halogenated internal standards were used, since the electron capture detector shows a favorable strong sensitivity to it.

More refined and increasingly more practiced is the so-called stable isotope labeling in the internal standard addition process (method C) [4].

In the following, this procedure will be exemplarily discussed.

Aflatoxins are nowadays a critical issue for food inspection. They are formed as hepatocarcinogenic secondary metabolites by fungal species such as *Aspergillus flavus* growing on a variety of foods under humid and warm storage conditions. Due to their hepatotoxic effect, a strict control of mycotoxin presence is regulated worldwide. Usually, high-performance liquid chromatography with fluorescence or tandem MS detection is used for separation and quantification.

Table 3.1: Examples for chemically modified compounds as internal standards (method B).

Analyte	Matrix	Internal standard
Nitrazepam	Blood serum	*N*-Desmethyldiazepam
Trimethoprim	Blood serum	Methyl analog
Theophyllin	Blood serum	Hydroxyethyltheophylline
Morphine	Blood serum	Ethylmorphine
Nicotine	Blood serum	Propylnornicotine
Progesterone	Blood serum	Testosterone
Polyamine	Urine	1,6-Hexanediamine
Benzo[*a*]pyrene	Air	Benzanthrone
Zearalenone	Maize	Zearalenone

Aside from difficulties to get complete recovery from various sample matrices, for example, nuts or grains, the matrix-dependent ionization depression in MS causes additional problems. So, it has become necessary to apply internal standard addition with stable isotope-labeled analytes. As these labeled substances show virtually the same chemical behavior (e.g., adsorption during sample cleanup or incomplete ionization in MS leading to signal suppression), they can be used for data compensation and calibration in one step. For the latter, this procedure is called stable isotope dilution assay, since the non-labeled analyte becomes "diluted" by addition of a known amount of labeled analyte. By data recording of the different stable isotope mass spectral intensities from the separated mass tracks, even a quantitative result can be deduced from such dilution. Therefore, such isotope labeling yields much more robust and valid data compared to other analytical methods.

In the case of aflatoxin B_2, therefore, it was necessary to synthesize deuterated aflatoxin B_2 and G_2 (D_2-aflatoxin B_2 and G_2; see Figure 3.2).

The food samples were ground and homogenized thoroughly. To 10 g of food sample, 40 ng D_2-aflatoxin B_2 was added and mixed again. The concentration of the spiked D_2-labeled standard matched exactly the actual EU legal limit for the aflatoxin. Next, an extraction step (2 h, 40 mL methanol/water (80:20; v/v)) was performed. After filtration, 20 mL of the filtrate was added to 180 mL buffered saline. Then, an immunofiltration step, which uses anti-aflatoxin antibodies for separation and enrichment, was applied. After desorption from the immunoaffinity column, an aliquot became subjected the liquid chromatography-mass spectrometry (LC–MS/MS) analysis. The ratio of unlabeled and labeled aflatoxins, ranging between 0.1 and 10, was at a total aflatoxin level of 50 µg/L. Figure 3.3 shows the selected reaction monitoring chromatograms of the labeled and unlabeled species.

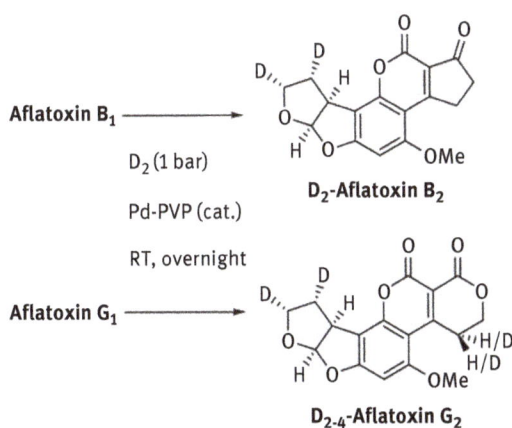

Figure 3.2: Synthesis of deuterated aflatoxin B_2 and G_2. Reproduced with permission from the American Chemical Society.

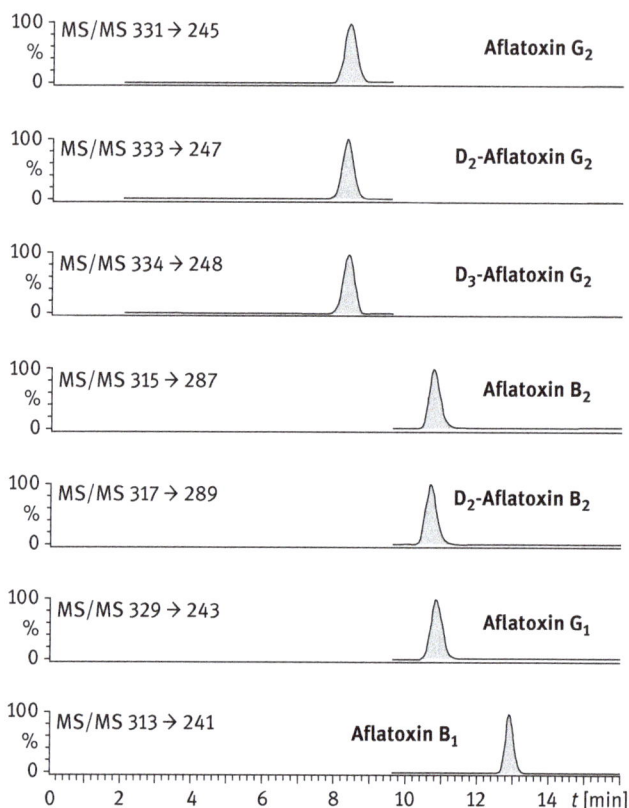

Figure 3.3: LC–MS/MS-selected reaction monitoring chromatograms of the labeled and non-labeled aflatoxins. Reproduced with permission from the American Chemical Society.

Linear calibration curves were constructed from the standard/analyte peak area ratios versus standard/analyte molar ratios. Additionally, calibration in the presence of matrix was performed by adding aflatoxin mixtures with the respective molar ratio of analyte to D_2-labeled standard ranging from 0.1 to 10 (40 ng total amount of aflatoxin spiked) to 10 g of ground sample.

All calibration curves constructed from the peak area ratios versus the molar ratios (labeled/unlabeled) showed excellent linearity. The limit of detection was determined to be 0.09 µg/kg for aflatoxin B_2 in wheat grains, which is well below the valid maximum permissible limits.

Sometimes, the use of radioactive isotope labeling is helpful. Despite the enormous circumstances connected to the application of a radioactive compound (safety means, waste disposal, licenses needed), this is the direct way to trace back analyte losses within a multistep separation scheme. One advantage of such a method is the easy and complete balancing of all subsequent extraction, cleanup and analysis steps after the matrix has been fortified with the pollutant.

In all cases (methods A, B or C), complete homogenization after standard addition remains the most crucial point to be observed. In favorable cases, such isotope labeling is achieved by adding isotope-labeled amino acids to cell cultures.

As mentioned at the beginning of this chapter, standard addition is only viable if the analyte is not included within non-accessible matrix cavities (e.g., montmorillonite in soil), where only long-lasting extraction procedures like supercritical fluid extraction open such aggregates. Especially, biological tissues with ongoing enzymatic activities within the cellular material may show nonlinear effects, even with isotope-labeled surrogates, due the different reaction rates when a high proportion of different isotopes is used (e.g., substitution of –H by –D changes mass by 100%!). Also, retention time in chromatographic separation can deviate due to this substitution.

3.4 Validation by Applying Independent Analytical Methods

An extremely high confidence can be attributed to those analytical procedures which involve the use of independently based determination mechanisms. These "orthogonal" determination techniques can be used in parallel after having split a sample into representative parts. In the following, this will be discussed.

PAHs represent a class of organic compounds with high relevance because of the known and accepted carcinogenicity of some of its representatives. Therefore, benzo[a]pyrene (BaP) is a trace analyte often requested in characterization of food, water, air and soil. As a consequence of this, hitherto the lowest threshold limit for drinking water is for BaP (10 ppt). Violation of such a legal threshold limit poses serious consequences, for example, closure of the public water distribution. Same is with BaP-polluted soil, classification as hazardous waste asks for expensive waste treatment. Food, exceeding such limits, would become detached from the shelf.

To assure the presence of such trace contaminants, different molecular properties of the same analyte are used. For BaP, these properties could be
- fluorescence emission (screened by fluorescence emission spectroscopy) [5];
- UV absorption (measured by UV absorption spectroscopy) [6];
- mass spectral features after different ionization (identified by MS) [7];
- antigenic properties for specific interaction with antibodies (analyzed by immunoassays) [8];
- distribution within non-miscible phases (e.g., by gas chromatographic separation) [9];
- Raman emission (observed by Raman or surface-enhanced Raman spectroscopy) [10].

From a juridical point of view, the application of at least two orthogonal techniques is seen as a sufficient measure. But even then, there is still a risk, when sample pretreatment does not gain the complete content of the analyte from a drawn sample. In case of PAHs, the most critical part in the whole sequence of a selected analytical procedure is the PAH separation from the sample matrix. Especially, those samples derived from combustion contain PAHs intertwined with soot. Depending on the soot particle structure, very strong adsorption interaction prevents an easy desorption, and hence high recovery rates. So, even the subsequent application of different quantification techniques will not give the true value, since the bottleneck is given by the sample treatment before.

3.5 Validation by Reference Material

One way to check the in-house procedure is the analysis of certified reference materials (CRMs), ideally in a similar sample matrix as the real sample. Such reference materials provide a kind of benchmark for a measurement. They are therefore used in, for example, method development and validation, calibration and quality assurance. The enormous advantage of a reference material is its transferability to everywhere around the world.

The amendment to ISO Guide 30 [11, 12] defines a reference material as a "material, sufficiently homogeneous and stable with respect to one or more specified properties, which has been established to be fit for its intended use in a measurement process."

The term "reference material" is a generic term, that is, it comprises materials that are investigated and documented at different levels:

> Certified reference materials (*CRM*): as defined in the ISO Guide 30 as "reference material, characterized by a metrologically valid procedure for one or more specified properties, accompanied by a certificate that provides the value of the specified property, its associated uncertainty, and a statement of metrological traceability."

Non-CRMs (not accompanied by a certificate, sometimes called "reference materials" in the sense of materials qualified only to a limited extent compared to "CRMs").

Frequently, reference materials are also classified according to their use, for instance by calling them "calibrants/calibrators", "QC materials," "proficiency testing materials" and so on.

From the metrological point of view, a kind of hierarchy can be established based on the uncertainty of the certified values and in relation to their position in the traceability chain:
- Primary reference materials (a material having the highest metrological qualities and whose value is determined by means of a primary method). Highest level is given by the use of different orthogonal determination methods to the same analyte.
- Secondary reference materials (reference material whose chemical composition is assigned by comparison with a primary reference material of the same chemical composition, or with several such primary reference materials).

–   "In-house" reference materials, QC materials and so on. This can be, for example, a huge homogenized sample, where a part of it repeatedly becomes introduced into the same analytical procedure. This can be used to check instrumentation function on a daily basis.

The uncertainty associated with the property values usually increases from reference material types 1–3 as does the length of the traceability chain. In several cases, neighboring material types may be situated on the same level of hierarchy.

Such CRMs are now offered by many national or supranational organizations, as there are

–   Bureau of Analysed Samples Ltd. (UK),
–   National Institute of Standards and Technology (NIST/USA),
–   National Research Council Canada (Canada),
–   Federal Institute for Materials Research and Testing (BAM/Germany),
–   Institute for Reference Materials and Measurements (BCR-IRMM/European Union),
–   National Metrology Institute of Japan.

Difficulties remain with biological or aqueous samples. Because of their inherent instability, special attention has to be paid to this. Standard reference material has been offered by NIST in former times, which was based on controlled leaching rate, for example, PAHs adsorbed to glass spheres and filled in a column. Leaching of PAHs at a constant rate happens under fixed flowrate of pure water through the column and with a fixed temperature applied. Sometimes, freeze-dried standards may be helpful for preparing a water reference material [13].

The number of available CRMs for organic trace species is steadily increasing, mainly for usage in food inspection and environmental analysis. Table 3.2 shows some examples.

Table 3.2: Selected CRMs, offered in different matrices.

Analyte	Matrix	Identification number	Organization
PAH/nitro-PAH	Diesel particulate matter	1650b	NIST
PCB	Organics in marine sediment	1941b	NIST
2,3,7,8-TCDD	Methanol	3063	NIST
PBDE, PFOS, pesticides	Lake superior fish tissue	1946	NIST
T-2 and HT-2 toxin	Oat	ERM®-BC720	BAM
Acrylamide	Crispbread	ERM®-BD274	BAM
Phthalate esters	Polystyrene resin pellet	3406-d	NMIJ
SF_6 and CF_4	N_2 (emission level)	4403-a	NMIJ

One has to be aware that calibration materials, which have been produced for more than 100 years in many varieties of inorganic materials (mainly for titrimetric determinations), will be rarely available for organic trace analysis. To produce a gravimetric standard, the crystal water has to be removed. Organic material cannot be liberated from adsorbed water just by heating for a while to temperatures above 100 °C. Pyrolysis or other metamorphic changes may happen. Therefore, standards from organic compounds remain a difficult task. Oxidation of reactive C=C bonds has been observed frequently (e.g., pyrene under air). Stabilization by adding reductive compounds is often practiced, but unwanted, since it may cause new interferences in the analytical determination process.

So-called SI-traceable *primary standards* are those which can be controlled in their production by a primary measurement. These are measurements of the basic SI units, like mass, size, time, electric current, temperature, molarity and light intensity at a given wavelength.

Gas mixtures are relatively easy to produce (just by handling partial pressures or volumes of the respective gases); the generation of trace amounts in the ppbv range or below remains challenging.

A good example to prepare such primary standard is for volatile organic compounds (VOC) trace gas (BTEX). This can be achieved by using a controlled leak of a container wherein the pure compound is stored. Usually, a refillable permeation tube (see Figure 3.4) is the method of choice [14, 15]. Such permeation tubes likewise consist of a container with a non-permeable wall (e.g., glass) and a stopper made of Polytetrafluoroethylene (PTFE), through which the gaseous analyte (e.g., toluene, benzene or xylene) can leak out at a constant permeation rate. The compound to turn into gas is stored within the reservoir as condensate, positioned vertically even with shrinking filling height.

Only the vapor phase above the liquid phase is in contact to the porous PTFE. At a constant temperature, the vapor pressure above the liquid is constant (according to Henry's law), and hence the permeated amount per unit of time is too. The only measurement necessary is the repeated gravimetric determination of the permeation loss of VOC over time (typically one determination per 14 days). Depending on fixed air flow, well-defined trace gas concentrations can be produced. Changing the storage temperature will change the vapor pressure and hence the permeation rate too. Since only weight and gas flow rate have to be determined, which can be achieved by basic unit measurement, it represents a primary standard.

Secondary standards are those which need chemical analysis for estimating the amount of analyte present. For this, many procedures are in use, for example, photochemical reagent formation or desorption under the flow of a gas or liquid. Sometimes, electrochemical synthesis helps, with the advantage of coulometric measurement.

It is important to realize the difference between a CRM and a SI-traceable primary standard. The primary standard serves as matrix with the true content. In

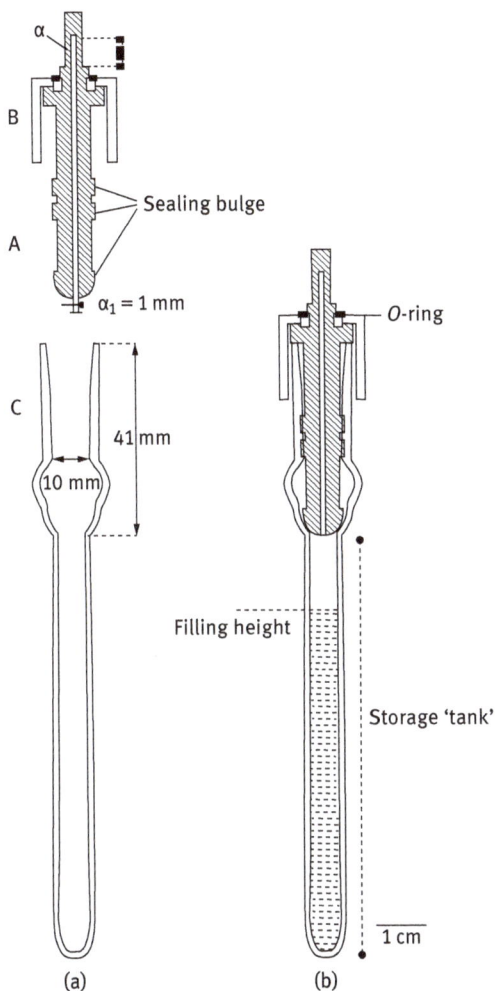

Figure 3.4: Refillable permeation tube as a primary standard for trace gas generation. (a) Open and (b) closed. Reproduced with permission from the American Chemical Society.

contrast to this, a CRM is a material whose content of a certain analyte is based on decision-making by consensus, reached by the best laboratories by using the three different determination methods. It is assumed to approach the true value (without applying basic measurement techniques).

Many standards of high purity for organic trace analysis can be obtained from various suppliers. The production of such materials is typically based on weighing the purified analyte and dissolution within pure solvent. From such master standards, working standards for daily use have to be prepared by the analyst himself.

Again, as said before, losses due to a change of state (e.g., due to degradation under air, light or temperature, adsorption or biodegradation by microorganisms, etc.)

need to be checked regularly. Especially in case of very poisonous materials like ultra-toxins, this becomes very difficult.

3.6 QC by Applying Good Laboratory Practices

Nowadays, regulation in some fields concerning organic trace analysis requires rigorous quality management in analytical laboratories.

The term "good laboratory practices" (GLP) is synonymous with what can be done to produce the best possible data by practicing a transparent scheme of operations [16–18]. Numerous regulations exist for this:
–   good clinical practices in clinical diagnostics,
–   bioavailability in pharmacology,
–   Toxic Substances Control Act for toxicity analysis.

GLP nowadays is exactly regulated and comprises means which have to be performed, for example, by manufacturers of pharmaceutical drugs or pesticide producers.

A GLP-practicing unit has to make all steps in planning, performing and reporting analyses transparent. This ranges from the organization, lab structure, chemicals storage, waste removal and treatment, to protocolling everything from the moment one enters a GLP area until the moment one leaves. *Standard operation procedures* (SOP) with detailed description of the whole analytical process have to be available. The head of a GLP unit can't work as the acting analyst or technician. Internal quality means have to be set and followed. Data handling and storage has to be regulated too.

The chemical substances, and the incoming samples, have to be labeled and characterized in an unambiguous way. Large amounts of chemicals must be retrievable to their origin, production date and purity. Shelf-life time needs to be controlled.

The lab rooms where GLP has to be practiced must be well separated from each other, in order to avoid any cross-contamination. So, no storage of larger solvent containers is allowed in areas where the analytical determination is carried out. Labs where physical or chemical operations are performed, have to be separated. Same is with sample treatment areas. Waste deposit and chemicals storage must also be well separated from each other.

Archiving of samples and data storage is of special importance. Especially in those cases where high economic risk is given or juridical consequences seem possible, archiving is of crucial importance, for instance the archive must be locked and safe in case of fire or flooding.

Aside from these requirements, the highest level of occupational hygiene has to be practiced since organic trace analysis in most cases is linked to a non-negligible toxicological risk.

Further Reading

Bulska E. Quality assurance and quality control of analytical results. In Namiesnik J, Szefer P, editor. Analytical Measurements in Aquatic Environments. Boca Raton: CRC Press, 2010;389–97.

Heydorn K. On the development of quality assurance in analytical chemistry. TRAC Trends Anal Chem 2015;64:xii–xiii.

Konieczka P, Namiesnik J. Quality assurance and quality control in the analytical chemical laboratory: a practical approach. Boca Raton: CRC Press, 2009.

Ni H-G, Zeng H, Zeng EY. Sampling and analytical framework for routine environmental monitoring of organic pollutants. TRAC Trends Anal Chem 2011;30:1549–59.

Olivares IR, Lopes FA. Essential steps to providing reliable results using the analytical quality assurance cycle. TRAC Trends Anal Chem 2012;35:109–21.

Bibliography

[1] Ess MN, Bladt H, Muhlbauer W, Seher SI, Zollner C, Lorenz S, Bruggemann D, Nieken U, Ivleva NP, Niessner R. Reactivity and structure of soot generated at varying biofuel content and engine operating parameters. Combust Flame 2016;163:157–69.

[2] Itoh N, Inagaki K, Narukawa T, Aoyagi Y, Narushima I, Koguchi M, Numata M. Certified reference material for quantification of polycyclic aromatic hydrocarbons and toxic elements in tunnel dust (NMIJ CRM 7308-a) from the National Metrology Institute of Japan. Anal Bioanal Chem 2011;401(9):2909–18.

[3] Jensen KM. Determination of nitrazepam in serum by gas-liquid-chromatography – application in bioavailability studies. J Chromatogr 1975;111(2):389–96.

[4] Cervino C, Asam S, Knopp D, Rychlik M, Niessner R. Use of isotope-labeled aflatoxins for LC-MS/MS stable isotope dilution analysis of foods. J Agric Food Chem 2008;56(6):1873–9.

[5] Lewitzka F, Niessner R. Application of time-resolved fluorescence spectroscopy on the analysis of PAH-coated aerosols. Aerosol Sci Technol 1995;23(3):454–64.

[6] Sun FS, Littlejohn D, Gibson MD. Ultrasonication extraction and solid phase extraction clean-up for determination of US EPA 16 priority pollutant polycyclic aromatic hydrocarbons in soils by reversed-phase liquid chromatography with ultraviolet absorption detection. Anal Chim Acta 1998;364(1-3):1–11.

[7] Letzel T, Poschl U, Rosenberg E, Grasserbauer M, Niessner R. In-source fragmentation of partially oxidized mono- and polycyclic aromatic hydrocarbons in atmospheric pressure chemical ionization mass spectrometry coupled to liquid chromatography. Rapid Commun Mass Spectrom 1999;13(24):2456–68.

[8] Pschenitza M, Hackenberg R, Niessner R, Knopp D. Analysis of benzo[a]pyrene in vegetable oils using molecularly imprinted solid phase extraction (MISPE) coupled with enzyme-linked immunosorbent assay (ELISA). Sensors 2014;14(6):9720–37.

[9] Rojas NY, Milquez HA, Sarmiento H. Characterizing priority polycyclic aromatic hydrocarbons (PAH) in particulate matter from diesel and palm oil-based biodiesel B15 combustion. Atmos Environ 2011;45(34):6158–62.

[10] Chen J, Huang YW, Zhao YP. Characterization of polycyclic aromatic hydrocarbons using Raman and surface-enhanced Raman spectroscopy. J Raman Spectrosc 2015;46(1):64–9.

[11] ISO Guide 30, Terms and definitions used in connection with reference materials. Geneva, International Organization for Standardization, 1992.

[12] ISO Guide 30, A. Revision of definitions for reference material and certified reference material. Geneva, International Organization for Standardization, 2008.

[13] Martin-Esteban A, Fernandez P, Camara C, Kramer GN, Maier EA. The preparation of a certified reference material of polar pesticides in freeze-dried water (CRM 606). Fresenius J Anal Chem 1999;363(7):632–40.

[14] Teckentrup A, Klockow D. Preparation of refillable permeation tubes. Anal Chem 1978:50(12):1728.

[15] Nelson G. Gas mixtures: preparation and control. Chelsea: Lewis Publishers, Inc., 1992.

[16] Merz W, Wittlinger R. Is good laboratory practice necessarily good analytical practice? Is GLP necessarily GAP? Mikrochim Acta 1991;3(1-3):11–16.

[17] Vantklooster HA, DeckersHA, Baijense CJ, Meuwsen IJ, Salm ML. Quality assurance in analytical chemical laboratories: towards harmonization and integration of current standards. TRAC Trends Anal Chem 1994;13(10):419–25.

[18] Andola H. Good laboratory practices for high performance liquid chromatography (HPLC): an overview. Int Res J Pharma Appl Sci 2012:2:46–52.

4 Sampling of Organic Trace Contaminants

4.1 General Remarks

The simple statement that different environmental media – air, water, sediment, soil, plant, animal and human tissues – require different sampling strategies and equipment hides the complexity of the manifold techniques and tools nowadays available for deriving reliable analytical information. It is important to notice that sampling is the main source of potential errors in deriving analytical results, often in the range of several 100%, whereas subsequent steps in sample preparation and cleanup and the final analysis contribute much less to erroneous results if systematic errors are excluded (see Chapter 2). Therefore, a good sampling strategy is of utmost importance. Obviously, when we consider that we will always collect only small parts of an environmental field or water body, in terms of errors, it is easier to sample homogeneous liquid or gaseous media than heterogeneous solid media for subsequent analyses.

We differentiate batch (or grab) sampling from composite sampling. Batch sampling means the collection of numerous samples at a certain time and place for analysis on-site or in the laboratory to obtain the individual analyte concentration to derive either median or average concentrations, for example, in a soil, or for getting a picture of the distribution pattern of the analyte in the media. Composite sampling means homogeneously mixing different grab samples at the sampling area in order to obtain an average concentration of a pollutant and, if analyzed at different times, to learn about the time pattern of the concentration profiles. The disadvantage of composite sampling is that no information on the individual concentrations is obtained. Continuous sampling on the other hand will provide average concentrations, for example, by passive samplers exposed for a certain time period (see Section 4.6) in a river or in a sediment or by use of continuous measuring devices by flow-through systems.

Already one century ago, Taggart determined a correlation between sample sizes and the homogeneity or heterogeneity of the analyte distribution in solid matrices (Figure 4.1). The principal concept, that with increasing heterogeneity of the analyte distribution in the solid matrix higher masses of the medium need to be sampled, is true also for the analysis of trace pollutants for instance in soils or sediments.

In a static system, for instance, a field contaminated with a persistent organic pollutant, that is, a chemical resistant to environmental degradation through chemical, biological and photolytic processes, and rather constant concentrations over time, samples could be selected
- along a (triangular or rectangular) grid pattern,
- randomly (e.g., by use of a random number generator) or
- judgmentally (source-oriented but being aware of the potential bias)

DOI 10.1515/9783110441154-004

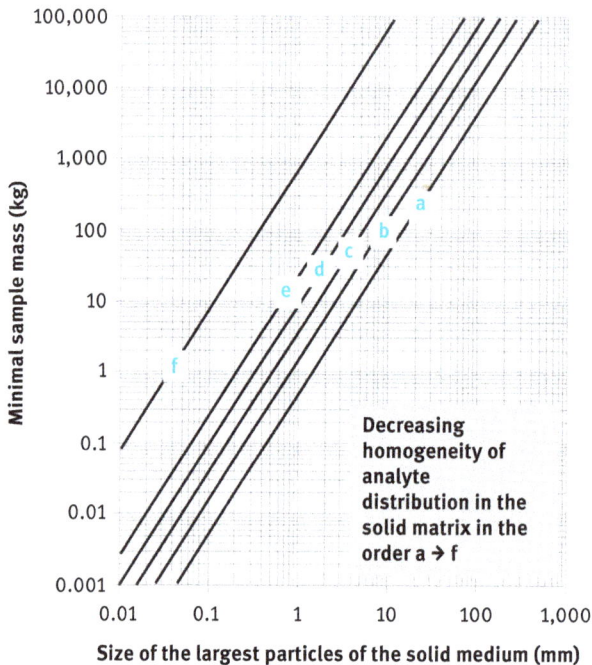

Figure 4.1: Nomogram according to Taggart [1] developed originally for ore analyses, cited, for example, by Kienitz [2].

–   in a stratified sampling scheme (e.g., if a sampling area contains parts with distinct differences),

to assess the concentration and distribution of the analyte (Figure 4.2). In a dynamic system, that is, with changing concentrations over time, either in space (for instance in a river) or time (for instance, due to degradation) or both, samples have to be taken at different random or regular time intervals in order to learn about the dynamics of the system.

A sampling strategy should be well prepared to answer the questions of where, when and how many samples should be taken. For preparation, the goal of the study should be clearly described, the area of interest identified and the environmental conditions such as climatic data, soil or water properties, and, if necessary, data on the site history collected. The analytical procedures should be outlined in order to delimit the necessary sample amounts. The spatial and temporal frequency of sampling considering the analytical problem and the statistical power of the campaign needs to be fixed. Also, most importantly, potential contamination sources have to be considered, for example, excess of a pollutant due to input from dust in the atmosphere, sampling devices, hands/gloves and clothing, glass- and plastic-ware (the latter often containing additives such as phthalates [3]) or loss of a pollutant by sorption to sampling materials and containers.

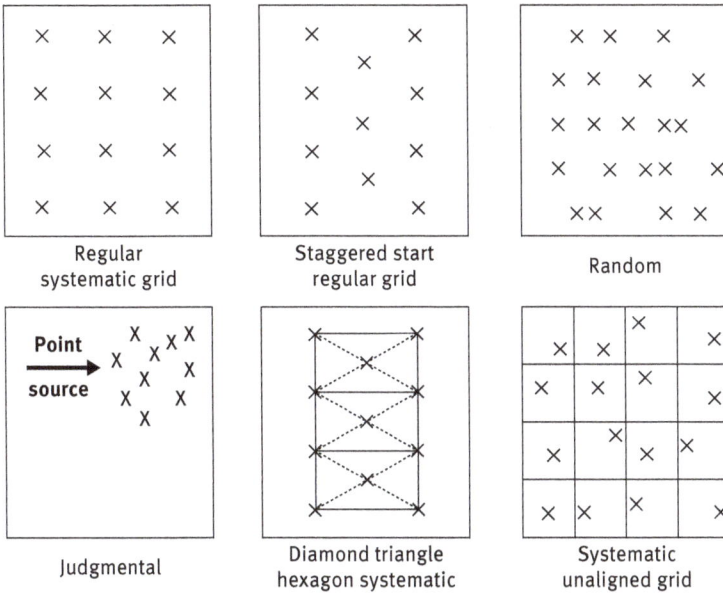

Figure 4.2: Sampling strategies for analyzing the concentration of a pollutant in a field.

4.2 Sampling from Organisms

4.2.1 General Remarks

Organic contaminants may be accumulated in the tissues of organisms. Therefore, chemical analysis of the appropriate biological tissues can reveal that the organism has been exposed to contaminants. The contamination of biological communities or of individual organisms can be monitored in a variety of ways to compare the contamination level with potential effects on the ecosystem such as the decrease of abundance of certain species at particular locations or behavioral, physiological and morphological changes.

Chemical analyses can identify that contaminants are present if the limit of quantitation is surpassed, but biological methods can integrate responses to combinations of all contaminants. Thus, if a biological response of a suited test system indicates the presence of pollutants, the corresponding samples can be analyzed by appropriate methods to identify and quantify the contaminants. The combination of biological methods and chemical analysis is a straightforward and comprehensive way to identify pollutant hotspots (see Chapter 10, Section 10.5) [4]. It must also be considered that chemical analyses indicate the pollution level that is valid only for the point in time when the sample was collected, whereas some biological methods reflect the effects of the physical and chemical conditions to which the organisms were exposed over a period of time. Of course, a disadvantage of biological methods is that other stressors, for example, anthropogenic or natural variations like changes in temperature or pH, may lead to a biological response. Thus, biological methods do not provide information on the identity of a contaminant unless supplementary information from chemical analyses is available.

4.2.2 Animal Tissue Samples

4.2.2.1 Aquatic Organisms

A variety of different sampling strategies have been used to collect aquatic organisms, summarized in Table 4.1. Depending on the research question, it has to be decided how many and which type of organisms have to be collected and from which part of the water-sediment system. Simple (hand picking) and more expensive alternatives (grab or dredge sampling, electrofishing) are available.

As an example, fish and sediment in the Three Gorges Reservoir of the Yangtze River have been sampled to monitor the pollution status regarding organochlorine pesticides, polycyclic aromatic hydrocarbons (PAHs), polychlorinated biphenyls (PCBs), brominated flame retardants and other halogenated environmental pollutants. Catfish was chosen as monitoring fish species due to its demersal lifestyle and low-to-medium migration – thus being suitable to indicate local contamination over longer periods of time – as well as its distribution all along the reservoir and other sections of the Yangtze River and due to its major economic importance. At each site, 10 fishes of comparable size (total length 165 ± 15 mm; weight of 40 ± 11 g; indifferent sex) were sampled by use of fishing nets. The fishes were kept in water of the local water body in an aerated tank to minimize alterations in the exposure conditions, till each fish was euthanized with clove oil, followed by an additional blow to the head. Bile samples were obtained from the gall bladder with a syringe for subsequent chemical and ecotoxicological analyses. Activities of phase I (ethoxyresorufin-O-deethylase) and phase II (glutathione-S-transferase) biotransformation enzymes, chemical analyses of pollutant metabolites and histopathological alterations were studied in situ in the fish. The PAH metabolite 1-hydroxypyrene was detected in bile of fish from all sites. All end points in combination with the chemical data suggest a pivotal role of PAHs in the observed ecotoxicological impacts [6].

Benthic organisms are sensitive indicators for contamination because of their long life history and relative stable residence and thus can provide an estimate of the bioaccumulation of organic pollutants. *Bellamya aeruginosa* (gastropoda) and *Corbicula fluminea* (mollusc) were collected using a Peterson grab sampler in East Taihu Lake (China). The organisms were washed with distilled water and classified into five groups according to weight. All samples were stored in a deep freezer at $-20\,°C$ until extraction. *B. aeruginosa* and *C. fluminea* were defrosted, and the flesh of each size grade was Soxhlet extracted and freeze-dried for organochloride pesticide analysis.

Table 4.1: Sampling techniques for aquatic organisms [5].

Sampling technique	Organisms	Advantage	Disadvantage
Collection by hand	Macrophytes, attached organisms	Cheap – no equipment necessary	Only specific organisms collected
Hand net on pole (ca. 500 mm mesh)	Benthic invertebrates	Cheap, simple	Mobile organisms may avoid net
Plankton net	Phyto- and zooplankton (ca. 60 mm mesh)	Cheap and simple. High density of organisms per sample	Selective according to mesh size. Damage to organisms possible
Bottle samples	Phytoplankton, zooplankton (incl. protozoa), microorganisms	Enables samples to be collected from discrete depths. No damage to organisms	Low density of organisms per sample. Small total volume sampled
Water pump	Phytoplankton, zooplankton (incl. protozoa), microorganisms	Rapid collection of large volume samples. Integrated depth sampling possible	Sample may need filtration or centrifugation to concentrate organisms. Damage to organisms possible
Grab (e.g., Ekman, Van Veen)	Benthic invertebrates living in, or on, the sediment. Macrophytes and associated, attached organisms	Little disturbance to sample	Requires winch for lowering and raising
Dredge type	Mainly surface living benthic invertebrates		Mobile organisms avoid sampler
Fish net/trap	Fish	Cheap. Nondestructive	Not selective
Electrofishing	Fish	Nondestructive	Selective technique according to the current used and fish size. Safety risk to operators

Concentrations of the contaminants and lipids in the organisms increased with body weight. Hexachlorocyclohexane analogs (HCHs), Dichlorodiphenyltrichloroethane analogs (DDTs), chlordanes and heptachlors were the dominant compounds detected in the organisms [7].

Zooplankton samples can be analyzed for the presence of pollutants after collection by use of a plankton net sampler (mesh 126 μm; Figure 4.3), sinusoidally filtering the water column of 0–50 m corresponding to the layer where the majority of zooplankton lives.

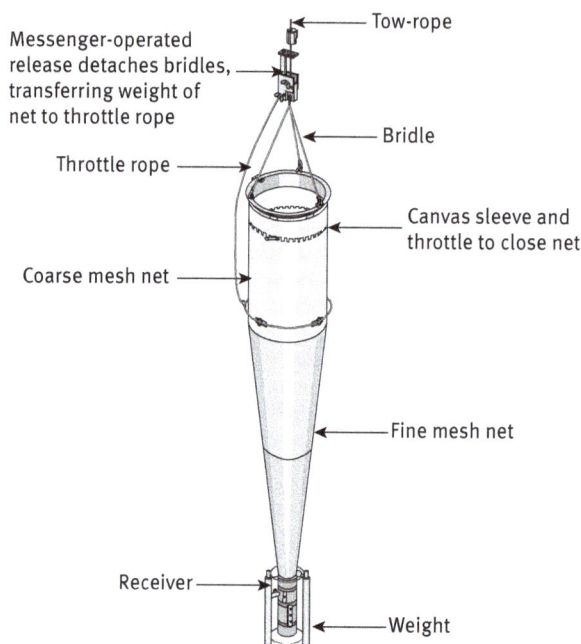

Figure 4.3: A plankton net sampler.

Microscopic analysis of zooplankton caught by this method showed that the samples were mainly composed of animals, although some large algal species and algal aggregates were also present. Freeze-dried zooplankton samples were introduced into a glass fiber thimble and Soxhlet extracted. The zooplankton extracts were characterized for lipid content, and chlorinated compounds were recovered by several *n*-hexane washings. Since most of the pollutants were undetectable in particulate matter samples collected at each depth separately, all the extracts of the entire water column were combined and the solvent volume was reduced to one-tenth. The detection limit for zooplankton was 1 ng/g lipid weight for each DDT residue. Chemical analysis revealed that the lipid normalized concentrations of Dichlordiphenyldichloroethylene (DDE) and Dichlorodiphenyldichloroethane (DDD) in zooplankton exceeded the levels of the same compounds in zooplanktivorous fishes [8].

4.2.2.2 Terrestrial and Aerial Organisms

Mammals Sampling of terrestrial and aerial organisms is performed either by collecting and analyzing dead animals or by different catching tools. As an example for the first strategy, Lopez-Perea et al. sampled 344 individuals representing 11 species of predatory wildlife that were found dead in the Mediterranean region of Spain. The diet of many wild predators is based on rodents that are often controlled or eradicated with rodenticides, which makes these predators susceptible to secondary poisoning by contaminated prey. The animals were found dead or moribund. Liver samples were taken and immediately frozen at −20 °C until chemical analysis using gram quantities of liver. Rodenticide analysis was performed by liquid chromatography coupled to mass spectrometry. Six different rodenticides were found in the liver of more than 50% of the animals [9]. Birds and mammals feeding on such small organisms will accumulate these residues.

Insects   The transfer of applied pesticides to predatory arthropods needs to be considered in the risk assessment of such chemicals. More specifically, transfer of pesticides to predatory arthropods may occur by ingestion of contaminated food: Gray garden slugs were sampled and kept in covered plastic boxes lined with moist potting soil and fed by organic cabbage. Adults of the ground beetle *Chlaenius tricolor*, an important slug predator, were collected from an agricultural farm using dry pitfall traps (Figure 4.4) and hand collection. Chemical analysis confirmed that neonicotinoid residues travelled up the food chain from soya beans to slugs to beetles; however, the pesticide residues declined exponentially along the food chain, at a similar rate in laboratory and in field studies [10].

Other techniques to sample terrestrial insects are briefly described: beat sampling, Berlese leaf litter sampling and light-trap sampling. In shake or beat sampling, a plastic cover or a plastic plate is placed below a shrub or tree from which insects will be analyzed. After heavy shaking of the branches or by beating with a stick, most insects will fall into the collection receptacle beneath where they can be separated and gathered for chemical analysis. Berlese funnels (Figure 4.5) are used to extract insects

Figure 4.4: A pitfall trap to collect terrestrial insects. A container is buried in the soil, covered by a lid leaving a space so that insects will fall into the container. Dry traps contain no liquid in the container and have to be collected daily. Wet traps contain a liquid for rapidly killing the insects. Since the insects will be analyzed later for chemical contaminants, only aqueous media should be used, eventually containing low amounts of detergents for rapid drowning. For biological studies, for example, to study the abundance and diversity of insects, the traps contain liquids as formalin (10% formaldehyde), alcohol, ethylene glycol or picric acid.

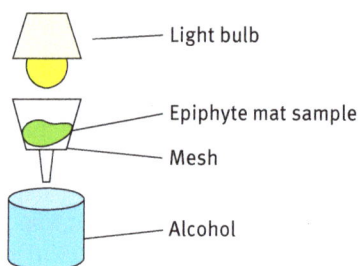

Figure 4.5: Berlese technique to collect insects from soil or litter.

from soil and litter samples. Since insects and other arthropods living in soil or litter avoid light, the principle is to force them to move downward because of a light source (light bulb) placed above the sample, from which they fall into a container with a killing agent or to sample them alive.

In light-trap sampling, flying nocturnal insects are collected. Many insects perceive light from yellow-green (550 nm) to ultraviolet (340 nm), the latter being used in private homes to attract annoying insects which will be killed on an electric grid. For sampling insects for subsequent chemical analysis, different light sources are available. The attracted insects enter into a collecting device, for example, a funnel with a large enough mouth to let them drop through easily, but small enough to prevent flying out again.

It was shown that systemic, soil-applied neonicotinoid insecticides are translocated to pollen and nectar of flowers and thus may reduce survival of flower-feeding beneficial insects. Imidacloprid residues in flowers of *Asclepias curassavica* from treated soil reached several thousands of ppb levels. These high imidacloprid concentrations caused significant mortality in lady beetle species feeding on pollen and nectar. The authors did not analyze the pesticide residues in the insects but only in the plants [11]. Imidacloprid was shown to impair the olfactory memory of foraging bees even after short exposure even at low concentrations (e.g., 10 nM) [12].

Ivermectin residues in cowpats reduced colonization by typical dung fauna. Animals were hand collected from the cowpats and especially Diptera (two winged insects like flies and midges) were present in significantly lower numbers. The ivermectin level in the feces from pour-on treated cattle 1 day after treatment with the pesticide averaged 22 mg/kg dry weight and leveled off after about 20 days to 1–22 mg/kg dry weight [13].

The stingless bee species *Scaptotrigona* aff. *depilis* has been investigated as a potential indicator of environmental pesticide contamination. Using LC–MS, the authors developed a protocol for residue extraction and detected neonicotinoids in the bee bodies. One hundred bees were sampled from each colony that returned to the colony after foraging. Low concentrations of the pesticides were detected in part of the sampled bees [14].

Honeybees are also good indicators of PAH-contaminated environments. Forager bees were caught at the entrance of the hive. Phenanthrene showed the highest mean concentrations in honeybees up to about 2 ng/g wet weight [15].

Birds   Usually, analysis of contaminants in birds is performed on their carcasses after recent accidents like kite-flying injuries and not by trapping living animals for reasons of animal welfare. Manifold capture techniques are available, most of them involving the use of food, water or decoys to attract the birds: clap traps (spring-loaded frames with nets), funnel traps (birds are lured or driven through a narrow entrance hindering the way back), cannon nets (large net which is closed by "cannons" or rockets that shoot a weight that drags a net behind it), mist nets (fine net hanging in the flight path of a bird across trees) or spot light trapping (stunning birds with bright light beams). Such catching techniques are mainly used for labeling the birds with stamped metal rings or with radiobearing devices to learn about their preferences in the habitat with residence times, feeding behavior and so on. From such studies, the potential exposure of birds can be deduced, for instance those living in agricultural landscapes with respect to pesticides. Any catching technique must be performed in a way to avoid foreseeable risks of injury, for instance by disturbing nesting birds at breeding sites or enhancing vulnerability to nest site predation following human intrusion. Birds captured should not remain in traps or nets any longer than is necessary which means that the traps should be checked as often as possible. The checking times are capture technique and weather dependent and range from every quarter of an hour to few times a day.

The population of many farmland bird species is declining due to the overall agricultural intensification and its effects on habitat and food resources. Pesticides may affect farmland birds either directly by poisoning, or indirectly [16, 17], especially by reducing weeds and invertebrates as important food resources for granivorous and insectivorous farmland birds.

A radiotracking survey of gray partridges was performed in an agricultural landscape to study their potential exposure to pesticides during the movements and feeding of the birds. Female partridges were caught at night with a hand-held net attracted and dazed with a strong light and marked with a metal leg ring fitted with a radiotransmitter (less than 5 min handling time). Residue analyses of pesticides were performed on all available carcasses (liver, gizzard/stomach contents and encephalon) to compare the actual exposure (pesticide residues detected in a tissue of a partridge) with the estimated exposure (calculated basically by the amounts of pesticides applied and the times and amounts that the birds were feeding on the agricultural site). 77% of crop types ($n = 30$) were sprayed at least once in the season by pyrethroids and 53% by triazoles but often the pesticides were used repeatedly. Extraction of the organs was performed by use of organic solvents. At least one active ingredient was detected in 29% of the carcasses analyzed (20 chemical pesticide classes were screened). The concentrations detected varied from 0.01 to 0.11 µg/g.

Modeling estimations suggest that the lethal pesticide-related poisonings decreased the population growth rate by less than 1% [18].

Amphibians Declining amphibian populations due to habitat loss in agricultural landscapes, diseases and potentially exposure to toxic chemicals have been observed worldwide [19]. Amphibians are not specifically considered in the pesticide risk assessment but assumed to be covered by the risk assessment for aquatic invertebrates and fish, or mammals and birds, respectively. However, the sensitivity of these animals after direct (pesticide overspray and drift) or indirect contact (exposure to soil containing pesticides) might be different compared to other organisms since it is known that the percutaneous passage of chemicals is higher in amphibians. Dermal absorption of chemicals is considered therefore the main exposure pathway of amphibians [20]. Also, the potential uptake of chemicals in amphibians is more likely since their skin-surface-to-body ratio is maximized to allow percutaneous gas, water and ion exchange with the environment, whereas it is minimized in mammals to avoid temperature loss. Literature data and specific exposure experiments by overspray suggest that amphibians can be sublethally or even lethally affected by field-relevant terrestrial pesticide application rates. Monitoring programs of amphibian migration should be performed combined with adjusted pesticide management to reduce temporal coincidence and thus potential exposure [21].

4.2.3 Human Samples

The uptake of environmental pollutants by dermal absorption, inhalation and ingestion will lead to a body burden of the chemical which is determined by the physical and chemical properties of the chemical, the exposure time and probably the individual susceptibility of the exposed organism. Analytical methods for human biomonitoring – the direct measurement of people's exposure to environmental contaminants by measuring them or their metabolites in human specimens – have been described and often reviewed (see Further Reading section). We here will refer to pollutant analysis in blood and plasma (invasive sampling), hair and urine (noninvasive sampling). Other often used matrices for noninvasive human biomonitoring are nails, breast milk, saliva and meconium (the neonate's first feces) with none of them being useful in every situation. Criteria for ideal analyses are the accessibility of samples in sufficient amounts, no health risk for the donor, pollutant levels detectable by the techniques available and ease of collection and storage.

4.2.3.1 Hair
The use of hair as a matrix for environmental analysis has a long history and many applications such as forensic, historical and clinical analysis and biomonitoring of the chronic exposure of humans to inorganic and organic pollutants. Main advantages

are the easiness of sampling and storage because this matrix is rather stable. A disadvantage is the limited weight of material often in the range of 50–500 mg compared to other matrices for human biomonitoring such as urine or blood and serum or breast milk which can be sampled in gram quantities. Higher amounts of hair in gram quantities for pollutant analyses have been reported [22]. Prior to analysis, hair samples may be cut into mm pieces or even powdered by a ball mill to optimize the recovery of the target analytes.

Often the concentrations of pollutants in hair are correlated to other human matrices [23] (fluids and tissues) showing that hair analysis reflects the internal dose of pollutants in humans although some doubts exist because of the possibility of external deposition of chemicals on the hair surface. However, differences in pollutant profiles in hair and other human matrices have also been reported: The DDT metabolite profiles in hair were different from those in breast milk and placenta [24]. It is assumed that biologically incorporated (inhalation, oral consumption) molecules are located inside hair shafts (cortex and medulla) while external contamination is likely to be located externally on the hair cuticle. The concentration of polybrominated diphenyl ethers (PDBE) increased from proximal (root end) to distal along the hair shaft which may be correlated to dietary and hair care [25]. Anyway, hair can be decontaminated from externally deposited pollutants by gentle washing with water and shampoo or even organic solvents, although it is somewhat unsecure whether externally deposited chemicals are removed, and on the other hand, no internally incorporated chemicals are completely removed by washing.

Appenzeller and Tsatsakis reviewed the available literature on hair analysis of organic pollutants and estimated the median concentrations of organochlorines, organophosphates, pyrethroids, PDBE, PAHs and PCBs in the range of below 1 to several 1,000 pg/mg hair, for dioxins and furans 0.001–0.1 pg/mg hair [26]. Pesticides and particularly organochlorine pesticides are by far the pollutants most often investigated in human hair. As for PAH, biomonitoring of human exposure often relies on the detection of urinary metabolites [27, 28] or those in white blood cells [29], but hair has also been investigated for the detection of PAH [30] and their monohydroxy metabolites [31].

4.2.3.2 Blood and Plasma

Biological fluids such as blood, plasma or urine contain many endogenous components that may affect the subsequent separation and characterization of contained pollutants. Therefore, sample preparation steps are often the major source of error in the overall analytical process. Typical cleanup steps comprise liquid–liquid extraction, liquid–liquid micro-extraction, solid-phase extraction (SPE), solid-phase micro-extraction and molecularly imprinted SPE. The main disadvantage of using blood in human biomonitoring is that it is invasive and the need for qualified staff. In general, about 1–5 mL of blood is necessary for analysis, and an extensive cleanup is required. Blood or its components (serum or plasma) are preferred when the internal exposure to persistent pollutants with long biological half-lives is analyzed. By contrast, nonpersistent compounds, which are rapidly transformed to polar metabolites, are found in low concentrations in blood, and urine is the matrix of choice.

Paracetamol was analyzed with high sensitivity in human blood plasma, obtained from a blood transfusion service, by magnetite nanoparticles which were coated with an imprinted methacrylate polymer to which the compound partitioned and from which it can be extracted for chemical analysis [32].

Trace levels of pesticides can be analyzed by help of the addition of internal standards labeled with stable isotopes (for instance deuterated compounds): As an example, endosulfan residue concentrations in blood were in the range of sub-µg/L to several µg/L [33].

Analysis of polybrominated diphenylethers in hair and serum of workers involved in the recycling of electronic waste revealed positive relationships in tri- to hepta-congeners between hair and serum concentrations, but this relationship was not found for octa- to deca-congeners, probably due to the longer persistence of highly brominated congeners compared to less brominated congeners in the human body [34].

4.2.3.3 Urine

Urine is the classical matrix for the assessment of human exposure to organic pollutants. Although this matrix allows for the detection of several compounds, mainly excreted metabolites, the kinetics of urinary compounds makes it representative of only the recent exposure for many chemicals. Typical enrichment of nonpolar contaminants can be achieved by SPE, solid-phase microextraction and of polar substances by mixed liquid–liquid extractions [23, 35].

Urine and hair samples were taken from agricultural workers involved in terbutylazine application during the application season. Each person was provided with instructions for urine collection and sampling tubes and was instructed to record the sampling times. The collected samples were chilled and stored at 4 °C and picked up by the investigators within 48 h. The samples were delivered to the laboratory and stored at −20 °C until analysis. In urine samples taken before going to bed on the day of application, the levels of the metabolite desethylterbutylazine were three- to ninefold higher (in the µg/L range) than those detected in samples from the next morning before the work shift, indicating fast absorption, distribution, biotransformation and excretion of the herbicide. The herbicide itself was only detected in part of the samples. Hair samples of the individuals only contained the parent compound in sub-ng/mg concentrations [36].

4.2.4 Plant Sampling

Chemical residues in food and feed commodities are gaining a lot of attention due to the potential of contamination of humans and animals with subsequent health problems [37]. Thus, the European Food Safety Authority and other authorities world-wide evaluate the risk of food and feed contamination in compliance with the defined maximum residue levels which are believed to ensure uncritical nourishment but are different in the various legislations. Analysis of contaminants in vegetation is a challenge due to the complex and highly different biological matrices, often rather low concentrations of the analytes and the presence of co-extracted matrix constituents that might interfere with the identification of the pollutants [38].

High variations of pollutant concentrations are due to different exposure (imagine spray drift of pesticides reaching bunches of grapes or leaves of apple trees where the interior grapes are less contaminated than the outer part), to different transport mechanisms within the plant, for instance higher concentrations in the roots compared to shoots and leaves, different seasons and growth stages. This necessitates a high number of replicates, often in field studies up to about 100 per hectare each with at least about 0.5 kg (see Table 4.2). For sampling leaves or grass, the use of pruners is indicated or hand-plucking (wearing gloves); for branches, saws or corers can be used. Wrapping the product in aluminum foil immediately after harvest is indicated since plastic bags, for example those made of polyethylene, might interfere with the integrity of the sample.

Diseased or under-sized crop parts or those at a stage when they would not normally be harvested should not be sampled; rather, sampling should be representative of typical harvesting practice. Loss of surface residues during handling, packing or preparation should be minimized. Individual fruits and vegetables should not be cut or divided.

Fruits from all segments of the plant, high and low, exposed and protected by foliage should be taken to determine the variance of contamination. For small fruits grown in a row, fruits from both sides should be selected. Bulb, root and tuber vegetables should be sampled from all over the field except at the very edges. Adhering soil should be removed by brushing or rinsing with cold water. Notice that part of the roots will be lost when removing the attached soil too vigorously.

Table 4.2: Typical amounts of samples needed for trace analysis, for example, of pesticides.

Commodity	Ca. amounts, sampled from different places, plants
Citrus fruits, pome fruits, large stone fruits	10–15 fruits
Small stone fruits	1 kg
Grapes, small berries, strawberries	0.5–1 kg
Fodder beets, sugar beets	10–15 plants
Potatoes, other root crops	2 kg
Brassica crops	2 kg
Cereal grains	1 kg
Herbs, teas, hops	0.2–0.5 kg
Maize cobs	2 kg

During shipment to the laboratory, the samples should be cooled. For analysis, the samples should be made up of individual subsamples taken randomly throughout the lot and of a similar size. Subsamples can consist of a single fruit or vegetable for larger items, or a single bundle for grapes or a small scoop for small items (peas, berries, etc.). For cereals, a number of subsamples (at least 10) from different places should be taken.

In anticipation of Chapter 6, where such techniques are described in more detail, common procedures to determine pesticide residues from plant samples comprise solvent extraction, liquid–solid extraction using Soxhlet, shake-flask, sonication, microwave-assisted extraction, pressurized liquid extraction and supercritical fluid extraction. Besides the QuEChERS extraction method, eventually modified, extractions with sorbents, that is, solid-phase micro-extraction, stir bar sorptive extraction and matrix solid-phase dispersion, have been used (see Further Reading section an some specific examples [39, 40]). Pine needles may be analyzed to assess the atmospheric presence of semi-volatile organic contaminants. A multicomponent protocol to quantify brominated flame-retardants, PCBs, organochlorine pesticides, polynuclear aromatic hydrocarbons and synthetic musks' fragrances has been developed. Pine needle extracts were obtained by ultrasonic solvents extraction of 5 g of pine needles cut into 1-cm pieces, and solid-phase extraction employing combinations of sorbents and solvents as well as gel permeation chromatography was tested [41].

4.3 Trace Gas Sampling

4.3.1 General Remarks

One common property of smaller organic molecules is their high volatility, in comparison to inorganic compounds of similar molecular mass. Therefore, we have to anticipate the presence of many analytes in both the gas and liquid states, or even in the solid phase. Depending on the accompanying matrix constituents, such partition equilibria are changing under the influence of external forces. Of course, the usual dependency on physicochemical variables, such as temperature and pressure, will influence these equilibria too.

A nice example for this is the presence of alkyl aldehydes (e.g., formaldehyde) under a photo-chemically triggered situation in the atmosphere. Without any stabilizing compound, we will expect this analyte under ambient conditions to the full extent in the gaseous state. But, if small fog droplets with dissolved sulfur dioxide are present, a chemical complex formation starts and methane sulfonic acid will be formed. Within this complex, the volatile molecule formaldehyde is now protected and caught within the liquid phase. Similar complex formations are observed with amines, terpenes or carboxylic acids. We see that sampling must follow these different forms of appearance. Other cases are semi-volatile compounds, for example, PAHs may cover many orders of magnitude in vapor pressure. So, if someone is looking to collect a wide variety of PAHs, he has to meet the requirements for such ambiguity in appearance (see Chapter 11.1).

Nowadays, typical fields of application are sampling of small gas volumes from working places or pestering malodorous situations for further evaluation by subsequent chemical analysis or an olfactory panel. Upcoming is the usage of exhaled breath air analysis for rapid monitoring of the health status of patients.

4.3.2 Gas Grab Sampling

Very often trace gas sampling has to take precautions in selecting the appropriate inlet configuration and the materials to be in contact with the gas phase. Especially when longer tubing is needed to transfer the gaseous analyte of interest from an air intake above roof of a house or from top of a meteorological observation tower to a lab container below, the nature of the respective molecule has to be considered. The same holds true for bag sampling as non-enriching recipient, where a certain gas volume is fed into an evacuated bag of polymeric material (see Figure 4.6) or a "glass mouse" for grab sampling. Often these containers are equipped with septa for taking a subsample for gas chromatography. In special cases, evacuated and electropolished stainless steel canisters provide optimal storage conditions for later chemical analysis. Polar molecules have a high tendency to stick to hydrous surface films formed on stainless steel capillaries or glass tubings. In contrast, PVF (polyvinyl fluoride: Tedlar™) is a very good choice for polar gases but offers a sink for nonpolar substances. Both situations result in severe losses in the transfer rate. A good solution is the usage of metal-layered bags, consisting of several polyethylene and aluminum layers, thus also preventing permeation losses. In Table 4.3, the recommended selections of layered or non-layered bags are listed. A storage time of several days should not be exceeded but has to be checked by separate storage experiments.

Figure 4.6: Plastic bags for gas grab sampling.

Table 4.3: Recommended gas sampling bag material.

Method/Application	Compounds	Tedlar™	Supel-Inert™ Foil
EPA 18	Gaseous organic compounds; VOCs by GC	X	NR
EPA 0040	Volatile organic compounds (VOCs)	X	NR
EPA TO-3	Volatile organic compounds (VOCs)	X	NR
EPA TO-12	Non-methane organic compounds (NMOC)	X	NR
EPA TO-14A/TO-15 mod.	Volatile organic compounds (VOCs) by GC/MS	X	NR
ASTM D5504-01/-08	Sulfur compounds in natural gas & fuels by GC	X	NR
NIOSH 3704	Perchloethylene (tetrachloroethylene)	X	NR
NIOSH 6603	Carbon dioxide	NR	X
OSHA ID-172	Carbon dioxide in workplace atmospheres	NR	X
OSHA ID-210	Carbon monoxide	NR	X
Landfill/Biogases (LFG)	Hydrogen, hydrogen sulfide, methane, nitric oxide, oxygen	NR	X

NR, not recommended; X, recommended.

It is a must to check any sampling configuration with test gases before. To apply test gas concentrations within the expected range is also mandatory.

Important is the parallel registration of the physical conditions during gas collection: pressure p_a [hPa], and temperature t_a [°C]. This is needed to calculate and correct the measured analyte concentration c to c_0 under standardized conditions (20 °C, 1,013 hPa):

$$c_0 = c \times \frac{273 + t_a}{293} \times \frac{1,013}{p_a} \tag{4.1}$$

This is necessary since all gas standards and threshold values given are always independent of pressure and temperature.

4.4 Rain, Surface Water and Particulate Matter Sampling

4.4.1 General Remarks

Sampling of rain and the many types of surface waters (run-off water, rivers, lakes, ditches, sea, industrial effluents, etc.) needs different devices that are adequate for the various situations. Some examples of the methods that may be used are given below.

4.4.2 Rain

Rainwater can be collected in automatic wet-only samplers which open only during rain events and close in dry periods, first to reduce evaporation and second to minimize contamination by dust and other matter. So-called bulk collectors, in the easiest version a bottle with an inserted funnel (Figure 4.7), accumulate both water and solid matter over a certain period of time. The collection of solid matter in such devices can be minimized by restricting the sampling time on rain events and/or using a mesh screen. Since bulk collectors do not need electric power, they are preferred in remote places.

Using bulk samplers, organochlorine pesticide residues in rainwater have been analyzed: Sampling was performed in stainless steel containers, glass bottles and funnels. Each rain event was sampled separately. After collection, the samples were transferred to the laboratory for sample cleanup. DDT, DDE, DDD, HCHs, dieldrin, heptachlor, chlordane, endrin and hexachlorobenzene have been detected in ng/L–μg/L ranges depending on the vicinity of contaminated sites [42]. Wet-only samplers were used to analyze rain in the Mississippi Delta comparing concentrations over a decade between 1995 and 2007: Concentration ranges were similar between years, however,

Figure 4.7: (Left) Rain bulk sampler for collecting rainwater and particulate matter. (Right) Wet-only rain water sampler with several polyethylene sampling bottles (to be switched motor-driven), motor-driven precipitation-controlled funnel lid, ambient temperature sensor and programmable climatization (heating/cooling).

even though the 1995 sampling site was 500 m away from active agricultural fields whereas the 2007 sampling site was within 3 m of a field. Atrazine, metolachlor and propanil were detected in >50% of the rain samples in both years. The total herbicide flux in 2007 was slightly greater than in 1995 and was dominated by glyphosate, malathion, methyl parathion and their degradation products [43].

4.4.3 Surface Water, Deep Water, Groundwater

For planning the sampling in water, it has to be considered that the concentration of a pollutant may change over time, for example, seasonal by growth of plants, biofilms, weekly by changes in emission from a chemical plant (eventually no emission during the weekend), or daily by photolytic or biotic degradation. Depending on the size of the sample – which is triggered by the analytical technique and its limit of detection/limit of quantification to be used – different types of containers are available, often made of glass or polyethylene. Sorption of the analyte to the container material and/or to the container lid is possible, so the material must be washed with an organic solvent after the sample has been removed. If volatile analytes are sampled from aqueous media, the container should not contain a head space but should be filled completely. On the other hand, a partly filled sampling bottle (for analysis of nonvolatile substances) can easily be shaken to homogenize the sample which is difficult to achieve in a completely filled bottle.

When taking water samples, large particles like leaves and detritus should be removed. To avoid contamination by pollutants concentrated in the thin (≤ 1 mm) surface films, often containing lipids, microorganisms and oil contaminants, devices that open under water should be used. The concentration of organic pollutants in surface films may differ significantly from that of the water below [44, 45]. If focused on the chemical composition of the surface layer, special devices are available for sampling [45]. Benson et al. [46] investigated the concentration of PAH in the sea surface microlayer by using a screen sampler (consisting of a wire mesh of about 10–16 wires per inch); two handles were attached to a framework, which supports the screen having a dimension of 65 cm × 45 cm. Surface microlayer water samples were collected by holding the rectangular screen by its handles in a horizontal position and parallel to the ocean surface for about 5 s after which the screen was withdrawn and then replaced. This was done several times before taking each sample in order to condition the screen wires to the chemical substance in the water. The screen was then withdrawn from beneath the surface of the water through the sea–air interface while still maintaining the horizontal position. As the screen is raised through the water below the surface, seawater merely flows between the wires and when passing through the surface, thin segment of the surface layer between the wires was then immediately tilted toward one rear corner of the frame so that the collected surface water can drain into a glass collection container [46].

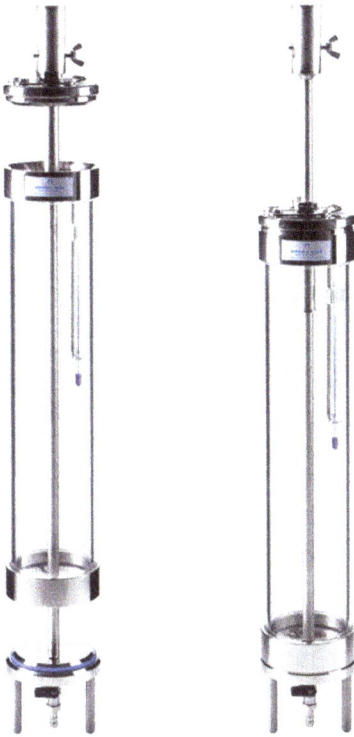

Figure 4.8: Ruttner sampler (left: open, right: closed) for collecting water from different depths suitable for taking samples in lakes, sea, water borings, wells and so on. Sampler volumes cover a range of about 1–5 L.

Several types of sampling devices are commercially available to collect samples from water at different depths. Samples in small streams at limited depth can be taken by immersing a bottle from a boat with the open side downward and turning it at the desired depth. To reach deeper water layers, the flask can also be held by a telescope bar. The Ruttner sampler (Figure 4.8) consists of a hollow PVC cylinder and two interconnected rubber valves at the ends. The two valves are pulled out and the strings connecting the valves fixed outside of the cylinder. The open cylinder is immersed into the water by use of a rope with an impregnated waterproof meter ruler. At the desired depth, a heavy metal cylinder is released by hand along the rope which will hit the lock fixing the valve strings. The valves will immediately close the Ruttner cylinder which is then retracted from the water. Other devices such as the Kemmerer or van Dorn sampler have a similar principle (see Further Reading section). Alternatively, samples may be taken at different depths using pumps and a plastic tube that is lowered to the appropriate depth.

Groundwater is usually sampled through monitoring wells that were installed for that purpose. If new boreholes at a site without monitoring wells are necessary, several drilling methods may have to be applied to perforate the below-ground strata.

Strategies for groundwater monitoring have been reviewed (see Further Reading section). For collection of a sample, the well must be purged (if the well is narrow with a diameter below about 5 cm and with a length below about 5 m). Otherwise the water must be pumped.

The Geoprobe groundwater sampler (Figure 4.9) is one example that allows using a bailer, bladder pump or other devices for sampling the aqueous phase.

4.4.4 Particulate Matter Sampling

Suspended particulate matter (SPM) in water is an important component of water-sediment systems regarding inorganic and organic pollutant transport. It has been shown that SPM, for instance, after resuspension of pollutant containing sediment particles during flood events or during dredging – sediments are the ultimate sink of pollutants in rivers, lakes and sea – may exert toxic effects on exposed fish [47]. The quantity and quality of SPM are specific for each water body and depend on the respective catchment area (geology, land use, urbanization, etc.), season, water flux and weather. SPM is therefore crucial in assessing the contamination of surface waters. The European Water Framework Directive (Directive 2000/60/EC) requires therefore explicitly an investigation of the whole water sample including SPM, because many priority and priority hazardous substances can sorb substantially to SPM.

SPM is sampled using mobile and stationary centrifuges, glass fiber or membrane filters (<0.7µm), stationary sedimentation tanks or mobile passive sedimentation boxes. An example for the latter has been described as a stainless steel box with dimensions of 400 mm × 250 mm × 300 mm (length × height × width) (see Figure 4.10) [48].

The box can be exposed directly in the surface water mounted at piers, sheet pile walls, pontoons or buoys using chains. Dynamic hanging points like pontoons or buoys are preferable to avoid changes in the depth of hanging of the sedimentation box (0.5–2.0 m below the water surface). The surface water is flowing through inlet openings into the box, baffles reduce the velocity and the SPM settles down into the sedimentation basins inside the box. After a certain exposure time, for example a week, the sedimentation basins are taken out, overstanding water is decanted cautiously to minimize loss of fine SPM. SPM is transferred through a stainless steel sieve (<2 mm) into a stainless steel freezing basin to preserve the contaminant status and homogenized using a Teflon™ spatula. Using this device, multiple analyses have been performed to analyze the SPM loads of PAH, chlorohydrocarbons, and biocides, described by the Environmental Specimen Bank of the German Environmental Protection Agency UBA (http://www.umweltprobenbank.de/en/documents).

Others filtered water samples with a glass fiber filter (0.7 µm) after being transported to the laboratory in order to analyze the PAH content of SPM [49, 50].

1

Drive Cap

Probe Rod

Drive Head

Sampler sheath

Expendable drive point (Steel or Aluminum)

2

Operator holds down on extension rods to keep screen in place as tool string is retracted

Extension rods pass through sampler to bottom of screen

Stainless steel or PVC screen

Screen Push Adapter

Expendable drive point remains downhole

Drive Head Top O-Ring

Screen Head O-ring

Drive Head

Drive Head Bottom O-Ring

Screen

Sampler Sheath

Expendable Drive Point O-Ring

Expendable Drive Point

PE Grout plug

3

Operator oscillates tubing up and down to bring groundwater sample to surface

Poly tubing extends through sampler to bottom of screen

Exposed Screen

Check Ball

Tubing

Check Valve

4

Nylon tubing connects to the Grout machine

Tool string is retracted as grout is pumped from lower end of screen

Nylon tubing carries grout to bottom of tool string

Internal threads attach grout nozzle to nylon tubing

Grout Nozzle

Grout is discharged below screen

Grout plug is knocked from bottom of screen before grouting

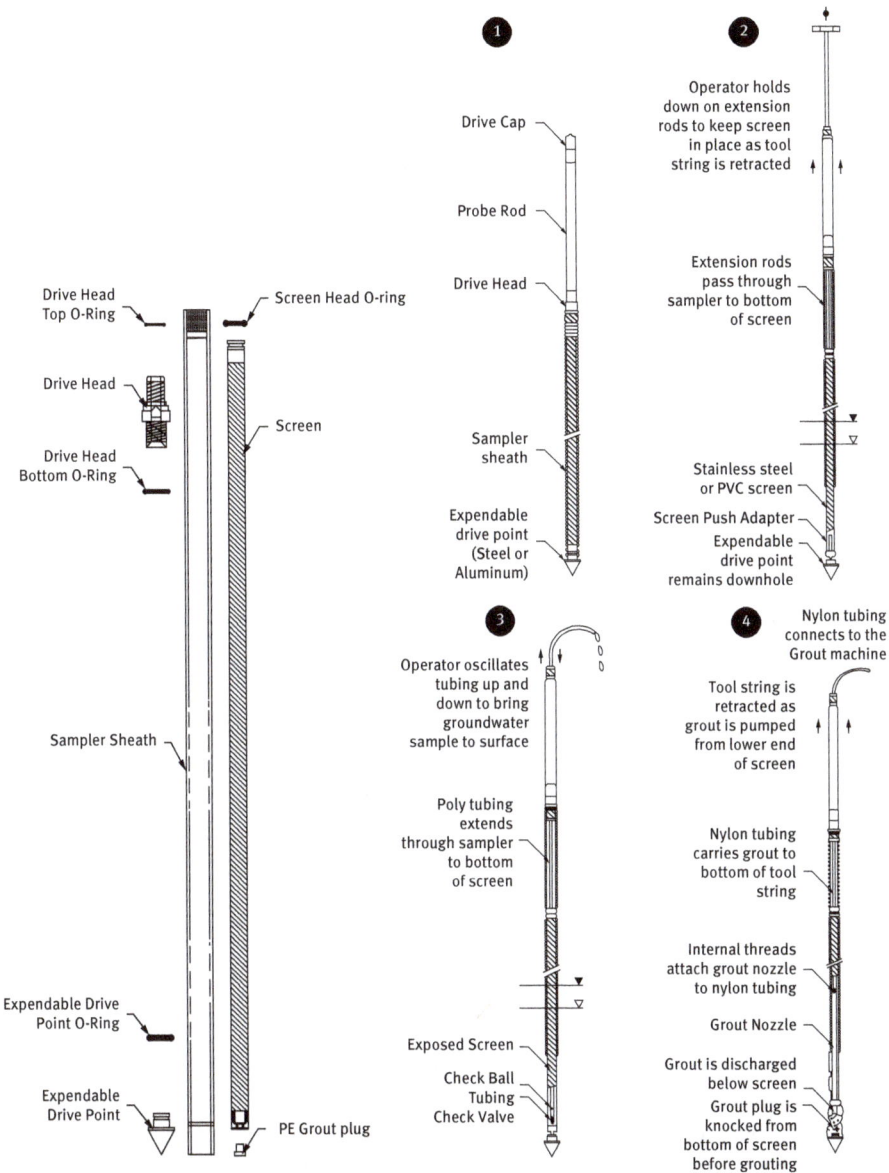

Figure 4.9: The Geoprobe groundwater sampler. **1.** Driving: The assembled Screen Point 16 Groundwater Sampler is driven to the desired sampling depth. **2.** Deployment: Extension rods are used to hold the screen in position while the rods and sheath are retracted. The screen sheath forms a mechanical annular seal above the screen interval. **3.** Sampling: The tubing check valve can be used to sample and measure non-aqueous phase liquids (NAPLs) within the screen interval as well as sample groundwater. A small bladder pump may be used to collect high-integrity volatile organic compound (VOC) samples. **4.** Grouting: Abandonment grouting can be conducted to meet ASTM guidelines. A high-pressure grout pump is used to pump grout into the probe hole as the screen and rods are extracted using the rod grip pull assembly.

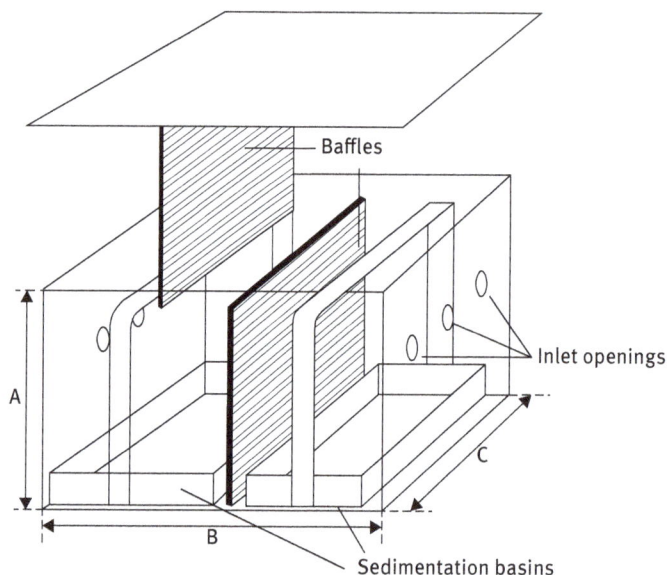

Figure 4.10: Sedimentation box for sampling suspended particulate matter in water sediment systems [48]. Reproduced with permission from Springer.

An SPE disk method was developed for the determination of 54 hazardous pollutants including PAHs, PCBs, PBDE, organic chlorinated pesticides and other pesticides in surface water containing SPM. The method allows analysis of 1 L surface water containing up to 1,000 mg SPM without prior separation of SPM [51].

4.5 Soil and Sediment Sampling

4.5.1 General Remarks

Soils and sediments are the most important solid environmental sinks of chemical pollutants. Chemicals entering soils deliberately (e.g., pesticides) or unintended (e.g., by accidents or by addition of polluted sewage sludge) often bind to the first centimeter layers by interacting with the inorganic and organic soil components. The highest residue amounts are usually found in the organo-mineral fraction of soil. Depending on the physicochemical properties of the pollutants parts will leach to lower soil layers or to the groundwater table, a process facilitated by preferential flow through macropores in the soil made by earth-dwelling organisms (e.g., earthworms), drying cracks, freeze–thaw cycles or plant roots.

Binding to soils and sediments stabilizes chemicals against microbial degradation; therefore, soils do not only act as a sink for pollutants but at the same time also as a long-term source of pollutants upon their release. Remobilization of type I NER (see Chapter 6, Section 6.3.4) may lead to the slow release of entrapped chemicals and/or metabolites [52]. Choi and Wania [53] modeled the long-term emission of chemicals from soils to atmosphere and water resources, a process described also by others [54].

Also chemicals with rather short dissipation half-lives as determined in classical lab and field degradation studies may reveal persistence: The herbicide atrazine (dissipation half-life < 100 days) was banned in Germany in 1991. However, atrazine and the primary metabolite hydroxy-atrazine are still, 25 years after the ban, present in groundwater-monitoring sites in concentrations, which are not in line with the dissipation half-life [55]. Soils at geologically and hydrologically predestined sites obviously play a role in the stabilization of herbicides [56]. Persistence, thus, is not only an inherent property of a chemical but is also affected by environmental processes. Also, the binding of pollutants to sediments leads to their long-term persistence [57].

4.5.2 Soil

The depth of sampling depends on the purpose of the analysis: For pesticides, the concentrations in the uppermost few centimeters (1, 2.5, 5 cm) are important parameters used for monitoring the "predicted environmental concentration (PEC)" that has been derived by modeling [58, 59]. The PEC is one of the two basic parameters to be used in environmental risk assessment of pollutants, that is, the ratio of PEC and PNEC ("Predicted No Effect Concentration") which will be multiplied by various safety factors depending on the complexity of the experimental study: The more complex the experimental design, that is, the more realistic environmental conditions have been applied, the lower is the safety factor. Of course, also other pollutants are deposited on the top soil layer, for example, by sedimentation of polluted dust particles, application of fertilizers and sewage sludge.

The top 20–25 cm is often adequate for soil sampling because this is the layer in which roots are exposed to the chemicals and may adsorb and absorb the pollutants. However, in typical lysimeter studies, soil sampling is extended up to about 1–1.5 m depth (see Further Reading section).

At least 3–5 soil samples from an area of 10–20 m^2 should be taken. Leaves, vegetation and other large particles should be removed before sampling. Since soils often have a high spatial heterogeneity, both horizontally and vertically, the number of samples needs to be increased, for instance 15–30 samples from heterogeneous plots. For different sampling strategies, refer to Figure 4.2. Large aggregated soil particles can be broken down by grinding, crushing, milling and so on, easiest by use of a mortar and pestle, before passing to a sieve after thorough mixing. Better (and easier) comminution is obtained by professional equipment, for example, Pulverisette mills

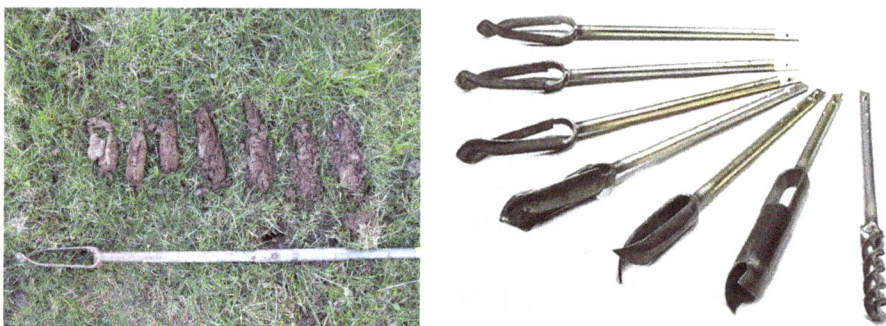

Figure 4.11: (Left) Soil samples from different depths by using a hand-driven soil auger. (Right) Different soil auger types specifically used depending on the soil conditions (e.g., hardness, dryness, texture).

(provided by Fritsch), and sieving devices such as Analysette types (Fritsch). An international standard is to sieve the mixed soil (after comminution) through a 2-mm sieve before analysis.

Surface samples can be taken by use of spoons, spades or shovels. Borers (augers) need to be used when sampling deeper in soil for which different types are available (Figure 4.11). The borer is screwed or hammered into the soil and the soil core is removed on withdrawal. It can be placed into a glass vial or left in the (tube) borer and shipped to the laboratory.

Another type of equipment is soil tube samplers, thin-walled, hollow steel tubes, which are driven into the ground to extract a relatively undisturbed soil sample for use in laboratory analysis (Figure 4.12). Each tube has one end that is chamfered to form

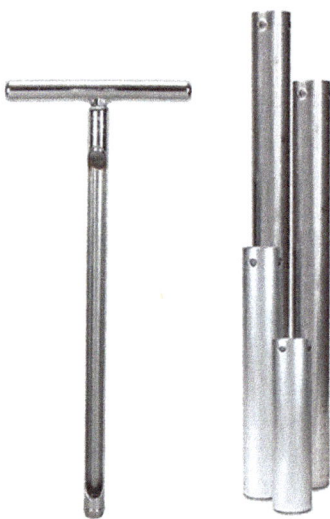

Figure 4.12: Soil tube samplers.

a cutting edge and the upper end includes holes for securing the tube to a drive head. Tube samplers are useful for collecting soils that are particularly sensitive to sampling disturbance, including fine cohesive soils and clays. The tubes can also be used to transport samples back to the laboratory.

4.5.3 Sediment

Samples from the surface of sediments can be taken by use of Van Veen or Ekman grab samplers. The Van Veen clamshell–bucket sampler (Figure 4.13) can grab up to about 20 cm deep sediment from about 0.1 m^2 areas of soft sediment bottom. The sampler is immersed into the water in the open scissor-like position with the two arms and buckets extended with a rope. When it hits the sediment, the arms are unlocked and the buckets with sharp cutting edges close when the rope is retracted. Up to a few kilograms of sediment can be sampled in this way depending on the size of the device. Disadvantages are a significant disturbance to sediments and the impracticality to sample overlying water and living organisms. A similar device is the Ponar sampler which has center-hinged jaws and a spring-loaded pin which releases when the sampler hits the bottom. The Ponar sampler can be used for all types of hard bottoms such as sand, gravel and clay. It can be used in streams, lakes reservoirs and the ocean and can grab 2–8 L of sediment.

The Ekman grab sampler (Figure 4.13) is designed for sampling in soft bottomed lakes and rivers composed of muck, mud or fine peat. As the sampler is lowered, two hinged upper lids swing open to let water pass through and close upon retrieval preventing sample washout. When the sampler reaches the bottom, a heavy metal peace

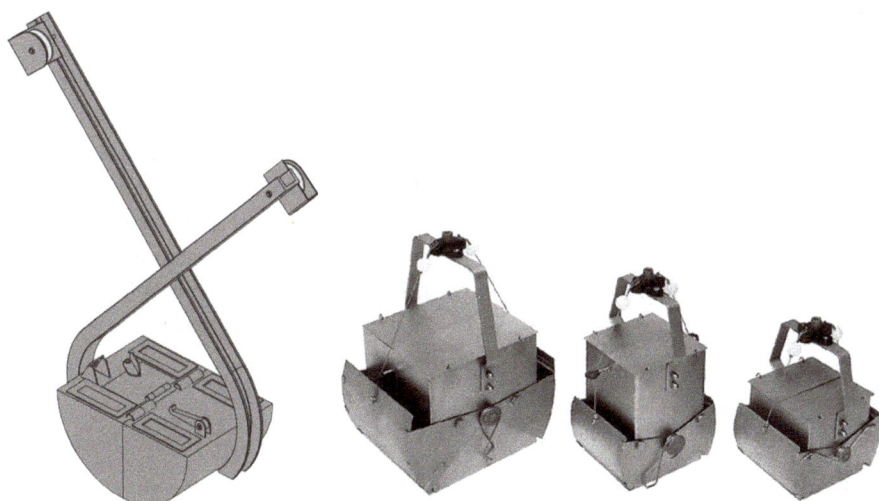

Figure 4.13: Van Veen (left) and Ekman (right) sediment grab samplers.

Figure 4.14: Ballcheck sediment core borer.

is sent down the line as a messenger tripping the overlapping spring-loaded scoops. With this device, sediment volumes between about 2 and 12 L can be sampled.

Sediment core samplers for undisturbed samples are based on gravity corers or spring-loaded piston corers. They are suitable especially for clay, silt or sand bottom and are typically used in lakes rather than in streams. Hand corers designed for manual operation can be used in shallow water as much as several meters in depth. Deeper water requires devices such as the Ballchek corer which depends on gravity to drive them into the sediment (Figure 4.14); the sample is prevented from leaving the core tube during raising by a sealing mechanism. Other corers work with a drop hammer to drive the sampler into the river or lake bottom. The corers are designed to retain the sample as the instrument is withdrawn from the sediment and returned to the surface. Sections of the core may be retained in the tube until it is shipped to the laboratory. Intact cores are best preserved by freezing until extraction for analysis.

4.6 Diffusion Sampling (Passive Sampling)

Diffusion sampling is becoming popular during the last decades because it represents a simple means to combine sampling with enrichment. Since such devices do not need active pumping of the respective sample medium (water or air), they can be produced in small sizes and at low cost, as well. The driving force, instead active pumping the analyte to a recipient, is the diffusion gradient toward a sink within the sampling unit. Prerequisite for this is a high Brownian motion, expressed by the diffusion coefficient D, of the analyte molecules or particles of interest. On the other hand, a passive sampler is relying on the flux of analyte per unit area and time unit. One could say that it mimics the natural deposition to a sink (soil, building surfaces, tree leaves, etc.). Hence, the dimension of D is $[g/cm^2 \, s]$.

Figure 4.15 depicts the basic scheme for such passive sampling devices.

First presented by Palmes in the 70s of last century, nowadays, many variations of this principle of non-steady-state diffusion as acting force for gas sampling are known. As the most simple model, it is an open tube of length z and a face opening area A. At the inner part of the closed end of the tube, an analyte-adsorbing material is fixed as diffusion sink. If outside the tube an analyte concentration $C_{analyte}$ is nonzero, at the distant end above the sorbent, we have $C_{analyte} = 0$. This holds true as long as there exists an irreversible adsorption.

Fick's first law of diffusion gives the relation between flux J of diffusing species and the concentration gradient, in general

$$J = -D \left(\frac{dc}{dz} \right) \tag{4.2}$$

The flux of analyte J toward the sink in distance z is represented by the analyte diffusion coefficient D times the analyte concentration gradient dc/dz from the opening area A of the sampling device to the surface of the irreversibly acting sorbent. This relationship is modified to include the area A of the opening of the diffusion sampler

$$\frac{dM}{Adt} = -D \left(\frac{dc}{dz} \right) \tag{4.3}$$

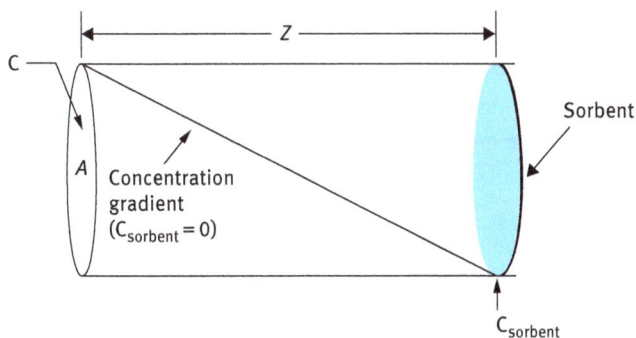

Figure 4.15: Schematic view of an open diffusion sampler.

where dM is the analyte mass passing through a cross-sectional area of a tube, A, for a time increment dt. This amount is proportional to the concentration gradient in the sampler (dc/dz) and the analyte diffusion coefficient D.

For a perfect sink (=irreversible sorption), the differential equation can be expressed by

$$M = R \int_{t_1}^{t_2} C_F \, dt \qquad (4.4)$$

M is the totally adsorbed analyte mass, which has to be determined after terminating sampling. C_F is then the ambient analyte concentration. R represents the diffusion sampling rate constant, which has to be determined by calibration experiments for a given sampler geometry and analyte. For a steady-state situation,

$$C_F = \frac{M}{Rt} \qquad (4.5)$$

with t the sampling time interval applied.

In principle, these considerations are valid for sampling within a liquid fluid, such as water, too. Within recent years, several applications, like time-averaged collection of hydrogeological tracer substances or other dissolved xenobiotics from borehole wells, are published. Monitoring of VOCs is a typical application for water or air monitoring.

Passive sampling means time-averaged collection. Only in case of constant ambient temperature and constant flow conditions around the sampling device, matching results with active sampling can be expected. This is a serious restriction when in no case, a threshold limit can be exceeded. Temporarily occurring concentration peaks can't be realized by passive sampling.

Typical applications for such passive samplers are in the field of occupational hygiene, where personal sampling asks for small, lightweight and wearable devices during a working shift of a worker. Indoor air monitoring, with moderate changes in ambient temperature, seems to be the most appropriate application. Many studies on material and even human (lung) exhalation of VOCs have been communicated over the last years.

A different application of passive samplers is the time-averaged collection of ultrafine particles, for instance from ambient aerosols. Here, the Brownian motion expressed by the diffusion coefficient of small particles becomes less dominant in contrast to surrounding flow conditions. Therefore, this application has to be taken with caution.

Table 4.4 gives selected applications for passive sampling.

Table 4.4: Selected applications of passive sampling.

Analyte (matrix)	Sink	Sampling duration
Formaldehyde (air)	2,4-Dinitrophenylhydrazine	Up to 10 days
BTEX (air)	Poly(dimethylsiloxane)	89 h
Alkanes (air)	Poly(dimethylsiloxane)	89 h
1,3-Butadiene	Carbograph S	24 h
Particulate matter $PM_{10-2.5}$ (ambient air: indoor and outdoor)	Standard stub as used in a scanning electron microscope image analysis	1 week
PAHs and PDBEs (seawater)	Polyethylene	Up to 3 months
Rhodamine WT (groundwater)	Amberlite XAD-7 resin	168 h

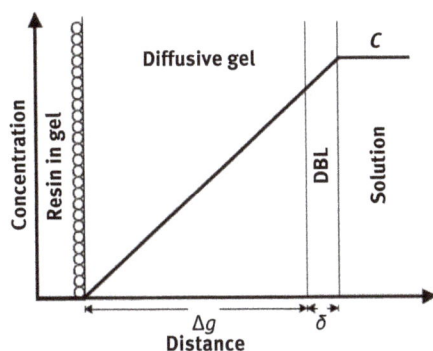

Figure 4.16: Scheme for diffusive gradient sampling in thin films (DGT). According to DGT theory, the concentration of an analyte, C, becomes 0 as the analyte approaches the binding layer as sink, permeating through the diffusive boundary layer (DBL; thickness δ) and the diffusive gel (thickness Δg).

An interesting application of this principle is the method establishing a diffusive gradient through thin films (DGT). This allows, for example, the assessment of soil pore water mobility of contaminants. Figure 4.16 shows the principle.

In front of an irreversibly acting recipient, for instance XAD resin or molecularly imprinted polymers, pore water-dissolved soil contaminants will exhibit a directed diffusion to the sink (strongly binding resin in a gel matrix: binding gel) through a defined diffusion barrier (e.g., agarose gel, polyimide membrane: diffusive gel). The analysis of the binding gel can be used to determine the bulk solution *concentration* of the element or compound. Once the mass of an analyte at the binding gel has been determined, the time-averaged concentration of the analyte in the bulk, C_{DGT}, can be determined by application of the following equation:

$$C_{DGT} = \frac{M\Delta g}{DtA} \tag{4.6}$$

where M is the mass of the analyte on the resin, Δg is the thickness of the diffusive layer and filter membrane together, D is the diffusion coefficient of the analyte, t is the exposition time and A is the area of the sampler entrance.

This allows to determine the rate constants of sorption and desorption for the mobile analyte. Several reports are published on mobility of organometallic complexes, antibiotics and others. Such application enables discrimination of differently mobile species in soil and sediment.

4.7 Sampling of Colloidal Matter (Hydrosol, Aerosol)

4.7.1 General Remarks

Finely dispersed insoluble material, suspended within a carrier medium, often represents a difficult matrix to be probed. Nowadays, many samples contain such colloids with particle diameters roughly between 1 and 1,000 nm. Depending on the prevailing flow conditions of the carrier medium, even larger particulate objects may represent part of a colloidal system.

A *hydrosol* contains hydrocolloids (solid or liquid) dispersed within water. A prominent example is milk, where the dispersed material is hydrophobic fat constituents. The particle size is in a range causing the "milky" turbidity created by light scattering. Surface water contains a wide variety of humic substances, polymeric matter of indefinite composition with size ranges between 2 and 10 nm. From physicochemical size distribution measurements, a number of concentrations typically up to 10^6 particles per mL in drinking water are known. The difficulty with such hydrosols stems from their dynamic behavior. Depending on number concentration and particle size, rapid agglomeration is observed. This leads to formation of new agglomerated particles with different composition. Depending on the pH, these colloids are charged, hence showing sudden collapse in number concentration due to coulombic attraction forces.

Hydrosols serve as transportation vehicles for adsorbable compounds, thus changing the migration properties of compounds within an aquifer (=water-containing underground layer in subsoil). This means that not the physicochemical property of the attached molecule governs the residence time and flow direction, rather the whole construct (particle plus attached molecules) is now determining the migration behavior. This is known for pesticides, toxins, PAHs and so on. Aside this, hydrosols may contain pathogenic microorganisms too.

An *aerosol* is the dispersion of solid and/or liquid particles (synonymous also to droplets) of a size measure of about 1 nm (at least for one dimension) up to several μm. Again, the surrounding flow conditions determine whether a particle/air system is an aerosol. Under high wind speeds, very big objects up to about a μm–mm size can be part of an aerosol. Clean air masses above the continental surfaces are characterized by particle number concentrations between 10^3 and 10^6 per mL. Particle sizes of freshly produced aerosols are in the lower nanometer range but subjected to rapid changes due to Brownian diffusion and subsequent coagulation.

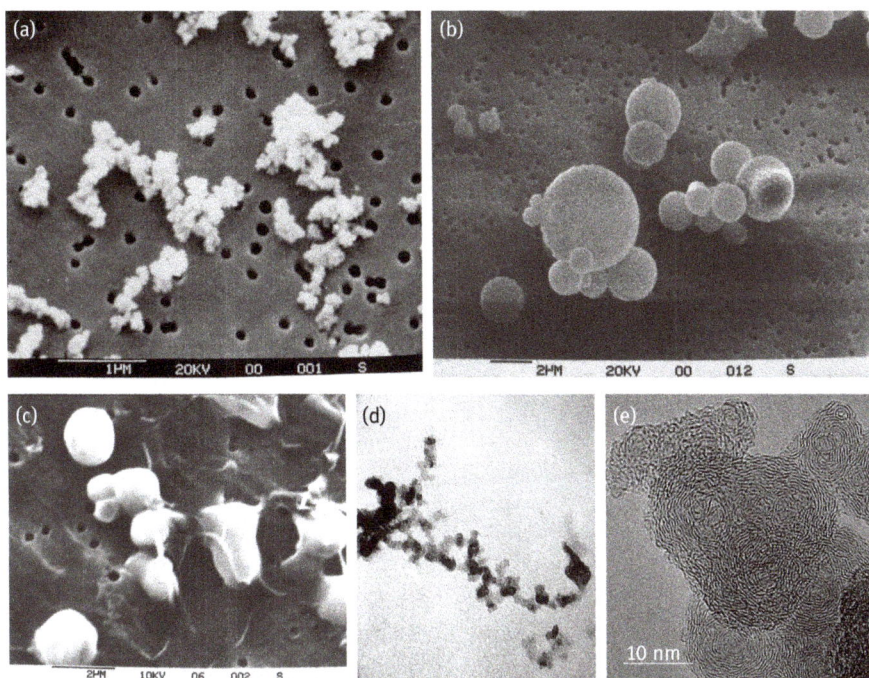

Figure 4.17: SEM/TEM microphotographs of selected aerosol particles. (a) Birch pollen particles; (b) polydisperse fly ash particles; (c) hexadecane-encapsulated ammonium sulfate particles; (d) TEM picture of agglomerated soot particles, magnification: 250.000; (e) High-resolution TEM picture of single propane soot particles. Source: Reinhard Niessner & Alexander Rinkenburger.

Also particle charging may have dramatic consequences in dispersion stability and residence time. Sometimes, one can observe long-range transport of sand grains and pollen from the Sahara desert across the Alps. Typical aerosol particles are shown in Figure 4.17.

Aerosols might be harmful to man when inhaled. This is especially the case for particle sizes between a few nanometers up to 10 μm due to high deposition masses in the smaller sized bronchial and alveolar lung systems (Figure 4.18).

Sources for toxic particulates are combustion residues (e.g., diesel exhaust) or the products of natural photochemistry. Also, working places are within the focus. Special attendance gained the intended application of toxic aerosols for biological and chemical warfare.

4.7.2 Hydrosol Sampling

Representative sampling of hydrosols is generally difficult. Once a partial water volume becomes withdrawn from a reservoir, dramatic changes of equilibria may happen. The instantaneous formation of calcite particles when lifting water subsamples

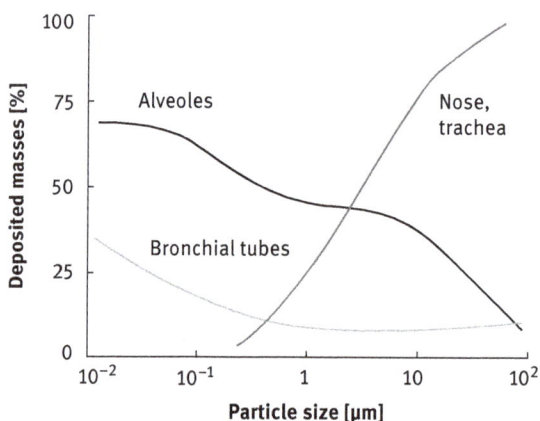

Figure 4.18: Deposition of airborne particles in the respiratory tract depending on their size.

from deep water wells or hot geothermal water wells has been observed many times. The reason is the change of dissolved CO_2 gas concentration due to different pressure and temperature, hence shifting the natural carbonate equilibria. Same happens with the sulfide system. The presence of newly formed small nuclei changes the dissolution equilibria of concomitant organic trace compounds too. Therefore, the usage of inert gas atmospheres cabinets for handling such water samples is sometimes recommended. Currently, the increased interest in the use of geothermal energy resources triggers research on these phenomena. For example, the hot water springs of the Molasse basin around Munich in Germany contain enormous amounts of PAHs and surface-active substances.

Hydrosols are important for many reasons. Environmentally, they act as carrier of many trace constituents in water. The size range of the particles varies from a few nanometers up to millimeters, depending on flow velocity of the aqueous phase as carrier. Despite the fact of the usually small amounts of nano- and micrometer colloidal matter, the large specific surface area of suspended matter exhibits the sink for most hydrophobic organic compounds. Also, polar substances might become fixed onto ion exchange sites of hydrocolloids. These hydrocolloids are moving by the various flow forces to different locations within a water–air–soil pore system. Depending on contact angle, which determines wettability, they will stick at different interfaces. Figure 4.19 depicts the situation.

Consequently, xenobiotics partition between colloids, aqueous phase or soil particle surfaces. Since the particles have several orders of magnitude lower diffusion coefficients than dissolved molecules, migration properties are very different from each other. Hydrosol collection often requires size-resolved sampling and enrichment.

This is achieved by usage of membrane filters with a well-defined monodisperse pore distribution (see chapter 6.4.4). Typically, these filters consist of polycarbonate thin sheet (thickness approx. 10 µm). The pore sizes can vary between lower nanometer up to micrometer ranges. Such filters are often applied in a sequential cascade arrangement (see Figure 4.20).

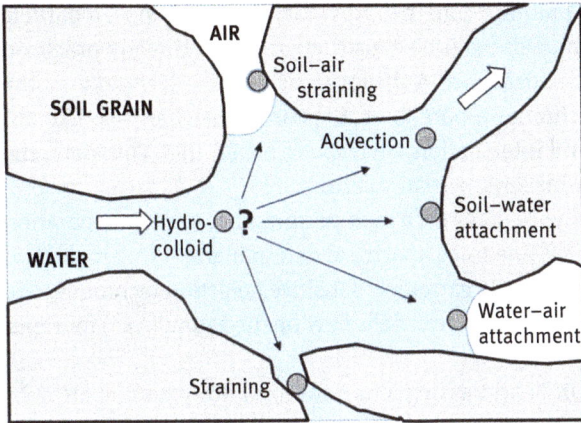

Figure 4.19: Transport and deposition routes for hydrocolloids.

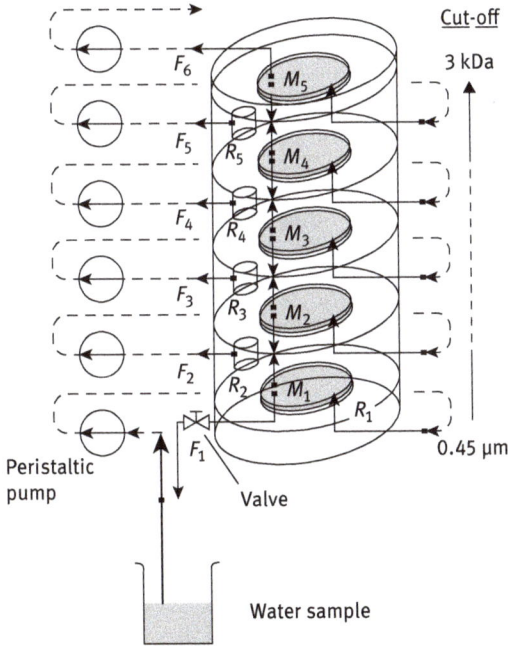

Figure 4.20: Flow scheme of cascade filtration (M_n: nanopore membranes of decreasing pore size, F_n: fractions obtained, R_n: flow-through reservoirs for fraction collection).

The hydrocolloids enter first a membrane filter with the largest cutoff size. Ideally, all particles larger than the cutoff size are collected by this filter. The next smaller colloid fraction will be collected on the next subsequent filter, now at a smaller cutoff size, and so on. Hence, the deposited particle fractions on all these membrane filters will represent the whole particle size distribution. Following sampling, they can be analyzed by an appropriate technique. From the individual mass fractions, a particle mass distribution as a function of size can be constructed. This is already practiced for metal speciation by humic substances. A disadvantage of this technique is the not very sharp separation function of a pore filter. Deposition mechanisms like diffusion deposition, impaction and interception are responsible for this. Therefore, the resulting size distribution contains large statistical errors.

A different approach is the hyphenation of a high-performance particle separation method with a subsequent on-/off-line trace analytical determination. Due to the low particulate matter content in water, only extremely sensitive detection techniques are applicable. Exemplarily, the use of asymmetric field flow fractionation (AF^4) for rapid particle separation is shown in Figure 4.21.

Particles dispersed within a liquid and passing a narrow-sized flow channel are subjected to a cross flow of particle-free solvent in a microchannel. Figure 4.22 shows this in more detail. The crossing flow depresses the particles toward the lower part, but depending on particle size, especially the smaller particles will show lateral diffusion. By this on time average during transition of the separation channel, the smaller particles will stay more within the center part of the channel cross section and hence elute first. The larger particles follow later. This particle separation is no chromatographic separation. A typical separation of a mixture of spherical polystyrene particles is shown in Figure 4.23.

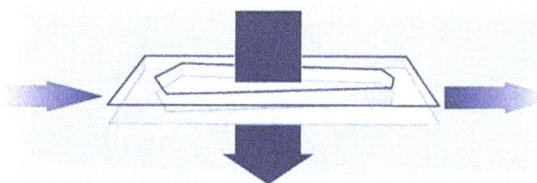

Figure 4.21: Force field in an AF^4 channel.

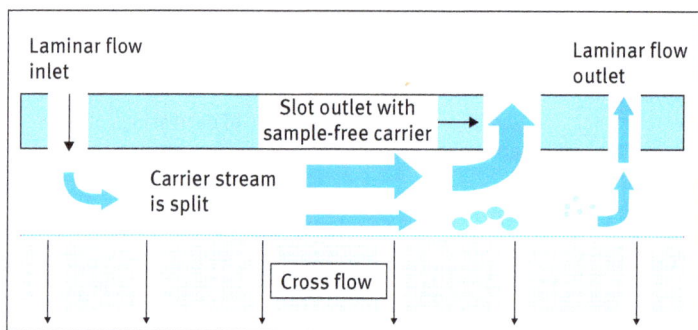

Figure 4.22: Flow splitting and particle movement in an AF^4 separator.

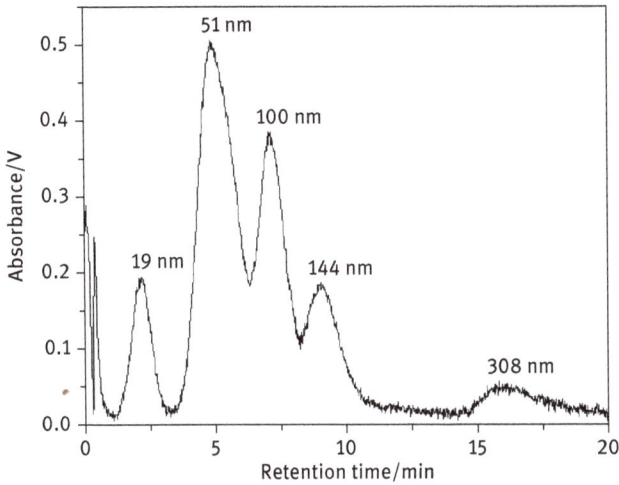

Figure 4.23: AF⁴ separation of a polystyrene particle mixture.

As can be seen from Figure 4.23, the separation of differently sized nm particles is achieved within a few minutes. Direct hyphenation to a mass spectrometer or other spectroscopic tools has been published, yielding size-related analysis of organic trace compounds.

4.7.3 Aerosol Sampling

What has been said for gas sampling (Section 4.3) is generally valid for aerosol sampling too. As distinctive feature and in contrast to gas sampling, an aerosol collector has to be adapted to the existing air flow conditions (wind speed, flow direction). Otherwise, dramatic under- or overestimation of collected material will be the result. Representative particle sampling is difficult.

Figure 4.24 depicts the optimum and the most deviating situations. In case of directly facing the particle streamlines and without difference in incoming particle velocity to the existing face velocity at tube inlet, no change and hence no particle

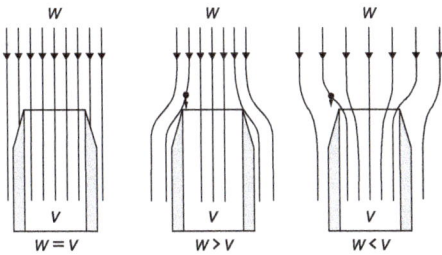

Figure 4.24: Particle entrance into a suction tube under differing inlet velocities. w = face velocity; v = velocity inside inlet.

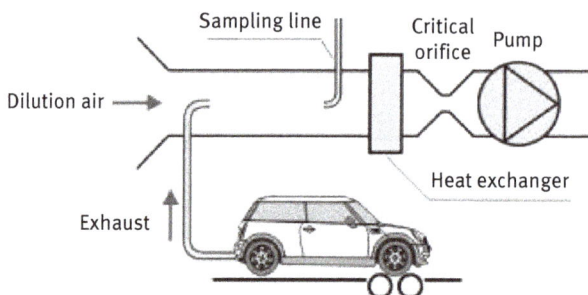

Figure 4.25: Schematic view of exhaust aerosol sampling under constant dilution.

losses have to be expected. This is the optimal configuration for aerosol collection, called *isokinetic* sampling. All other situations are named *anisokinetic* sampling, yielding strong deviations in collection rate due to formation of a stagnation or suction zone head-on to the tube inlet.

Figure 4.25 shows a schematic view of an inline aerosol sampling system, for example, for engine exhaust sampling. The car exhaust becomes injected into a particle-free air flow of constant flow rate, where isokinetic sampling can be arranged.

Depending on the volatility of emitted compounds, losses due to particle evaporation of semi-volatile compounds during residence time within the dilution tunnel have to be tolerated. The actual dilution is estimated by parallel continuous NO_x measurement at the car exhaust tailpipe and within the dilution tunnel.

Ambient aerosol sampling usually applies circularly formed collector heads (e.g., Figure 4.26). This minimizes influence of wind direction.

Also sampling of larger particles (>20 μm) is challenging, because of fast gravitational settling and inertial forces due to the actual flow conditions. Because these particles contribute most to the total sampled aerosol mass, fatal sampling errors are sometimes unavoidable.

Size-related aerosol sampling is under permanent discussion among inhalation toxicologists. Whereas total suspended dust particle sampling was in the focus some decades before, nowadays, fine (>0.1 μm), and especially ultrafine particles (<0.1 μm) are considered to be the most relevant size fractions to be studied and evaluated.

Legally important is sampling of the particulate matter below 2.5 μm (the so-called PM2.5 fraction). This is the direct consequence of observed increased particle deposition (of this size fraction) within the tracheoalveolar and alveolar tract of the lung apparatus of man (see Fig. 4.18). Once aerosol particles become deposited there, the usual routes of detoxification (mucociliary clearance, macrophage digestion of deposited debris, etc.) may become limited, thus increasing the toxic dose (=exposure time × effective concentration of deposited substance).

Plenty of purpose-oriented sampling arrangements have been published. Each analytical task of chemical aerosol sampling has its own peculiarities and needs to be optimized individually.

Figure 4.26: Photography of a high-volume aerosol collector unit with circular sampling head. Reproduced with permission from Roy A. Armstrong, Ph.D., Bio-optical Oceanography Laboratory, University of Puerto Rico.

4.8 Sources for Artifact Formation

4.8.1 General Remarks

Measurement artifacts are numerous but unfortunately rarely reported. The reason for this, aside from it effectively being the confession of having made a fault, is often ignorance of possible artifact formation. To reveal such artifacts, periodic checks of any step of the total analytical process are necessary. To do so, the use of reference material, or the introduction of independently labeled analytes, can help a lot (see also Chapter 3).

4.8.2 Sources of Artifacts

Artifacts may have different origins, as there are
- contamination through external media or instrumentation
- instability of analytes due to radiation, photolysis, temperature and pressure changes, or presence of reactive chemical environments during or after sampling (the latter due to improper storage conditions).

Contamination by reaction vessels, media and instrumentation is a severe restriction for trace analysis in general. Depending on the concentration level, already the laboratory environment may cause problems. Carryover of substances has to be minimized by separation of working areas, as described in the rules for good laboratory practices (GLP guidelines). Sometimes, the analyst himself is a source of contamination, for example, through his exhalation. Smokers shouldn't perform a PAH analysis. Skin acts as a sink and source for many organic compounds. Clean-air hoods with active charcoal air purification are an option.

Recycling of used solvents should be avoided. Reuse of pipette tips also sometimes creates problems. Rigorous cleaning of glassware by strongly oxidizing agents (e.g., "Piranha" solution, a mixture of sulfuric acid, H_2SO_4, and hydrogen peroxide, H_2O_2) is needed. Ultrapure water for rinsing has to be checked for organic impurities. Usually, only conductance is used for monitoring water purity. Even ultrapure solvents reveal contaminants leading to interferences in fluorescence spectroscopic analysis. Gas impurities are generally underestimated. A purity of a bottled gas at 99.999% level means that there is still plenty of unwanted contaminants. Septum materials have been reported as contamination sources. Filter membranes for gas/particle separation may for instance contain melamine resins as binder.

Destruction during storage has numerous reasons. Improper physical conditions (temperature, sun light, etc.) may alter its constituents (bond breaking or isomerization). Cross-contamination by neighbored leaky sample containers isn't rare. Sometimes, the catalysts used for polymer production are still active at inner wall surfaces of plastic bottles or seals of screw caps leading to degradation of the analyte. Freezing a liquid sample often leads to demixing and/or precipitation of previously homogeneously distributed compounds. Chlorinated water samples are subject to further chlorination reactions within the sample.

We must not forget microbial impurities (bacteria and protozoa) which are especially thankful for feeding them with organic chemicals.

Artifact formation during sampling will happen when an analyte dispersed before within the matrix is separated/enriched and subsequently contacted with other reactive chemicals present in the same matrix but sampled later. A prominent example is PAH sampling from ambient atmosphere. During a 24-h sampling period, ozone concentrations are often changing due to photochemical intensity. When particulate PAHs like benzo[a]pyrene are enriched by membrane filtration, this isn't only prone to volatilization losses; in fact, oxidation by ozone destroys the PAH. NO_2 in hot exhaust gas easily produces nitro-PAHs to an enormous amount. The same happens with amino functionalities which are completely destroyed.

All these imponderabilia are the motivation for searching for in situ and *online* measurement technologies.

Further Reading

Agarwal SK. Environmental monitoring. New Delhi: APH Publishing Corporation, 2005.

Alves A, Kucharska A, Erratico C, Xu F, Den Hond E, Koppen G, Vanermen G, Covaci A, Voorspoels S. Human biomonitoring of emerging pollutants through non-invasive matrices: state of the art and future potential. Anal Bioanal Chem 2014;406(17):4063–88.

Balderacchi M, Benoit P, Cambier P, Eklo OM, Gargini A, Gemitzi A, Gurel M, Klove B, Nakic Z, Preda E, Ruzicic S, Wachniew P, Trevisan, M. Groundwater pollution and quality monitoring approaches at the European level. Crit Rev Environ Sci Technol 2013;43(4):323–408.

Beceiro-Gonzalez E, Gonzalez-Castro MJ, Muniategui-Lorenzo S, Lopez-Mahia P, Prada-Rodriguez D. Analytical methodology for the determination of organochlorine pesticides in vegetation. J AOAC Int 2012;95(5):1291–310.

Bruzzoniti MC, Checchini L, De Carlo RM, Orlandini S, Rivoira L, Del Bubba M. QuEChERS sample preparation for the determination of pesticides and other organic residues in environmental matrices: a critical review. Anal Bioanal Chem 2014;406(17):4089–116.

Conclin R. Field sampling: principles and practices in environmental analysis. New York: Marcel Dekker Inc., 2004.

Esteban M, Castano, A. Non-invasive matrices in human biomonitoring: a review. Environ Int 2009;35(2):438–49.

Fenik J, Tankiewicz M, Biziuk, M. Properties and determination of pesticides in fruits and vegetables. TRAC Trends Anal Chem 2011;30(6):814–26.

Haroune L, Cassoulet R, Lafontaine M-P, Belisle M, Garant D, Pelletier F, Cabana H, Bellenger J-P Liquid chromatography-tandem mass spectrometry determination for multiclass pesticides from insect samples by microwave-assisted solvent extraction followed by a salt-out effect and micro-dispersion purification. Anal Chim Acta 2015;891:160–70.

Katagi T. Soil column leaching of pesticides. In Whitacre DM, editor. Reviews of environmental contamination and toxicology. Heidelberg, Springer, vol 221, 2013; 1–105.

Radojevic M, Bashkin, V. Practical environmental analysis. London: Royal Society of Chemistry, 2006.

Smolders R, Schramm K-W, Nickmilder M, Schoeters, G. Applicability of non-invasively collected matrices for human biomonitoring. Environ Health 2009;8:1–10.

Vereecken H. Mobility and leaching of glyphosate: a review. Pest Manage Sci 2005;61(12):1139–51.

Keith LH. Environmental sampling and analysis: a practical guide. Boca Raton: CRC Press, 1991.

Yusa V, Millet M, Coscolla C, Pardo O, Roca, M. Occurrence of biomarkers of pesticide exposure in non-invasive human specimens. Chemosphere 2015;139:91–108.

Yusa V, Millet M, Coscolla C, Roca, M. Analytical methods for human biomonitoring of pesticides. A review. Anal Chim Acta 2015;891:15–31.

Zhang C. Fundamentals of environmental sampling and analysis. New York: Wiley, 2007.

Bibliography

[1]   Taggart A. Handbook of mineral dressing, ores and industrial minerals. New York: Wiley, 1945:1905.
[2]   Kienitz H, Bock R, Fresenius W, Huber W, Tölg G. Analytiker-Taschenbuch, Band 1. Berlin: Springer-Verlag, 1980.

[3]   Guo Y, Kannan K. Challenges encountered in the analysis of phthalate esters in foodstuffs and other biological matrices. Anal Bioanal Chem 2012;404(9):2539–54.

[4]   Burgess RM, Ho KT, Brack W, Lamoree M. Effects-directed analysis (EDA) and toxicity identification evaluation (TIE): Complementary but different approaches for diagnosing causes of environmental toxicity. Environ Toxicol Chem 2013;32(9):1935–45.

[5]   Friedrich G, Chapman D, Beim A. The use of biological material. In Chapman D, editor. Water quality assessments. A guide to the use of biota, sediments and water in environmental monitoring. London: Chapman & Hall, 1996; 171–238.

[6]   Floehr T, Scholz-Starke B, Xiao H, Hercht H, Wu L, Hou J, Schmidt-Posthaus H, Segner H, Kammann U, Yuan X, RoSs-Nickoll M, Schaffer A, Hollert H. Linking Ah receptor mediated effects of sediments and impacts on fish to key pollutants in the Yangtze Three Gorges Reservoir, China – a comprehensive perspective. Sci Total Environ 2015;538:191–211.

[7]   Zhao Z, Zhang L, Wu J, Fan C. Distribution and bioaccumulation of organochlorine pesticides in surface sediments and benthic organisms from Taihu Lake, China. Chemosphere 2009;77(9):1191–8.

[8]   Bettinetti R, Galassi S, Guzzella L, Quadroni S, Volta P. The role of zooplankton in DDT biomagnification in a pelagic food web of Lake Maggiore (Northern Italy). Environ Sci Poll Res 2010;17(9):1508–18.

[9]   Lopez-Perea JJ, Camarero PR, Molina-Lopez RA, Parpal L, Obon E, Sola J, Mateo R. Interspecific and geographical differences in anticoagulant rodenticide residues of predatory wildlife from the Mediterranean region of Spain. Sci Total Environ 2015;511:259–67.

[10]   Douglas MR, Rohr JR, Tooker JF. Neonicotinoid insecticide travels through a soil food chain, disrupting biological control of non-target pests and decreasing soya bean yield. J Appl Ecol 2015;52(1):250–60.

[11]   Krischik V, Rogers M, Gupta G, Varshney A. Soil-applied imidacloprid translocates to ornamental flowers and reduces survival of adult Coleomegilla maculata, Harmonia axyridis, and Hippodamia convergens Lady Beetles, and Larval Danaus plexippus and Vanessa cardui Butterflies. Plos One 2015;10(3):1–22.

[12]   Wright GA, Softley S, Earnshaw H. Low doses of neonicotinoid pesticides in food rewards impair short-term olfactory memory in foraging-age honeybees. Sci Rep 2015;5:1–7.

[13]   Sutton G, Bennett J, Bateman M. Effects of ivermectin residues on dung invertebrate communities in a UK farmland habitat. Insect Conserv Diversity 2014;7(1):64–72.

[14]   Rosa AD, Price RI, Caliman MJ, Queiroz EP, Blochtein B, Pires CS, Imperatriz-Fonseca VL. The stingless bee species, Scaptotrigona aff. depilis, as a potential indicator of environmental pesticide contamination. Environ Toxicol Chem 2015;34(8):1851–3.

[15]   Perugini M, Di Serafino G, Giacomelli A, Medrzycki P, Sabatini AG, Oddo LP, Marinelli E, Amorena M. Monitoring of polycyclic aromatic hydrocarbons in bees (Apis mellifera) and honey in urban areas and wildlife reserves. J Agric Food Chem 2009;57(16):7440–4.

[16]   Koehler H-R, Triebskorn R. Wildlife ecotoxicology of pesticides: can we track effects to the population level and beyond? Science 2013;341(6147):759–65.

[17]   Gibbons D, Morrissey C, Mineau P. A review of the direct and indirect effects of neonicotinoids and fipronil on vertebrate wildlife. Environ Sci Poll Res 2015;22(1):103–18.

[18]   Millot F, Berny P, Decors A, Bro E. Little field evidence of direct acute and short-term effects of current pesticides on the grey partridge. Ecotoxicol Environ Saf 2015;117:41–61.

[19]   Mann RM, Hyne RV, Choung CB, Wilson SP. Amphibians and agricultural chemicals: Review of the risks in a complex environment. Environ Poll 2009;157(11):2903–27.

[20]   Quaranta A, Bellantuono V, Cassano G, Lippe C. Why amphibians are more sensitive than mammals to xenobiotics. Plos One 2009;4(11):1–4.

[21]   Lenhardt PP, Bruehl CA, Berger G. Temporal coincidence of amphibian migration and pesticide applications on arable fields in spring. Basic Appl Ecol 2015;16(1):54–63.

[22] Krol S, Zabiegala B, Namiesnik J. Human hair as a biomarker of human exposure to persistent organic pollutants (POPs). TRAC Trends Anal Chem 2013;47:84–98.

[23] Hardy EM, Duca RC, Salquebre G, Appenzeller BM. Multi-residue analysis of organic pollutants in hair and urine for matrices comparison. Forensic Sci Int 2015;249:6–19.

[24] Man YB, Chan JK, Wang HS, Wu SC, Wong MH. DDTs in mothers' milk, placenta and hair, and health risk assessment for infants at two coastal and inland cities in China. Environ Int 2014;65:73–82.

[25] Carnevale A, Aleksa K, Goodyer CG, Koren G. Investigating the use of hair to assess polybrominated diphenyl ether exposure retrospectively. Ther Drug Monit 2014;36(2):244–51.

[26] Appenzeller BM, Tsatsakis AM. Hair analysis for biomonitoring of environmental and occupational exposure to organic pollutants: State of the art, critical review and future needs. Toxicol Lett 2012;210(2):119–40.

[27] Chetiyanukornkul T, Toriba A, Kameda T, Tang N, Hayakawa K. Simultaneous determination of urinary hydroxylated metabolites of naphthalene, fluorene, phenanthrene, fluoranthene and pyrene as multiple biomarkers of exposure to polycyclic aromatic hydrocarbons. Anal Bioanal Chem 2006;386(3):712–18.

[28] Elovaara E, Mikkola J, Makela M, Paidanius B, Priha E. Assessment of soil remediation workers' exposure to polycyclic aromatic hydrocarbons (PAH): biomonitoring of naphthols, phenanthrols, and 1-hydroxypyrene in urine. Toxicol Lett 2006;162(2-3):158–63

[29] Godschalk RW, Van Schooten FJ, Bartsch H. A critical evaluation of DNA adducts as biological markers for human exposure to polycyclic aromatic compounds. J Biochem Mol Biol 2003;36(1):1–11.

[30] Toriba A, Kuramae Y, Chetiyanukornkul T, Kizu R, Makino T, Nakazawa H, Hayakawa K. Quantification of polycyclic aromatic hydrocarbons (PAHs) in human hair by HPLC with fluorescence detection: a biological monitoring method to evaluate the exposure to PAHs. Biomed Chromatogr 2003;17(2-3):126–32.

[31] Schummer C, Appenzeller BM, Millet M, Wennig R. Determination of hydroxylated metabolites of polycyclic aromatic hydrocarbons in human hair by gas chromatography-negative chemical ionization mass spectrometry. J Chromatogr A 2009;1216(32):6012–19.

[32] Azodi-Deilami S, Najafabadi AH, Asadi E, Abdouss M, Kordestani D. Magnetic molecularly imprinted polymer nanoparticles for the solid-phase extraction of Paracetamol from plasma samples, followed its determination by HPLC. Microchim Acta 2014;181(15-16):1823–32.

[33] Liu SM, Pleil JD. Human blood and environmental media screening method for pesticides and polychlorinated biphenyl compounds using liquid extraction and gas chromatography-mass spectrometry analysis. J Chromatogr B-Anal Technol Biomed Life Sci 2002;769(1):155–67.

[34] Zheng J, Chen K-H, Luo X-J, Yan X, He C-T, Yu Y-J, Hu G-C, Peng X-W, Ren M-Z, Yang Z-Y, Mai B-X. Polybrominated diphenyl ethers (PBDEs) in paired human hair and serum from e-waste recycling workers: source apportionment of hair PBDEs and relationship between hair and serum. Environ Sci Technol 2014;48(1):791–6.

[35] Magiera S, Hejniak J, Baranowski J. Comparison of different sorbent materials for solid-phase extraction of selected drugs in human urine analyzed by UHPLC-UV. J Chromatogr B Anal Technol Biomed Life Sci 2014;958:22–8.

[36] Mercadante R, Polledri E, Bertazzi PA, Fustinoni S. Biomonitoring short- and long-term exposure to the herbicide terbuthylazine in agriculture workers and in the general population using urine and hair specimens. Environ Int 2013;60:42–7.

[37] Syed JH, Alamdar A, Mohammad A, Ahad K, Shabir Z, Ahmed H, Ali SM, Sani SG, Bokhari H, Gallagher KD, Ahmad I, Eqani SA. Pesticide residues in fruits and vegetables from Pakistan: a review of the occurrence and associated human health risks. Environ Sci Poll Res 2014;21(23):13367–93.

[38] Gomez-Ramos MM, Ferrer C, Malato O, Agueera A, Fernandez-Alba AR. Liquid chromatography-high-resolution mass spectrometry for pesticide residue analysis in fruit and vegetables: Screening and quantitative studies. J Chromatogr A 2013;1287:24–37.

[39] Diez C, Traag WA, Zommer P, Marinero P, Atienza J. Comparison of an acetonitrile extraction/partitioning and "dispersive solid-phase extraction" method with classical multi-residue methods for the extraction of herbicide residues in barley samples. J Chromatogr A 2006;1131(1-2):11–23.

[40] He Z, Peng Y, Wang L, Luo M, Liu X. Unequivocal enantiomeric identification and analysis of 10 chiral pesticides in fruit and vegetables by QuEChERS method combined with liquid chromatography-quadruple/linear ion trap mass spectrometry determination. Chirality 2015;27(12):958–64.

[41] Silva JA, Ratola N, Ramos S, Homem V, Santos L, Alves A. An analytical multi-residue approach for the determination of semi-volatile organic pollutants in pine needles. Anal Chim Acta 2015;858:24–31.

[42] Mahugija JA, Henkelmann B, Schramm K-W. Levels and patterns of organochlorine pesticides and their degradation products in rainwater in Kibaha Coast Region, Tanzania. Chemosphere 2015;118:12–19.

[43] Majewski MS, Coupe RH, Foreman WT, Capel PD. Pesticides in Mississippi air and rain: a comparison between 1995 and 2007. Environ Toxicol Chem 2014;33(6):1283–93.

[44] Wurl O, Obbard JP. A review of pollutants in the sea-surface microlayer (SML): a unique habitat for marine organisms. Mar Pollut Bull 2004;48(11-12):1016–30.

[45] Hardy JT. The sea-surface microlayer – biology, chemistry and anthropogenic enrichment. Prog Oceanogr 1982;11(4):307–28.

[46] Benson NU, Essien JP, Asuquo FE, Eritobor AL. Occurrence and distribution of polycyclic aromatic hydrocarbons in surface microlayer and subsurface seawater of Lagos Lagoon, Nigeria. Environ Monit Assess 2014;186(9):5519–29.

[47] Brinkmann M, Hudjetz S, Kammann U, Hennig M, Kuckelkorn J, Chinoraks M, Cofalla C, Wiseman S, Giesy, JP, Schaeffer A, Hecker M, Woelz J, Schuettrumpf H, Hollert H. How flood events affect rainbow trout: Evidence of a biomarker cascade in rainbow trout after exposure to PAH contaminated sediment suspensions. Aquat Toxicol 2013;128:13–24.

[48] Schulze T, Ricking M, Schroter-Kermani C, Koerner A, Denner H-D, Weinfurtner K, Winkler A, Pekdeger A. The German environmental specimen bank – sampling, processing, and archiving sediment and suspended particulate matter. J Soils Sediments 2007;7(6):361–7.

[49] Li H, Lu L, Huang W, Yang J, Ran Y. In-situ partitioning and bioconcentration of polycyclic aromatic hydrocarbons among water, suspended particulate matter, and fish in the Dongjiang and Pearl Rivers and the Pearl River Estuary, China. Mar Poll Bull 2014;83(1):306–16.

[50] Qin N, He W, Kong X-Z, Liu W-X, He Q-S, Yang B, Wang Q-M, Yang C, Jiang Y-J, Jorgensen SE, Xu F-L, Zhao X-L. Distribution, partitioning and sources of polycyclic aromatic hydrocarbons in the water-SPM-sediment system of Lake Chaohu, China. Sci Total Environ 2014;496:414–23.

[51] Erger C, Balsaa P, Werres F, Schmidt TC. Multi-component trace analysis of organic xenobiotics in surface water containing suspended particular matter by solid phase extraction/gas chromatography-mass spectrometry. J Chromatogr A 2012;1249:181–9.

[52] von der Ohe, PC, Dulio V, Slobodnik J, De Deckere E, Kuehne R, Ebert R-U, Ginebreda A, De Cooman W, Schueuermann G, Brack W. A new risk assessment approach for the prioritization of 500 classical and emerging organic microcontaminants as potential river basin specific pollutants under the European water framework directive. Sci Total Environ 2011;409(11):2064–77.

[53] Choi S-D, Wania F. On the reversibility of environmental contamination with persistent organic pollutants. Environ Sci Technol 2011;45(20):8834–41.

[54] Bogdal C, Mueller, CE, Buser, AM, Wang Z, Scheringer M, Gerecke, AC, Schmid P, Zennegg M, MacLeod M, Hungerbuehler K. Emissions of polychlorinated biphenyls, polychlorinated dibenzo-p-dioxins, and polychlorinated dibenzofurans during 2010 and 2011 in Zurich, Switzerland. Environ Sci Technol 2014;48(1):482–90.

[55] Vonberg D, Vanderborght J, Cremer N, Puetz T, Herbst M, Vereecken H. 20 years of long-term atrazine monitoring in a shallow aquifer in western Germany. Water Res 2014;50:294–306.

[56] Vonberg D, Hofmann D, Vanderborght J, Lelickens A, Koeppchen S, Puetz T, Burauel P, Vereecken H. Atrazine soil core residue analysis from an agricultural field 21 years after its ban. J Environ Qual 2014;43(4):1450–9.

[57] Ricking M, Schwarzbauer J. DDT isomers and metabolites in the environment: an overview. Environ Chem Lett 2012;10(4):317–23.

[58] Hardy I, Gottesbueren B, Huber A, Jene B, Reinken G, Resseler H. Comparison of lysimeter results and leaching model calculations for regulatory risk assessment. J Verbraucherschutz Lebensmittelsicherh–J Consum Prot Food Saf 2008;3(4):364–75.

[59] Labite H, Holden NM, Richards KG, Kramers G, Premrov A, Coxon CE, Cummins E. Comparison of pesticide leaching potential to groundwater under EU FOCUS and site specific conditions. Sci Total Environ 2013;463:432–41.

5 Sample Treatment Before Analysis

5.1 General Remarks

Economic considerations usually prevent immediate analysis of freshly sampled material. In fact, collection of many samples is preferred, in order to have at least a series of analyzed samples after a lengthy and often cumbersome calibration procedure. Sometimes, instrumental equipment needs a considerable warm-up time to gain stable operational conditions too. So, the subsequent introduction of only a few samples for analysis would cause unreasonable expenditures of time, consumption of expensive chemicals (standards, isotopes, reference materials, etc.), and hence waste of money.

Depending on the circumstances of sampling and matrix, it might be also advisable to store the collected samples immediately until end of a sampling period is reached, particularly when absence of any lab structure (field measurement, expedition, dirty industrial environment, etc.) is asking for such. For general literature, see Refs. [1, 2].

5.2 Sample Pretreatment (Stabilization and Storage)

Sample pretreatment means preparation for a safe storage in analytically clean containers for a longer time period. This implicates that no change of state of the analyte must happen. Therefore, the selection of appropriate sample containers is essential.

Aqueous samples are best collected and stored in quartz glass vessels at a temperature of about 5 °C. Quartz glass is not only impermeable and free of organic material, it can also be rigorously cleaned and does not possess high surface roughness. Same is with electropolished stainless steel containers. Cleaning such containers is always necessary, especially when polar compounds need to be analyzed. They tend to be adsorbed readily and everywhere, even under ambient conditions. Examples are phthalates or lipophilic halogenated compounds (PCBs, PCDFs, PCDDs) or polyaromatic hydrocarbons. Baking out vessels under a heated (400 °C) nitrogen flow for some hours and storage in a clean bench are approved measures. PTFE containers, often preferred for inorganic trace analysis, have the disadvantage of acting as a reversible sink for organics. Hydrophobic organic traces become absorbed within the polymer frame and may cross-contaminate a subsequently introduced liquid sample by desorption. Many plastic containers still contain substantial amounts of pre-polymers and plasticizers.

Sometimes, especially for clinical samples like urine, the following measures have been approved:

DOI 10.1515/9783110441154-005

– pH control	By adding acid/base/buffer
– Additives to prevent nonspecific binding	By adding BSA (bovine serum albumin, good for compounds with high protein binding); surfactants (CHAPS, SDBS, Tween)

Initiation of photochemical postreactions can be easily prevented by wrapping the sample container with intransparent aluminum foil. Especially, the UV-B radiation may start dehalogenation reactions.

Freezing of liquid samples is not always recommendable. During thawing, precipitation of inorganic contaminants as metal hydroxides or proteins may happen, which in turn serve as sink for adsorbable organic traces. A good solution is removal of oxygen by inert gas (argon, helium) purging. Of course, this is not applicable to volatile trace compounds. Freeze–drying (lyophilization), the removal of water under vacuum, is an increasingly applied method to obtain long-term stable samples (e.g., for antibiotics in milk).

Soil samples are generally very complex. The microbial content of soil may continue to degrade organic substances and incorporate them into the frame of humic substances. But also acidic sites of alumosilicates can start hydrolysis of adsorbed analytes. Sterilization by hot steam, autoclaving, γ-irradiation methods can be helpful but pose the risk of strong alterations of the matrix and the analyte. Sometimes, the immediate intimate mixing with ultrapure sodium sulfate powder preserves the traces, since hydrolysis by water is excluded. This is, for example, an approved method for aflatoxin determination in soil samples. If analysis of soil samples has to be performed not immediately after sampling, they should be stored frozen (at least −20 °C) to reduce further transformation of the analyte.

Blood samples are per se complicated. Due to blood cells, proteins and inorganic constituents, the pretreatment has to be optimized for the respective analyte. Either centrifugation, or fixing the redox potential by complexing iron (through fluoride addition), or avoidance of emulsification/precipitation is necessary. Table 5.1 contains a collection of accepted stabilizers for various analytical tasks.

For lipid analysis, it might be useful to transform the blood constituents into a powder by addition of ultrapure anhydrous sodium sulfate. From such powder, usual lipid extraction procedures can be applied.

Particulate matter (e.g., aerosol particles) sampled by air filtration should be stored under dry conditions in small gas-tight filter boxes and further deposited in an inert-gas filled desiccator. It is known that under humid conditions, deposited particles may become dissolved and creep over the filter surface.

Organic gases, especially labile and reactive traces, need special consideration. Depending on inertness to oxidative degradation, hydrolysis or reductive transformation, masking of functional groups containing analytes during sampling can be used. Aldehydes and ketones are frequently prevented from autoxidation or condensation by bisulfite adduct formation. Similarly, carbonyls can be protected by hydrazone

Table 5.1: Blood-stabilizing agents.

Stabilizer class	Examples	Compound class
Esterase/Protease inhibitors	NaF, phenylmethane-sulfonylfluoride, BNPP, eserine, paraoxon, acetylcholine, dichlorvos, Ellman's reagent, DFP	Ester prodrugs, amides
Acids	pH ~ 3 formic acid, HCl pH ~ 4 o-phosphoric acid pH ~ 5 citric acid	Acylglucuronides, lactones, esters, amides, enantiomers
Antioxidants	Ascorbic acid, mercaptoethanol/propanol, Na-metabisulfate, L-cystein; EDTA as anticoagulant	Folates, catecholamines
Reducing agents	Dithiothreitol, pyrosulfite	Thiols
Alkylating agents	Methyl acrylate, N-ethyl maleimide	Thiols
Enzyme inhibitors	Aprotinine (DPPIV inhibitor)	Peptides/Proteins

BNPP, Bis(4-nitrophenyl)-phosphate; DFP, Diisopropylfluorophosphate; DPPIV, Dipeptidyl peptidase IV.

formation. Volatile amines can be protected from further transformation by protonation (e.g., by citric acid). The addition of halogenated aminosulfonic acids avoids nitrosamine formation. Adduct formation has been reported as means for stabilizing volatile PAHs by using Ni° surfaces. Here, the aromatic compounds form π-electron complexes with the Ni surface. Organic sulfides are loosely bound to gold wool, where they can be thermally desorbed by moderate temperature increase. In general, a variety of protection techniques known in organic synthesis are applicable.

The widely used gas sorption tubes filled with active charcoal or silica only needs to be closely capped and stored under nitrogen.

Bibliography

[1] Anderson R. Sample pretreatment and separation. Hoboken, NJ: John Wiley & Sons, 1987.
[2] Ballschmiter K. Sample treatment techniques for organic trace analysis. Pure Appl Chem 1983;55(12):1943–56.

6 Enrichment and Sample Cleanup

6.1 General Remarks

Many of the about 100,000 registered chemicals enter the environment intentionally, that is, pesticides, or unintentionally by use of the products, after their use and discharge or by accidental spills [1]. Often, these chemicals and their metabolic products are found only in trace amounts in the environmental media and organisms. Usually, the chemical of interest cannot be analyzed directly because of interfering compounds and matrix effects, except in cases of matrix-poor water, in which direct analysis may be possible if the limit of detection and quantification of the analytical method is sufficiently low. Therefore, physicochemical or bioanalytical enrichment methods must usually be applied in order to clean and concentrate the sample and to prepare it for the desired analytical technique. Chemicals at such low concentrations may still affect exposed organisms, and therefore knowledge about the concentrations in environmental media is needed for a proper risk assessment, which is based on the ratio of exposure and effect parameters. In the following, some techniques for analyte enrichment and cleanup for liquid, solid and gaseous samples are presented.

It is advisable to control the efficiency of enrichment and the recovery of each individual step by addition of defined amounts of a reference substance, ideally labeled by a stable isotope [2], which can be analytically readily followed, or by standard addition of a suitable trace substance. Certain examples for analyses where stable isotope labeling was applied will be included in the following.

6.2 Liquid Samples (Water, Body Fluids, Beverages)

Water samples may be derived from groundwater, rain, snow and ice, surface water such as from lakes, rivers, marine water, waste and industrial process water and drinking water, the latter containing probably the lowest concentrations of matrix components. Sometimes, water from solid matrices also needs to be analyzed, such as soil pore water, for instance to determine the bioavailable part of pesticide residues.

We will also give some examples for sample enrichment from biological fluids such as blood and urine of animals and humans, and morning dew drops of plants. Also, the sample preparation for the determination of organic contaminants in beverages will be presented.

6.2.1 Liquid–Liquid Extraction

Liquid–liquid extraction (LLE) is used to extract organic contaminants from aqueous samples by using an immiscible organic solvent as a second phase. Depending on the distribution coefficient K, that is, the partitioning of the analyte between both

DOI 10.1515/9783110441154-006

phases, concentrations in the organic solvent c_s and the aqueous sample c_a, will reach an equilibrium, often obtained by shaking the two phases for a certain period and determination of the concentrations in the two well-separated phases:

$$K = \frac{c_s}{c_a} \tag{6.1}$$

K is not always constant but shows some concentration dependency. The distribution coefficient does not depend on the volume ratio V_R of both phases, and the fraction F that is extracted into the organic solvent can be calculated by

$$F = \frac{1}{1 + (V_R/K)} \tag{6.2}$$

If a single extraction does not reveal a good enrichment of the analyte, repeated extractions may be performed. By using the same solvent volume in each of the successive extraction steps, the percentage of extracted analyte remains constant, and the concentration in the aqueous phase after n extractions (c_a^n) can be calculated from the initial concentration in the aqueous phase (c_a^0) and the volumes of aqueous (V_a) and solvent phases (V_s) by

$$c_a^n = c_a^0 \left[\frac{V_a}{KV_s + V_a} \right]^n \tag{6.3}$$

If the analyte is ionizable, variation of pH will affect its distribution between the organic and aqueous phase, for example, a carboxylic acid at pH below the pK_a will have a lower polarity than the deprotonated acid and, thus, the distribution coefficient K (Eq. (6.1)) will be higher.

The simplest test system for LLE is to use a separatory funnel (Figure 6.1) in which, depending on the densities of the solvent either the lower or upper phase contains the extracted analyte (solvent examples to be used for extracting aqueous samples are hexane or ether, both ca. 0.7 kg/L, or dichloromethane, 1.3 kg/L and chloroform 1.5 kg/L). Too vigorous shaking may lead to the formation of an emulsion and should

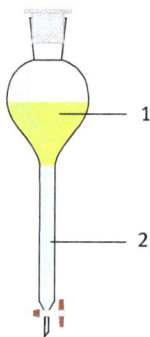

Figure 6.1: A separatory funnel for LLE. Depending on the density of the immiscible organic solvent, it will separate after shaking either to the upper (1) or lower (2) phase.

be avoided; in case an emulsion has been formed, either use an alternative organic immiscible solvent or gently centrifuge the two phases. Another emulsion-breaking technique is the addition of salt (NaCl or Na_2SO_4) or a few drops of methanol that may help to separate the two phases. Using pentane LLE, vinylchloride traces in groundwater were enriched for subsequent gas chromatography (GC)–mass spectrometry (MS) analysis [3].

Support-assisted LLE is a variation of standard LLE. An aqueous sample is applied to a high-surface area matrix such as purified diatomaceous earth, for instance in a syringe. A water-immiscible organic solvent is then passed over the matrix holding the aqueous layer, and the analyte may partition into the organic phase that can be collected and further prepared for analysis.

For analytes with a low volatility, continuous LLEs can be performed: The extracting solvent is vaporized and condensed in a cooling column (Figure 6.2), and droplets are continuously passing the aqueous phase. In both cases, with a density lower or higher than water, the organic solvent is recycled and can be used, after prolonged extraction, for further analysis.

6.2.2 Pre-concentration after LLE (Evaporation, Freeze-Drying)

The volume of the organic solvent after LLE needs to be reduced to get a concentration of the analyte high enough for the analytical method to be used. There are several techniques available that are presented in the following.

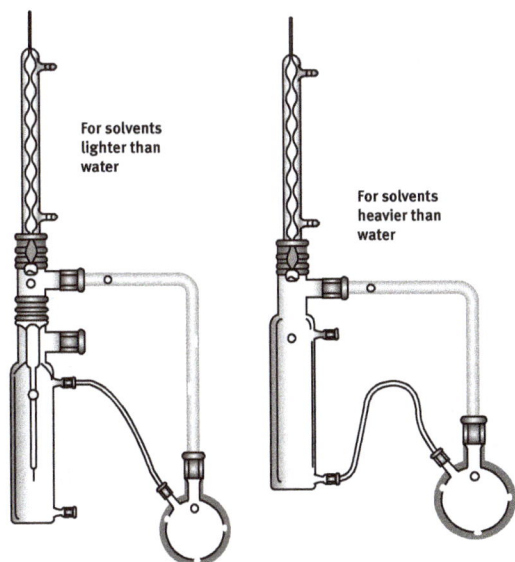

For solvents
lighter than
water

For solvents
heavier than
water

Figure 6.2: Glassware for continuous liquid–liquid extraction.

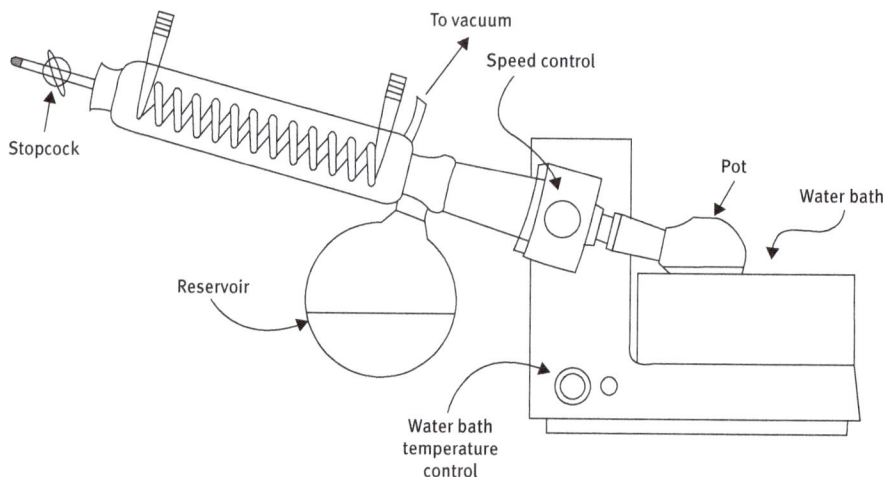

Figure 6.3: Rotary evaporator.

The volume of the extract may be efficiently reduced by means of a rotary evaporator (Figure 6.3), used now for many decades in all kinds of laboratories for analytical and preparative purposes. However, besides enrichment of the analyte, also other components, for instance from the extracted sample matrix, will be enhanced in concentration, that is, subsequent purification steps are essential.

Due to a reduced pressure in the system, the solvent can be evaporated under gentle heating that is also an advantage for the analyte if it is heat sensitive. However, in case of (semi-)volatile compounds, for example, PAHs of low-molecular weight, this enrichment method is not suitable, because losses will occur; the addition of a high boiling "keeper" solvent (for instance, n-octanol) may help one to reduce such losses, but other enrichment methods are preferred. The dissolved sample is rotated in order to reduce the tendency of bumping and to guarantee a homogeneous temperature in the sample and the formation of a solvent film with a large surface in the rotating flask. Even solvents with high boiling points above 100°C may be evaporated at manageable temperatures if the pressure is sufficiently low.

Somewhat related to this technique, with the advantage of preventing bumping but the disadvantage of allowing only smaller volumes to be evaporated compared to rotary evaporation, is the combination of reduced pressure, heating and centrifugation (centrifugal evaporation). Samples are placed in a centrifuge that can be evacuated and heated. Under such conditions, even solvents with high boiling points, for instance DMSO (boiling point 189°C), can be evaporated at temperatures below 50°C. Evaporated solvents are condensed at a cold trap.

The remaining sample, typically with a volume of a few mL, may be further reduced by gently passing nitrogen over the solvent surface until the desired volume is reached.

Freeze-drying (synonyms are lyophilization or cryodesiccation) is the removal of solvents from a sample in the frozen state under reduced pressure, that is, a phase transfer from the solid state directly into the gaseous state (sublimation). Generally, samples that contain water as the main liquid element can be freeze-dried, whereas those containing organic solvents with lower freezing points are usually dried using vacuum concentration (without freezing), although also freeze-drying is in principle possible [4]. First, the aqueous sample is frozen at temperatures below the triple point of the phase diagram, at which the solid and liquid phases coexist, usually by rotating the flask in a bath, filled with dry ice and an organic solvent, or with liquid nitrogen. Once the product is frozen, it is sublimated at a low temperature under reduced pressure. The free water (primary drying) and eventually that bound to the dissolved sample, for example, a protein (secondary drying) is recondensed at a cold trap installed in the freeze-drying apparatus.

The analysis of volatile organic compounds (VOC), such as di- or trichloroethane, bromodichloromethane, dibromochloromethane, BTEX (benzene, toluene, ethylbenzene and xylene) or trihalomethanes, needs specific extraction procedures. They all depend on the ratio of the concentrations of the analyte in the gas phase (c_{gas}) and the liquid phase (c_{liq}), and thus on the water solubility and the vapor pressure of the analyte and the temperature. K, the partitioning constant ($K = c_{gas}/c_{liq}$), is increased at higher temperatures or by adding salt to the aqueous sample that often decreases the water solubility of the organic analyte. A recent review on several VOC extraction methods has been published by Chary and Fernandez-Alba (2012) (see Further Reading). In the case of VOC analysis, gas–liquid extraction techniques can be applied or alternative enrichment procedures may be selected such as solid-phase extraction (SPE) (see below).

6.2.3 Solid-phase Extraction

SPE enables the extraction, cleanup and concentration of analytes prior to their analysis and quantification. SPE resembles the liquid column chromatography process as the analyte partitions between a solid phase, packed in a, for example, polypropylene syringe, and the mobile phase, for example, an aqueous sample that is passed through the solid phase, for example, by means of a vacuum.

SPE enrichment comprises five steps (Figure 6.4): First, the solid matrix is solvated by a solvent like methanol. Second, the matrix is washed (conditioned) by water or buffer and third, the aqueous sample is loaded. Then, the matrix containing the sorbed analyte is washed with water, buffer or a solvent, that does not elute the analyte (i.e., a nonpolar solvent for polar compounds and a polar solvent for nonpolar compounds), to remove impurities and co-contaminants in the sample. After drying the sorbent by drawing air through it as the last step, the analyte is eluted from the matrix by use of a suitable solvent. By these steps, it is possible to not only extract

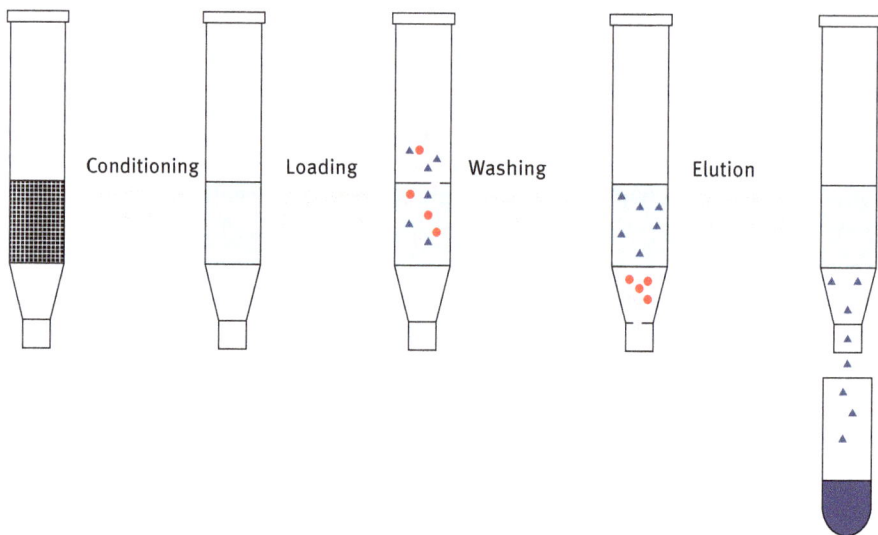

Figure 6.4: Principle of solid-phase extraction.

the analyte from the water sample but also to transfer it to a volume, much smaller than the original sample volume, for example, a few hundred milliliters of aqueous sample may be rapidly reduced to a few hundred microliters of a concentrated solution in an organic solvent that afterward can directly be analyzed. Not only flow rates in the individual steps of about 5 mL/min are usually used but also higher rates up to 30 mL/min (except the elution step that should always be performed with a low flow rate) have been successfully applied without loss in recovery (recovery = amount percolated/amount extracted).

With respect to the sorption of the analyte, the partition coefficient K equals the ratio of the concentrations in the solid (c_s) and mobile phase (c_m): $K = c_s/c_m$. The fraction F of the analyte extracted from the aqueous sample can be calculated by

$$F = \frac{c_s V_s}{c_s V_s + c_m V_m} = \frac{1}{1 + (KV_r)^{-1}} \tag{6.4}$$

with V_s and V_m representing the volumes of the solid and mobile phases and V_r as the ratio V_s/V_m.

Typical SPE sizes for trace analytical work are from 50-mg sorbent mass and 1-mL column volume up to 10-g sorbent mass and 60-mL column volume. A variety of solid phases, resembling those in liquid chromatography, are available for SPE and depending on the type hydrogen bridging, hydrophobic interactions, electrostatic and ion-exchange forces will contribute to the partitioning process. Cartridges with different fillings (particle sizes ca. 30–70 µm) and in different sizes are commercially available. The sorbent selection depends on the properties of the analyte:

polar compounds can be extracted using normal phase silica, Florisil, cyano-, amino-, diol-modified silica or alumina; moderately polar substances by phenyl-, ethyl- or cyclohexyl-modified silica; ionic compounds by ion exchange phases and unpolar substances by C_{18}- or C_8-bonded silica (reversed phase). Specific binding of analytes to molecular imprinted polymers (MIP) and immunoaffinity columns are alternative methods to specifically bind analytes for enrichment and subsequent analysis (see later). A thorough survey of available phases that are suitable for SPE of environmental pollutants was provided by Barcelo (see Further Reading) and are recommended by the manufacturers. For compounds adsorbed on reversed phase beds, suitable eluents are methanol, acetonitrile or ethyl acetate; ionic compounds adsorbed to ion-exchanger beds are eluted by adjusting the pH well below or above the pK_a or by buffers with high ionic strength.

For optimal SPE results, the proper combination of sorbent, washing solution and elution solvent is crucial. SPE has the advantages of less solvent waste being produced compared to LLE, and it is ready to use even in the field, operationally simple and rapid to perform. It can be used off-line, that is, the analyte solution is subsequently and independently analyzed by a suitable technique like GC or high-performance liquid chromatography (HPLC), or online by direct coupling to the analytical device, for example, HPLC or GC. Potential drawbacks in SPE are the non-efficient binding of the analyte to the sorbent – which can be solved by choosing an alternative packing – and the incomplete elution of the analyte from the sorbent. This may be solved by use of another eluting solvent, but anyway for the establishment of a quantitative recovery, corresponding analysis with a reference compound (as internal or external standard) must be performed. Problems that co-contaminants have a similar affinity to the sorbent as the analyte may be overcome by changing the composition of the liquid used for the partitioning, that is, by changing the selectivity of the sorbent. Breakthrough of analytes because of overloading the sorbent is unlikely for trace analysis of organic contaminants that are often present in the sub-µg/L range; 10 L passed through a typical 500-mg bed packing will thus contain only a few micrograms of analyte.

The usefulness of SPE cartridges for the determination of the beta-blockers acebutolol, atenolol, metoprolol, nadolol and propranolol and beta-agonists salbutamol and terbutaline in environmental aqueous samples using GC techniques has been tested by Caban et al. [5]. For most compounds, present in a mixture of each 2.5, g/L, the recoveries varied considerably depending on the applied extraction conditions by a factor of at least 10.

SPE cartridges can be obtained with many different matrices for adsorbing analytes depending on the physicochemical properties (Table 6.1).

Immunoaffinity SPE columns contain a matrix in which the stationary phase consists of an antibody or antibody-related reagent to selectively purify a target compound. A molecule that is capable of initiating antibody production is called an antigen. Common natural antigens include viruses, bacteria and foreign proteins from

Table 6.1: Solid-phase extraction cartridges and analyte properties.

Aqueous samples (water, aqueous extracts, biological fluids)		Organic samples (organic extracts)	
↓	↓	↓	↓
Reversed phase	Ion exchange	Ion exchange	Normal phase
↓	↓	↓	↓
Analytes: moderately polar to nonpolar compounds	Analytes: weak cations/anions	Analytes: strong cations/anions	Analytes: polar-to-moderately polar compounds
Example analytes: pesticides, PAHs, PCBs, phthalates	Example analytes: organic chemicals with carboxylate or ammonium groups, ionizable chemicals	Example analytes: organic chemicals with sulfonate or ammonium groups, ionizable chemicals	Example analytes: alcohols, aldehydes, amines, pesticides, PCBs, ketones, nitro compounds, organic acids and phenols
↓	↓	↓	↓
SPE cartridges: octadecyl (C$_{18}$), octyl (C$_8$), phenyl, cyanopropyl	SPE cartridges: for anions: matrix with quaternary amino groups. For cations: matrix with aliphatic sulfonic acid groups	SPE cartridges: for anions: matrix with quaternary amino groups, aminopropyl. For cations: matrix with carboxylic acid groups	SPE cartridges: florisil (magnesium silicate), silica, alumina (for different pH ranges), dioles, cyanopropyl, ethylenediamine-N-propyl (primary + secondary amines), graphite carbon based (Hypercarb™)

animals and plants that are capable of producing an immune response. Due to the large size of naturally occurring antigens, antibodies that bind to several different regions of the antigen with a range of binding affinities are often generated. In order for a substance to be recognized by the body's immune system and to lead to the production of antibodies, this substance must have a size corresponding to a mass of several thousand Daltons. However, antibodies can be produced also against smaller substances (haptens), which need first to be coupled to a larger species (e.g., a carrier protein like serum albumin) before antibody production can occur. A small substance that is used to produce antibodies after being linked to a carrier agent is known as a hapten (see Chapter 10) [6].

A specification of SPE is the use of MIP. The basic principle is the polymerization of monomers in the presence of a template molecule, for example, a pesticide or pharmaceutical agent, that can be extracted – or cleaved if covalently bound– from the polymer afterward, leaving a complementary cavity in the matrix to which the analyte will bind, ideally with high specificity. The technique can be used for SPE using the MIP as solid support in cartridges or in combination with solid-phase microextraction (SPME) [7]. Dai et al. reviewed recently the analytical applications of molecularly imprinted polymers on the surface of carbon nanotubes (see Further Reading).

6.2.4 Solid-phase Microextraction

A method for extracting organic chemicals from aqueous phases and biological tissues related to SPE is the SPME technique invented by Pawliszyn some 25 years ago [8]. A fused silica fiber, 1–2 cm long, to which a nonvolatile organic liquid or polymer is immobilized, acts as the sorbent. The most common coating material is polydimethylsiloxane (PDMS). Other phases like polyacrylate and mixed Carbowax™ and Carboxen™ coatings are available [9]. SPME can also be used for trapping volatile organic substances from the gas phase.

The coated fiber can be used as part of a syringe and placed within the needle from which it can be extended to the sample (Figure 6.5). After a certain sampling time, the fiber is retracted and used for analysis, either by extraction with organic solvents and subsequent injection into an HPLC, GC or capillary electrophoresis system or alternatively directly into a heated injection port of a GC from which the analyte is desorbed (Figure 6.6). For GC, some analytes need to be derivatized before. In situ derivatization

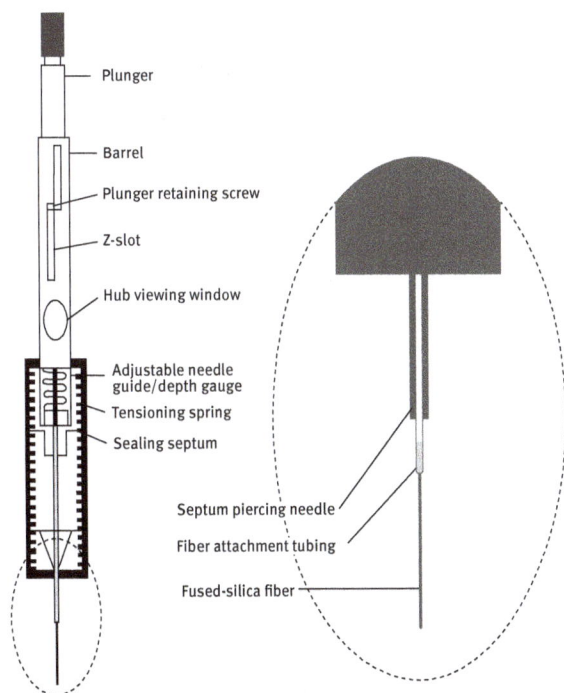

Plunger

Barrel

Plunger retaining screw

Z-slot

Hub viewing window

Adjustable needle
guide/depth gauge

Tensioning spring

Sealing septum

Septum piercing needle

Fiber attachment tubing

Fused-silica fiber

Figure 6.5: Details of a commercial SPME device.

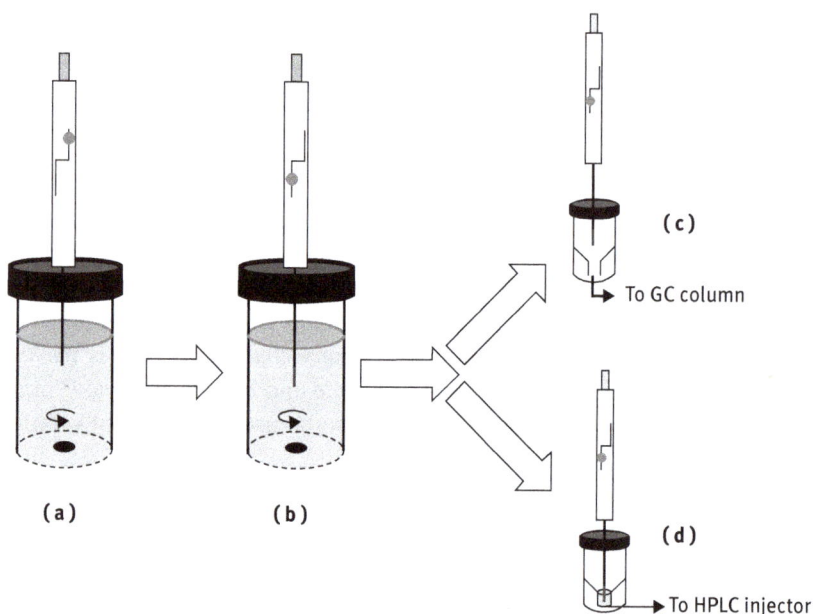

Figure 6.6: Principle of solid-phase microextraction. (a) Insert sampling device to the solution with the analyte; (b) expose fiber to the solution; (c) use fiber for GC analysis (either after extraction or directly); (d) extract fiber for HPLC analysis. Please refer to the text.

is one common technique in SPME: A derivatization agent is first added to the sample and the SPME fiber extracts the derivatized analytes that are injected to the GC. Also, direct coupling of the SPME technique to HPLC-MS is possible: The needle containing the fiber is placed into a modified Rheodyne or Valco valve, and the analytes are eluted by the mobile phase. The SPME technique is very well suited for GC and MS applications. The sampling fibers can be used multiple times.

Quantification of the analyte is possible, but it depends on many parameters such as the sampling (exposure) time, the partition coefficient of the analyte between aqueous and acceptor phase, the temperature and the length of the fiber used for the final analysis. MS detection is the optimal quantitation technique as it allows isotopically labeled standards to be added to the sample such as ^{13}C- or ^{2}H-derivatives.

Microwave-assisted headspace SPME with PDMS fibers was used for PAH analysis in environmental water. By GC–MS analysis, detection limits of 30–130 ng/L were obtained. Dissolved organic matter (DOM) was found not to affect the PAH analysis, but it is possible to determine the association constant of pollutants to DOM by this method. Numerous papers on SPME for analyzing pesticides, benzotriazoles, alkylphenols, bisphenol A, perfluorinated chemicals, hormones and pharmaceuticals, often with detection limits in the low ng/L range, have been published and were reviewed, for instance, by Vas and Vekey and recently by Padrón et al. (see Further Reading).

6.2.5 Dispersive liquid–liquid microextraction

Another more recent development is the micellar extraction (ME) or dispersive liquid–liquid microextraction (DLLME) that has been used in combination with many analytical techniques. It represents a simple and rapid method (minutes), which requires low volume of sample (5 mL).

In DLLME, analytes are extracted in an organic extracting solvent dispersed in water. To achieve dispersion, the solvent (the dispersing solvent, such as CCl_4 [10]) is rapidly injected to a water sample. A dispersion is formed, which facilitates fast extraction of analytes from the water sample. The solvent droplets are removed by centrifugation and the extracting solvent containing analytes is taken for analysis with a microsyringe (Figure 6.7).

For analysis of PAH in tap water, rain waters and river surface waters, good recovery was achieved using trichloroethylene as the extraction solvent and acetonitrile as the dispersive solvent. DLLME was used for the simultaneous extraction of morphine, 6-acetylmorphine, cocaine, benzoylecgonine and methadone from human plasma. The extractant solvent was chloroform and the dispersant solvent acetonitrile together with NaCl as a salting-out additive [11].

For instance, Berijani et al. analyzed organophosphorus pesticides from water samples [12]. A mixture of 12 µL chlorobenzene as extraction solvent and 1 mL acetone as disperser solvent was rapidly injected into 5 mL of the aqueous sample. The droplets of chlorobenzene into which the analyte is extracted from the water phase

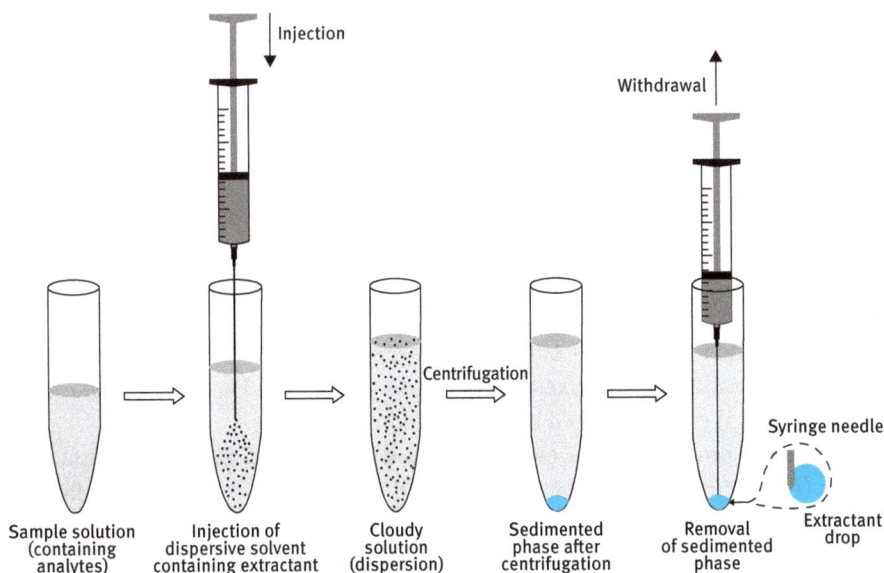

Figure 6.7: Principle of dispersive liquid–liquid microextraction.

give rise to a cloudy state of the sample. The fine droplets of chlorobenzene are sedimented by centrifugation and can be used for analysis. A novel ultrasound-assisted DLLME based on solidification of floating organic droplet method combined with GC was developed for the determination of pyrethroid pesticides in tea. 1-Dodecanol and ethanol were used as the extraction and dispersive solvent, respectively. Under optimal conditions, the enrichment factors were 100-fold with recoveries between 92% and 100% [13].

A similar approach as DLLME is the method of hollow fiber liquid–liquid–liquid microextraction. Here, also the analytes are extracted from, for example, biological fluids, into a few microliters of a solvent. Typical sizes of hollow fibers are 600 μm inner diameter, 200 μm wall thickness and 0.2 μm pore size. Cleaned hollow fibers are cut into few centimeter pieces and filled with the acceptor phase. For analysis of analgetical pharmaca from urine, the fiber pieces were filled with 0.5 M H_2SO_4 as acceptor phase by means of a Hamilton microsyringe. The fiber was immersed in isoamyl benzoate shortly to impregnate the pores of the fiber and washed afterward to remove excess solvent from the fiber surface. Then, the fiber pieces were placed into 5 mL urine sample under stirring. After extraction for about 20 min, the fiber was removed from the sample and the acceptor phase withdrawn and injected into the analytical device [14].

6.2.6 Micellar Extraction

Micellar extraction, often also named *cloud-point extraction*, was introduced by Hinze in the 1980s of the last century. ME offers the possibility to combine extraction (without solvents!) and enrichment within one step. It makes use of the phenomenon that surfactants within water may form micelles, when the so-called critical micelle concentration is exceeded. This can be followed, for example, by light extinction measurements. Depending on the nature of surfactant and solvent being dispersed, these micelles form a hetero-dispersed fraction within a liquid. Micelles own a nonpolar, hydrocarbon-like core and may solubilize lipophilic contaminants like PAHs (see Figure 6.8).

Nonionic surfactants form micelles, which lose their water solubility, if temperature is raised above a critical point, known as the cloud point (CP). Above the CP, the surfactant solution becomes turbid and separates into two phases, a surfactant-rich micellar phase (MP) and a surfactant-poor water phase (see Figure 6.9). Separation from each other is easily achieved by centrifugation.

The general procedure of ME can be described in the following way: to a surfactant solution (e.g., Triton X-100, Genapol 80) lower than 1% (w/w) in a calibrated flask, a specific amount of salt is added to assist phase separation. This solution is spiked with appropriate volumes of liquid sample. The flask is kept thermostated in a water bath until separation into MP, and surfactant-poor part is complete. The MP is then removed by pipetting for further analysis.

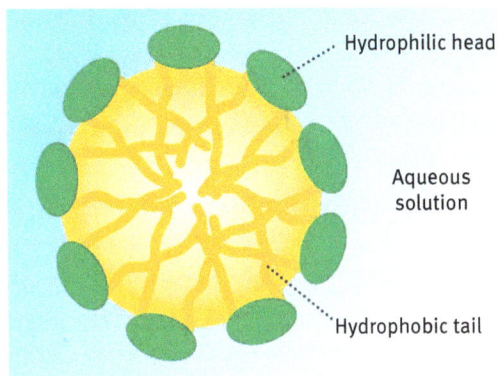

Figure 6.8: Micelle structure in water.

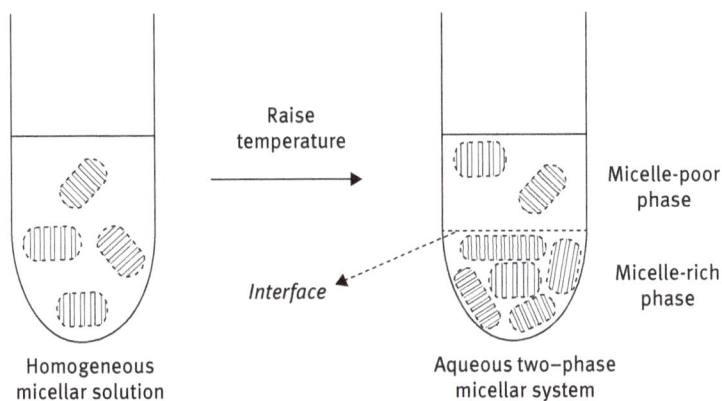

Figure 6.9: Formation of a micellar phase with micelle-enriched analyte.

The great advantage of this is that there is no longer a need for large volumes of inflammable and toxic organic solvents (as usual in LLEs). In general, the direct determination of the enriched analyte within the MP is not a trivial task: Large amounts of surfactants can cause enormous problems, for example, in the application of GC. Direct measurement of PAHs within the MP by synchronous luminescence for PAHs has been reported. Using a back-extraction, pesticides can be determined by GC. Especially, the trend to "green chemistry" favors the substitute of solvent extraction. Lipid extraction is currently in the focus.

Of special interest is the observed desorption capability for soil particle-bound analytes. More difficult matrices like PCB-containing landfill leachates can be extracted not only much more conveniently by ME than by LLE but also with a higher performance. Some examples on this technique is found in Further Reading.

6.2.7 Headspace Extraction

Headspace extraction is based on Henry's law describing the ratio of the concentration of an analyte in an aqueous medium (c_{aq}) divided by its partial pressure p in the gas phase above the water phase under equilibrium conditions.

$$H^c = \frac{c_{aq}}{p} \tag{6.5}$$

The concentration-based Henry coefficient H^c has a unit of mol/(m^3 Pa). The dimensionless Henry coefficient H equals the ratio between the aqueous-phase concentration c_{aq} of an analyte and its gas-phase concentration c_{gas}:

$$H = \frac{c_{aq}}{c_{gas}} \tag{6.6}$$

For headspace analysis, the aqueous sample is placed in a septum-sealed container, which may be heated to shift the equilibrium of the liquid–gas partitioning to the gaseous phase. Using a syringe, an aliquot of the gas phase is sampled and injected into a gas chromatograph. This type of analyzing volatile organic compounds (VOC) can be performed either directly, if the analyte concentration is sufficiently high, or by concentrating the VOC in a cold trap or in an adsorption cartridge (filled, e.g., with activated charcoal, Tenax™ or silica adsorbents) from which the analyte is desorbed afterwards, for example, by heating (Figure 6.10).

The detection limits of the static headspace technique are higher than those of purge and trap analyses [15], but the latter requires more complex instrumentation and may lead to a higher water concentration in the gas phase.

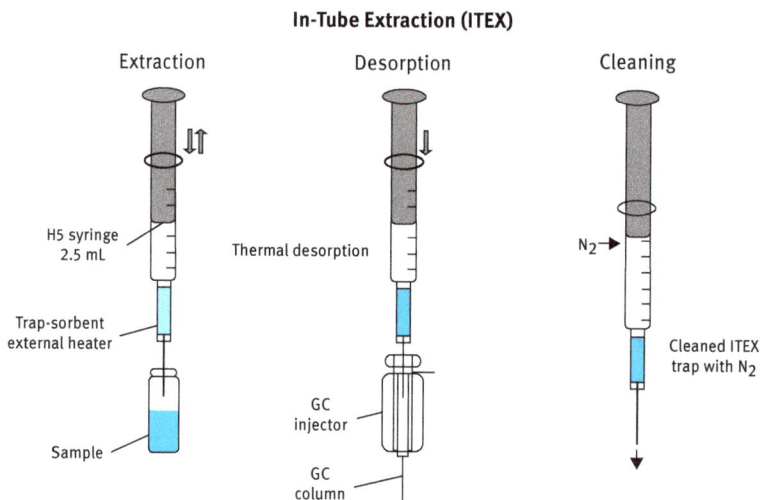

In-Tube Extraction (ITEX)

Figure 6.10: Principle of headspace extraction analysis.

Priority VOC from water and wastewater samples have been analyzed after headspace sampling and combining this with a needle-trap device filled with Tenax™ and Carboxen™ particles to concentrate the VOC, the LOQ could be pushed into the ng/L range. Kavcar et al. analyzed several trihalomethane species, BTEX and naphthalene by using the headspace method and found that VOC concentrations in Turkish megacities are in part of toxic concern, especially that of bromodichloromethane and dibromochloromethane. All trihalomethane species were detected in higher concentrations in tap water, whereas non-tap water contained more BTEX and naphthalene [16]. Various VOC have been analyzed in the field with microwave radiation as heating source for a more rapid headspace sampling and use of a field-portable GC–MS [17].

Gas–liquid-partition coefficients of, for example, BTEX, C_9–C_{10} aldehydes, toluene and ethylbenzene have been determined by the headspace method in combination with GC–MS. The authors studied the effect of solid particles on their partitioning behavior and found that for instance the presence of diesel soot and mineral dust expectedly reduced the K constants significantly ($K = c_{gas}/c_{liq}$) [18].

6.2.8 Purge and Trap Extraction

Purging the water by passing synthetic air, N_2 or argon through the aqueous sample, for instance using a glass frit placed below the liquid surface, and trapping the volatiles on solid adsorbents in a trap is an often used method for VOC analysis. The purge chamber may be warmed up in order to increase the VOC concentration in the gas phase. The VOC may be trapped on a Tenax™ or activated charcoal trap that is heated afterward to desorb the volatile to be analyzed by GC. Since desorption may last for some time, a second trap, for example, cold trap, may be used to collect the desorbed residues which is then heated to instantaneously feed the analyte(s) into the GC. Purge and trap analysis may be either static or continuous, the latter shown schematically in Figure 6.11. Commercial instruments provide a software-regulated analysis scheme regarding purge time and temperature and trapping parameters. If these parameters are not optimized, for instance choosing too long purging times, the analysis is jeopardized by too high water amounts trapped that may lead to badly resolved gas chromatograms.

Purge and trap extraction of VOC may be combined with isotope specific analysis (^{12}C/^{13}C) by using an isotope-ratio mass spectrometer: LOQ for BTEX and halogenated hydrocarbons by using this technique in groundwater samples were in the sub-µg/L range [19]. Vinylchloride traces in a groundwater monitoring study were directly analyzed by nitrogen purging of the water and directing the VOC into a cooled glass tube that was filled with Tenax™ followed by thermal desorption for GC–MS analysis [3]. The major contents of VOC measured by purge and trap extraction in the effluent of a wastewater treatment plant in Taiwan contained dimethylsulfide, acetone and

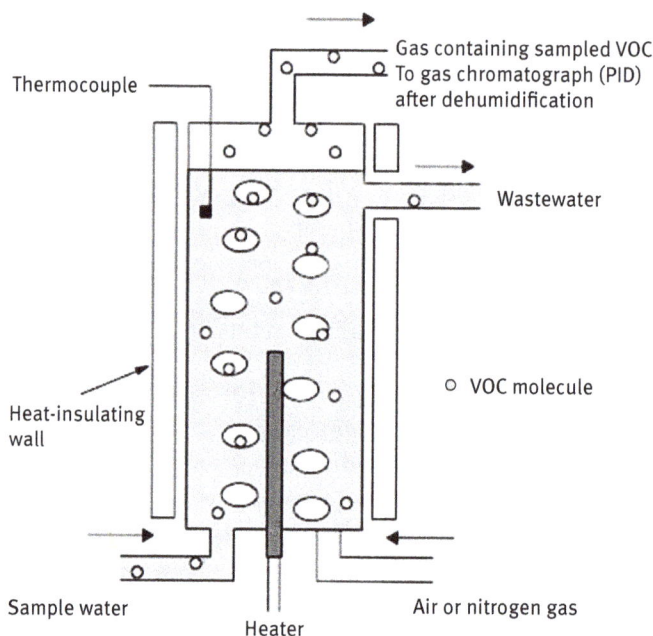

Figure 6.11: Scheme of a continuous purge and trap analysis of aqueous samples.
PID = Photoionization detector.

isopropyl alcohol, all in the range of 10–100 µg/L, as determined by purge and trap GC–MS [20].

A higher extraction efficiency than the conventional purge and trap technique has been obtained by a continuous gas–liquid counter-current flow extraction and a concentration system using a spiral tube combined with GC–MS analysis. They showed that the gas–liquid partitioning and extraction efficiencies are higher than those of purge and trap analysis. The methodology was also applied to an online monitoring system of trace volatile organic components in river water [21].

6.3 Solid Samples (Particulate Matter, Soil, Sediment, Plant and Animal Material)

In liquid samples, analytes can either be directly quantified by a suitable analytical technique, if the concentration is high enough, or, if present in trace amounts, need to be extracted and enriched. For solid samples, that is, particulate matter, soil, sediment and biological tissue such as plant and animal material, always efficient extraction methods are needed, followed by cleanup, for example, by chromatography, and enrichment procedures. Since the analytes may have a strong affinity to the solid matrix, more vigorous extractions are necessary, which lead to the penetration of the

solvents into the voids of the solid matrix. As a consequence, natural and xenobiotic co-contaminants may be released that might lead to interference in the analytical
method.

In any case, varying amounts of the analytes are so tightly bound to the matrix
that they cannot or not with a reasonable expenditure be released by nondestructive extraction methods. Such residues are called non-extractable residues (NER) and
methods to investigate NER are also shortly described below (Section 6.3.4).

Determination of the recovery when extracting organic contaminants from solid
samples is challenging. Spiking a soil or sediment, that is, addition of an analyte followed by immediate extraction, does not reflect the real extraction efficiency because
of the process of aging which leads for many contaminants to a decreasing extract-
ability with time. Since it is impossible to "age" a sample after addition of an analyte
for prolonged time, for example, years, a compromise would be an incubation time in
soils or sediments of not less than 24 h before the extraction is performed. An elegant
method to determine the recovery of extraction efficiencies is the isotope dilution MS
(see Further Reading and Chapter 3.3). An isotopically enriched substance, labeled
for instance either with ^{13}C, ^{15}N or ^{2}H, is added to the sample that leads to the dilution
of the enriched reference substance. By MS, the signal ratio of the added reference
is quantified. As an example, pesticides in soya beans have been quantified by this
method [22].

Prior to extraction, grinding of the sample (soils, sediments) may lead to a better
extraction recovery.

6.3.1 Solid–Liquid Extraction

6.3.1.1 Bioavailable Concentrations

Sometimes, especially from an ecotoxicological point of view, it is reasonable not only
to determine the total amount of an analyte in soils or sediments but also to quantify
the bioavailable part. For this purpose, mainly two methods are available: sequential
extraction and the determination of the analyte concentration in soil/sediment pore
water, either by physical separation of the pore water, for example by centrifugation,
and subsequent analysis or by SPE methods of the pore water such as Tenax^{TM}
extraction [23], polyoxymethylene SPE [24] or solid-phase microextraction, the latter
described in Section 6.2.3.

Sequential extraction starts with shaking the solid sample with an aqueous solution (e.g., 0.01 M $CaCl_2$), by which the potentially bioavailable fraction is removed,
followed by solvents with a higher elution potential, typically organic solvent/water
mixtures or pure organic solvents (see Further Reading). By this method, a pragmatic
scheme for the fractionation of PAH in soils correlating with their bioavailabilities has
been developed: Four PAHs were spiked into two soils, and the distribution of PAH
fractions was measured for up to 210 days. PAHs were fractionated into water-soluble,

organic acid-soluble, organically bound and residual fractions via sequential extraction supported by ultrasonic treatment. The water-soluble and organic acid-soluble fractions corresponded to the most bioavailable portions, as tested by a semipermeable membrane desorption assay. The bioavailable residues decreased significantly during aging and the remaining fractions correspondingly increased [25].

The concentrations of imazaquin in the pore water of five soils were determined to estimate the bioavailability of the pesticide in terms of phytotoxicity to sorghum. Pore water of the soil was obtained by use of a centrifugal filter device, centrifugation and filtration prior to HPLC analysis. Dose–response curves based on the pore water concentrations versus the used toxicity endpoint (plant biomass) revealed that the effective dose based on the pore water concentration was almost identical in the five investigated soils. Thus, the soil pore water concentration is a parameter for estimating the bioavailability of chemical pollutants [26].

6.3.1.2 Exhaustive Extractions to Release Bioavailable and Inert ("Not Bioavailable") Pollutants

In order to release as much as possible of a contaminant in solid samples, so-called exhaustive extractions are performed, either in the cold or, as described below, at elevated temperatures and pressure and other change in conditions, respectively. Samples are extracted for prolonged times using an overhead shaker or a horizontal shaker. From labeling experiments, for example, by using ^{14}C-labeled pesticides, it is known that the extractions must be continued several times until in the last step, for example, less than about 5% of the first extraction is released from the matrix. This is usually obtained after at least three to five extraction steps.

The selection of the proper organic solvents is a critical step. The physicochemical properties of the analyte, that is, its volatility, water solubility, the solubility in the organic solvent to be used, the pK_a and the stability, as well as the sample properties (such as the water and organic matter contents), must be considered. Some examples for liquid–solid extractions will be presented. Before extraction of solid dry matrices, it is common to add some water to the samples, to hydrate them and make the pores in the sample more accessible to the extraction solvent [27, 28]. A low water content when using organic solvents for extraction will also prevent shrinking of three-layer clay silicates potentially trapping the analyte in the interlayers that could occur if pure organic solvents like acetone were used.

6.3.1.3 QuEChERS Method

QuEChERS stands for a quick, easy, cheap, effective, rugged and safe extraction method. In the original approach, acetonitrile is added in a 1:1 v:w ratio to the sample to extract the analyte from the matrix. After shaking, anhydrous MgSO$_4$ and NaCl are added, and the mixture obtained is shaken again to dehydrate the sample. After centrifugation, an aliquot of the acetonitrile supernatant is recovered for a further cleanup step by dispersive SPE. The SPE matrix contains a weak anion exchanger, dispersed in the extraction solvent, to bind co-extracted acidic compounds (e.g., sugars, fatty acids and organic acids) and anhydrous $MgSO_4$. After shaking and centrifugation, the supernatant is analyzed, for example, by GC– or LC–MS. To control the partitioning of the analyte between water and the solvent (acetonitrile) and, because the potential lability of the analyte under acidic and/or alkaline conditions, the extraction should be performed by use of a buffer (e.g., acetate or citrate buffer). For lipid-containing samples, for instance food matrices, graphitized carbon black or silica–C_{18} may be used for cleanup instead of the anion exchanger. The extraction efficiency may be optimized by using an ultrasonic device.

A comprehensive overview on the use of the QuEChERS method for different analytes (mainly pesticides, phenols, perfluorinated chemicals, chlorinated hydrocarbons, pharmaceuticals) and matrices (soil, sediment, food, feed) is presented in the Further Reading section and in some specific examples [29–37].

6.3.2 Assisted Solid–Liquid Extraction (Soxhlet, ASE, Microwave Extraction, Ultrasonication, Supercritical Fluid Extraction)

LLE as described earlier is a laborious technique and requires large amounts of organic solvents. Therefore, improved methods have been developed such as Soxhlet extraction, accelerated solvent extraction (ASE), supercritical fluid extraction (SFE), microwave assisted extraction (MAE), which will be described in the following.

6.3.2.1 Soxhlet Extraction

Invented some 150 years ago, Soxhlet extractors are still in use in many laboratories. The solid sample, often mixed with a drying agent such as sodium sulfate because wet samples may repel lipophilic solvents, is loaded in a porous extraction sleeve made of thick filter paper. The solvent is distilled from a reservoir and is liquefied at a water cooled condenser from which the hot solvent drips into the sample sleeve. The apparatus is constructed in a way that the solvent is collected and, if the solvent level reaches a drain pipe (a siphon), it is directed back to the sample reservoir in which the solvent cycling starts over again. The continuous flux of the hot solvent through the sample allows for very efficient extraction results. A picture of a Soxhlet extractor is shown in Figure 6.12.

Care has to be taken to prevent loss of semi-volatile and volatile analytes; for such samples, Soxhlet extraction is not the proper choice. Also, the potential thermal degradation during the long extraction procedure (several hours) needs to be considered.

6.3.2.2 Accelerated solvent extraction, Pressurized Solvent Extraction

This method has been in use for about 20 years and is nowadays commercially available from many brands. The method uses elevated pressure and simultaneously

Condenser

Reactor
section

Vapor
bypass tube

Syphon
tube

Extraction
thimble

Side port for
pumped-mode
operation
[stopper in for
normal-mode
operation]

Reservoir

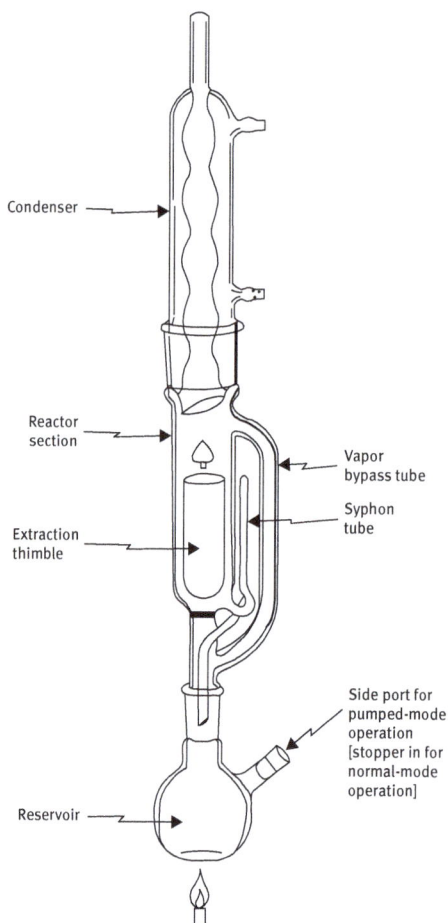

Figure 6.12: Scheme of a Soxhlet extractor.

elevated extraction temperatures and allows for rapid and thorough extraction of solid samples. The sample is filled into a stainless steel cartridge (3–30 mL) to which the solvent is added by a solvent pump. The cartridge is closed and valves control the increased pressure that will build up when the temperature is increased (50–200 °C). After a predefined extraction period, the extract is directed to a vial, and the sample purged with some fresh solvent and inert gas. A picture of an ASE unit is shown in Figure 6.13. The method can be used for extraction of all kinds of solid samples including soil, sediment, sludge, plant and animal tissues. The improved extraction properties are due to the lower viscosity of the solvent at the high temperatures and the increased pressure by which the solvent has a very efficient contact with the solid matrix. Regarding automation, common pressurized solvent extraction (PSE) instruments on the market allow 24 or 12 samples to be extracted simultaneously.

The disadvantage of thermal degradation of the analyte at the high temperatures applied is counterbalanced by the much shorter extraction times compared to

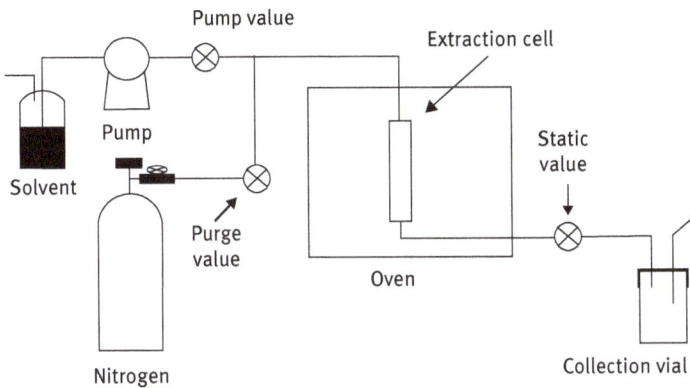

Figure 6.13: Scheme of accelerated solvent extraction (ASE).

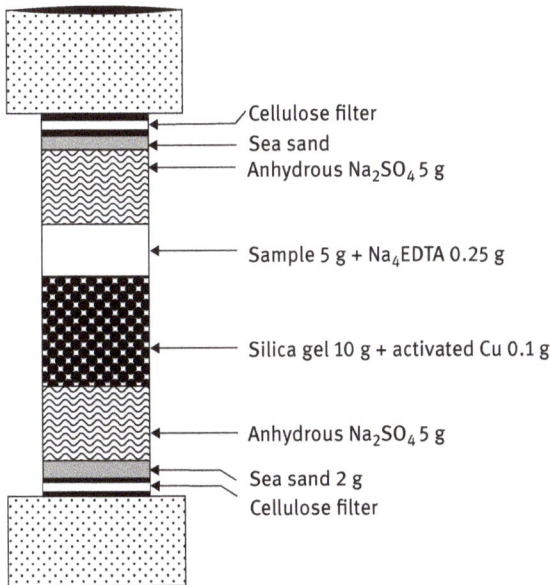

Figure 6.14: Extraction cartridge used for pressurized liquid extraction (PLE) of PAH in sediments according to Kim et al. [38]. Reproduced with permission from Elsevier.

the Soxhlet method. Another advantage is the small amount of solvent to be used. The solvent extraction properties in ASE can be improved by addition of a modifier. A sodium-EDTA-assisted pressurized liquid extraction (PLE) system was developed for the routine analysis of polycyclic aromatic hydrocarbons in sediments with a reasonable high extraction efficiency (around 90%) using the cartridge shown in Figure 6.14 [38].

The PSE method can even be applied with water as a solvent: Organic contaminants can be extracted from solid and powdered samples such as plant and food materials as well as soils and sediments. It has been shown that at certain temperatures and pressures, the polarity of water can be varied close to those of alcohols. Also here, the addition of an organic or inorganic modifier can enhance the solubility of the analytes in water and thus increase the extraction efficiency.

Organochlorine pesticides of the DDT type have been extracted from soils by use of ethanol, 1-propanol and three fractions of petroleum ether differing in boiling points. By testing different factors, that is, organic solvent concentration, washing time, mixing speed, solution-to-soil ratio and washing temperature, the optimal extraction parameters were a solvent-to-soil ratio of 10:1, an extraction time of 3 h, a mixing speed of 100 rounds/min and an extraction temperature of 50°C. These parameters were optimized for remediation purposes. For a quantitative analysis, ASEs at a temperature of 100°C and a pressure of 1,500 psi (10.345 MPa) with hexane/acetone (1:1, v/v) were performed. Recoveries above 90% of spiked soil were obtained.

Superheated but subcritical water was used to extract polycyclic aromatic hydrocarbons from soils and sediments. Decreasing the polarity of water by successive increase of the extraction temperature from 50 to 200°C at a pressure of 10.3 MPa enabled selective extraction of PAH proportional to their octanol–water partition coefficients. As the temperature of extraction increased from 50 to 100°C and subsequently to 150°C, there was an increase in the PAHs extraction extents from contaminated soil, but a further increase to 200°C did not result in significant increase in the better extraction efficiencies of PAHs.

PLE was developed for the simultaneous extraction of polychlorinated dibenzo-p-dioxins/dibenzofurans and dioxin-like polychlorinated biphenyls from contaminated sediments. The final method incorporated cleanup adsorbents (Florisil, alumina and silica) into the extraction cell in a 1:1 ratio of matrix to individual adsorbent (w/w). Sulfur, a common interference found in sediments, was successfully removed by placing activated copper in the extraction bottle prior to extraction. No additional post-extraction cleanup was required, and sample throughput was reduced to 2.5 h per sample [39].

PLE can also be applied in food and feed analysis (see Further Reading).

6.3.2.3 Supercritical Fluid Extraction

A supercritical fluid state means that a fluid is above its critical temperature (T_c) and critical pressure (P_c), that is, it combines conditions of a typical gas with its high diffusivity and of a liquid state with good extraction potential (see Figure 6.15 and Table 6.2). Manipulating the temperature and pressure of the fluid can solubilize the material of interest and selectively extract it.

The gas-similar diffusion properties and the liquid-similar solvent properties of supercritical media make them ideal for extraction purposes. At an industrial scale, this technique is used to decaffeinate coffee, to extracts spices and flavors and to

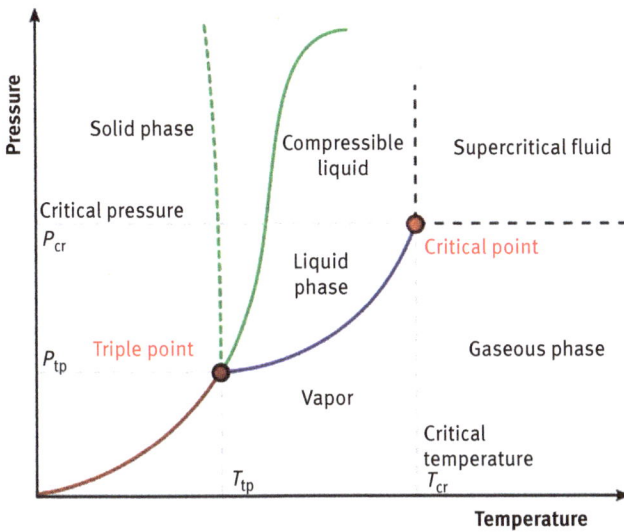

Figure 6.15: Phase diagram (the critical phase with gas and liquid-like properties starts above a certain pressure and temperature, the critical point).

Table 6.2: Typical parameters of gases, liquids and supercritical media.

Characteristics	Gases	Liquids	Supercritical media
Density [g/mL]	10^{-3}	10^{0}	0.2–0.8
Viscosity [g/(cm s)]	10^{-4}	10^{-2}	10^{-4}
Diffusion coefficient [cm^2/s]	10^{-1}	10^{-6}	10^{-3}

remediate solids such as charcoal used in water purification and polluted soils. As sketched in Figure 6.16, the supercritical substance, in most cases carbon dioxide but also supercritical water or methanol (see Table 6.3), is pumped through the sample in an extraction vessel. The extract is pumped into a fraction collector that is under atmospheric pressure and the solvent, for instance supercritical CO_2, is evaporated under such conditions. Heating must be supplied, as the adiabatic expansion of the supercritical liquid results in significant cooling. Otherwise, water or other extracted material present in the sample may freeze in the restrictor or valve and cause block-ages. The extracted pollutant can be dissolved in a suitable solvent and analyzed after further cleanup. Online systems allow direct transfer of the extract to GC, HPLC or SFC.

The efficiency of extraction is given by the density of the supercritical medium, which is a function of the temperature and pressure, and the solubility of the pol-lutant in the chosen media. For instance, supercritical CO_2 can extract analytes of

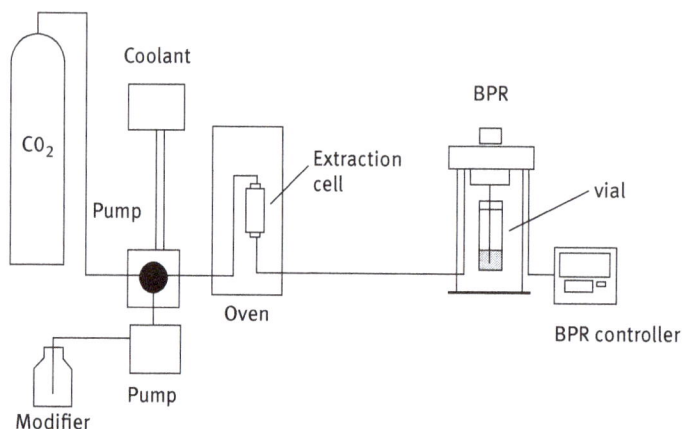

Figure 6.16: Scheme of an extraction device using supercritical carbon dioxide (backpressure regulator, BPR) according to Dean and Xiong [41]. Reproduced with permission from Elsevier.

Table 6.3: Supercritical (SC) temperature, pressure and density of some solvents used in supercritical fluid extraction.

Solvent	SC temperature [K]	SC pressure [MPa]	SC density [g/mL]
CO_2	304.1	7.38	0.47
N_2O	306.6	7.35	0.45
NH_3	405.7	11.40	0.24
H_2O	647.1	22.06	0.32

low polarity at low pressure, and high polarity analytes at high pressure, the latter however usually not performed for practical reasons. More polar compounds can be extracted by addition of a modifier solvent such as 1–10 vol% of methanol, acetone or acetonitrile, which change the polarity of the solvent and increases the extraction efficiency considerably.

The advantages of SFE are its speed and efficiency, the avoidance of large amounts of organic solvents, the use of lower temperatures and thus less thermal stress for the analyte compared to classical extraction techniques (e.g., Soxhlet) and the possibility of directly lining it up to chromatographic systems. Because of the broad application potential of SFE, many review articles are available describing different environmental pollutants and matrices (for reviews, see Further Reading).

Empirical correlations were proposed for solute solubilities in various supercritical fluids, but reliable predictive models are still desired [40].

Extraction with supercritical CO_2 at a flow rate of 3.0 mL/min, a density of 0.90 g/mL and an extraction temperature of 50°C was optimal for extraction of the herbicide cyanazine from soil after 35 days of incubation. After addition of methanol:water (1:1) as modifier at a concentration of 20% (v:w), high recoveries above 90% were obtained. Several organic modifiers at a concentration of 5 vol% of supercritical CO_2 were compared (n-hexane, methanol and toluene) to recover different PAH from soil under various temperature (50 and 80°C) and pressure conditions (23–60 MPa). With increasing pressure, as the supercritical fluid viscosity increased, the solvating power of the supercritical solvent increased as well up to a certain threshold pressure. The addition of methanol at 5 vol% led to higher recoveries than the use of n-hexane and toluene.

Despite the manifold advantages, the necessary high pressures in SFE increase the costs compared to conventional liquid extraction. Transport of the supercritical medium into the sample and of the analyte from the sample to the surface on the one hand, and dissolution of the analyte in the supercritical fluid, are two processes which need to be optimized. Diffusion of the supercritical liquid can be increased by increasing the temperature, while increase in pressure well above the critical pressure may enhance the solubility. As a compromise, the flow rate should be high enough for the extraction to be diffusion limited and, to minimize the amount of solvent used, low enough to be solubility limited. The optimal flow rate therefore depends on the competing factors of time and solvent costs.

6.3.2.4 Ultrasound-Assisted Extraction

Ultrasound-assisted extraction, usually performed at room temperature in an ultrasonic bath or using an ultrasound disrupter finger, combines chemical and physical extraction of analytes from solid matrices. Cavitation bubbles produced during ultrasonication may attain significant internal temperature and pressure that eventually collapse causing the extraction solvent to propagate outward with considerable velocity and on a collision course with matrix particles. These collisions break up the matrix and cause smaller particles to be produced, thereby exposing more surface area to the extraction solvent (see Further Reading). The ultrasound equipment should have a minimum power of 300 W, with pulsing capability. Extraction supported by ultrasonication therefore is an efficient method for extracting organic contaminants from solid matrices comparable to other commonly used methods such as Soxhlet, pressurized liquid extraction and microwave-assisted extraction (MAE).

The extraction recoveries in comparison with other extraction methods may not be higher, but the ultrasonic-supported process is faster and does not need any sophisticated equipment. Simultaneous extraction of seven pesticides was possible by ultrasonication with methanol or acetone. Approximately 2 g of soil was weighed into a 28-mL silylated vial and 8 mL of solvent added. Silylation of a glass surface replaces the H-atoms of the surface hydroxyl groups by covalently bound organosilyl groups, thus making the surface hydrophobic (see Figure 6.17). The vial was then sonicated for 20 min. Extracts were separated from soil by centrifugation and the solution

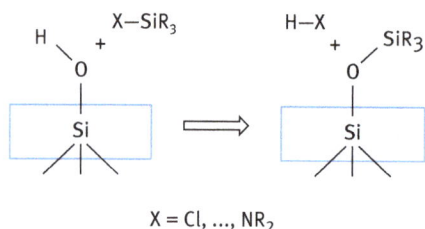

Figure 6.17: Silanization of glass surfaces.

removed. The ultrasonication step was repeated twice more resulting in a final volume of solution of 24 mL. This was then dried under a gentle stream of nitrogen for a minimum time period and stored ready for GC–MS analysis. Results showed extraction recoveries for various pesticides from soil between 70% and 114% [42].

A recent overview of applications of ultrasound for analysis of different environmental and biological samples such as food, soil and water is given in the Further Reading section.

6.3.2.5 Microwave-Assisted Extraction

Microwaves, that is, electromagnetic waves with frequencies of 0.3–300 GHz, can interact with solvents with a dipole moment (either permanent or induced in an electric field) which as a result are heated up. Heating occurs by ionic conduction, that is, the electrophoretic migration of ions under the influence of the changing electric field. Frictions of the migrating ions with the solvent will lead to the development of heat. Second, heat is produced by dipole rotation, that is, realignment of the dipoles of the solvent molecules with the rapidly changing electric field of the microwaves and thus collisions with surrounding molecules and the matrix. Unlike conductive heating methods, microwaves heat the whole sample simultaneously and very rapidly.

Some solvents for MAE are shown in Table 6.4. The choice of a suitable solvent will be based upon the property of the solvent and the solid matrix that will be extracted to absorb microwaves as well as the solubility of the target analyte and its thermal stability. The latter can be addressed by choosing mixtures of high and low microwave absorbing solvents and by using short extraction times or low-to-moderate power, which however extends the necessary extraction times. In order to optimize the extraction efficiency, the particle sizes of the sample should be small, if possible.

Closed systems with diffused (multimode) microwaves, allowing one to control pressure and temperature during extraction, and focused microwave (single-mode) ovens, in which only a restricted zone of the extraction vessel with the sample is irradiated, are available (Figure 6.18). A typical power used in closed systems ranges between 0.6 and 1 kW, and about 0.25 kW in open systems. The microwave extraction assembly comprises a wave generator (magnetron), a waveguide to direct the

Table 6.4: Dielectric constants and dipole moments of solvents.

Solvent	Dielectric constant at 20°C [–]	Dipole moment at 25°C [D]
Hexane	1.9	<0.1
Toluene	2.4	0.4
Dichloromethane	8.9	1.1
Acetone	20.7	2.7
Ethanol	24.3	1.7
Methanol	32.6	2.9
Water	78.5	1.9

Hexane is an example of a microwave-transparent solvent, whereas, for example, ethanol is an excellent microwave-absorbing solvent.

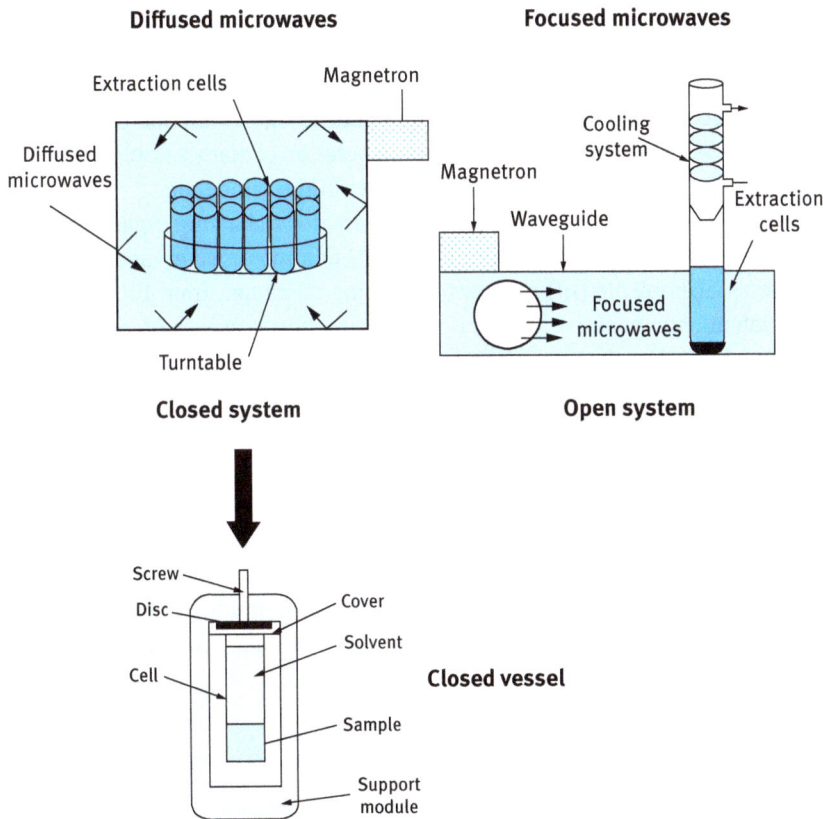

Figure 6.18: Closed/multimode and focused microwave extraction systems.

propagation of microwaves from the source to the microwave cavity, an applicator, on which the sample is placed, and a circulator ensuring that the microwaves proceed only in the forward direction. In the case of multimode systems, the applicator is a closed cavity inside which a random dispersion of microwaves is brought about (diffused microwaves) supported by beam reflectors and a rotating sample table to provide a uniform distribution of microwave energy inside the oven. In focused microwave ovens, the waveguide acts as the applicator and the extraction vessel is placed directly in the cavity. Only the lower part of the vessel is exposed to the microwaves and as glass is transparent to microwaves, the upper region of the vessel can be cooled. Advantages of the closed systems are potentially higher extraction efficiencies (boiling points of the solvents under pressure may be raised up to ca. 100°C above their atmospheric boiling point), less possibility of losing volatile substances and no need to fill up solvents during the extraction, whereas in open systems, part of the solvent will evaporate. Disadvantages of the closed systems are the possibility of thermal degradation of the analyte and the low amount of sample to be extracted. Open systems are operated at atmospheric pressure; they can process large sample amounts and reagents and solvents can be added during the extraction cycle. However, only closed systems allow the simultaneous operation of many samples; modern instruments can extract up to about 40 solid samples at the same time.

A typical protocol, as described in several papers, for extracting organochlorine pesticides from vegetables uses hexane–acetone (1:1) mixture or water–acetonitrile (5 + 95) or acetone–acetonitrile (1:1) includes a ramping step (e.g., from 100 to 800 W in a closed system), some minutes extraction at the highest power, a break for some minutes (0 W) followed by the same procedure again.

Some examples for microwave extractions are given in Table 6.5.

6.3.3 Enzymatic Digestion

Enzymes, either individually or in the form of cocktails, are used in analytical chemistry to specifically destroy matrix components in soils, plants and animal tissue in order to allow the release of bound pollutants under milder conditions compared to Soxhlet, accelerated solvent or microwave extractions. Enzymatic digestion

Table 6.5: Examples for microwave extractions of various analytes in different matrices.

Analyte	Matrix	Literature
Organochlorine pesticides	Vegetables	[43]
PAH	Soil	[44]
Pesticides	Soil	[45]
Biochemical metabolites	Biological tissues	[46]

is therefore a suitable technique for extracting thermolabile chemicals. Hydrolytic enzymes to be used for this purpose are

- lipases, which hydrolyze fats into long-chain fatty acids and glycerol,
- amylases, which split starch and glycogen to maltose and to residual polysaccharides,
- proteases, which attack the peptidic bonds of proteins and peptides. Proteases can be distinguished in proteinases (pronase E, proteinase-K, subtilisin) and peptidases (trypsin, pepsin).

When using the enzymes, their specific temperature and pH sensitivities have to be taken into account.

Some examples for enzymatic digestions are given in Table 6.6.

6.3.4 Non-extractable Residues

When analyzing concentrations of contaminant in solid matrices – soil, sediment, plants and animal tissue – always, after thorough extraction by one of the methods described above, a certain part of the analyte will remain unextracted. Each chemical will form such NER to a varying extent. According to a recent classification scheme, there are three types of NER: residues physically entrapped in the voids of the solid matrix but not covalently bound represent type I NER, those chemically bound to the organic material, for example, by ester, amide or ether linkages to soil humic matter, are type II NER, and a third fraction, type III NER, contains biogenic residues like amino acids or fatty acids that have been formed by complete microbial metabolization of the xenobiotic and formation of biomass [52]. Type I and type II NER contain xenobiotic residues, the parent substance and primary degradation products, whereas type III NER are of no environmental concern as they represent natural organic material. It has been shown that type I NER can be released from the solid matrix that therefore represents a long-term source of slowly released xenobiotics. A well-known example is the atrazine case: Even though banned in Europe in the beginning of the 1990s, in 2015, atrazine metabolites we still being found in monitoring studies in µg/L concentrations [53]. One of the reasons may be the long-term storage of such residues in the soil matrix which are slowly released. It was found that the

Table 6.6: Examples for enzymatic digestions to release analytes from different matrices.

Analyte	Matrix	Literature
Metals, metalloids	Biological tissue, food, blood	[47]
Pesticides	Soil	[48, 49]
Antibiotics	Animal tissue	[50]
Other pharmaceuticals	Animal tissue	[51]

concentrations though vary significantly within one field. Considering the typical atrazine loads on agricultural fields, long-term leaching can probably not sustain the atrazine concentrations found in monitoring programs in an entire aquifer, but a local increase of concentrations in hot spots is quite probable [54].

The differentiation of the binding mode of type I NER and type II NER is possible by a combination of a silylation derivatization and subsequent size exclusion chromatography. By silylation, the hydrogen atoms of functional groups in soil organic matter, for example, carboxylic or amino groups, are exchanged with trimethylsilyl groups. Thus, missing hydrogen bonding leads to disintegration of the humic substances into smaller fragments, which have before formed supramolecular aggregates by noncovalent interactions. If NER are entrapped in the humic matrix (type I NER), they are released after silylation, whereas NER formed by covalent binding (type II NER) remain bound to the fragmented humic matter. These two scenarios can be distinguished for instance by size exclusion chromatography [52].

Also in food, hydrophobic contaminants will form NER to various extents from which a significant amount can be mobilized in the gastrointestinal tract of animals by action of digestive enzymes as shown for a variety of contaminants. The same is true for soil NER that can be partly released in vitro and in vivo in earthworms and vertebrates.

6.3.5 Steam Distillation

By steam distillation, organic analytes can be extracted both from aqueous solutions and from solid matrices. If used for extraction from *aqueous samples*, the enriched organic substance has to form a separate layer from the water phase, that is, the substance must be immiscible with water. If a solution of such a substance at low concentrations with water is heated, both components exert their own vapor pressure dependent on the temperature, that is, the vapor pressure of the whole system increases. When the sum of the partial pressures of the two liquids exceeds the atmospheric pressure, the mixture starts boiling. Thus, the analyte may be distilled at a lower temperature compared to the boiling temperature of the analyte. In practice, water steam is introduced into a distillation apparatus. The vaporized compounds are condensed at a cooling device, and the separated phase of the immiscible analyte can be collected for analysis. If applied under reduced pressure, the boiling temperature is further reduced, and even thermolabile analytes can be enriched by this method.

Steam distillation is also employed on *solid samples*, for example, to isolate oils (e.g., not only as biocide alternatives, perfume essences or other valuable organic compounds but also for analytical purposes of contaminants in soils or plants). Steam distillation occurs in a so-called Bleidner apparatus (Figure 6.19) where the steam distillate is simultaneously extracted into an organic solvent. A typical description of the method: 60 g of sediment was transferred with 50 mL distilled water into a

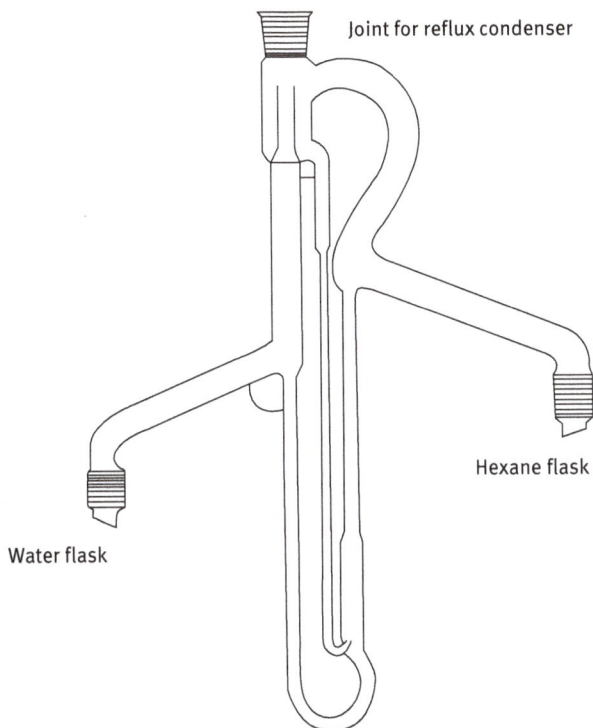

Figure 6.19: Bleidner apparatus for steam distillation of solid samples according to Brenner [55]. Reproduced with permission from Elsevier.

500-mL round-bottom flask to which a small amount of antifoaming agent was added. The flask was connected to the lower side-arm of a Bleidner apparatus and heated for 5 h while refluxing. Besides this, a 100-mL round-bottom flask containing 50 mL isooctane (as the organic solvent to later trap the analyte) was connected to the upper side-arm of the Bleidner apparatus and heated to 100°C. Within the head of the apparatus, both vapors mixed and condensed in a cooler mounted above. Ultimately, the water phase was collected within the 500-mL flask, isooctane containing the organochlorine trace compounds within the 100-mL flask. The steam distillation has also been successfully used for 2,3,7,8-TCDD analysis trapping the analyte in polyurethane foam plugs which can in addition be impregnated with, for example, ethylene glycol or methoxyethanol [55].

By Bleidner distillation, the extraction of the pesticide metabolite 3,4-dichloroaniline from humic complexes in soil was evaluated. In freshly spiked soil, the recovery after extensive extraction was literally 100%, but after 3 months' incubation in soil, the recovery efficiency declined to less than 70% [56].

A related method is called sweep co-distillation [57, 58] and based on evaporation of an analyte in an organic solvent in a steam of an inert gas such as nitrogen. The evaporated extracts are distilled in a glass tube containing either glass wool or glass beads, which trap the less volatile material. The distillate is frozen, followed by a second round of evaporation and analysis by GC.

6.3.6 Size Fractionation of Dispersed Solid Matter

Solid matter (soil, sediment, ash, dust, etc.) often consists of differently sized particulates. The interest in size-fractionation stems from the fact that different sizes may have different origin and history. Airborne dust or soil consists of particles, whose sizes reflect the different mechanisms of formation. Figure 6.20 shows this exemplarily for dust samples.

As can be seen from Figure 6.20, particle sizes between 0.1 and 1 µm are derived from accumulation of even smaller particles (nucleation mode). They are mainly formed from the gas phase by chemical supersaturation (due to formation of new products with lower vapor pressure) and subsequent agglomeration. On the other hand, particles found in the range of 1–50 µm are produced by mechanical stress on solids, like by shear forces or abrasion. Already a morphological characterization of the individual particles exhibits considerable differences. Coarse particles are in most cases irregularly shaped, opposite to those from combustion, which are agglomerates of spherical particles (e.g., shown in Fig. 4.17 d).

The chemistry of such modes therefore is distinctly different. In the coarse mode (>1 µm), we will find windblown dust and humic substances with attached organic fractions from soil erosion, in contrast to mainly soot within the carbon-rich part of particles <1 µm. Combustion might be the reason for this. Due to adsorption, all

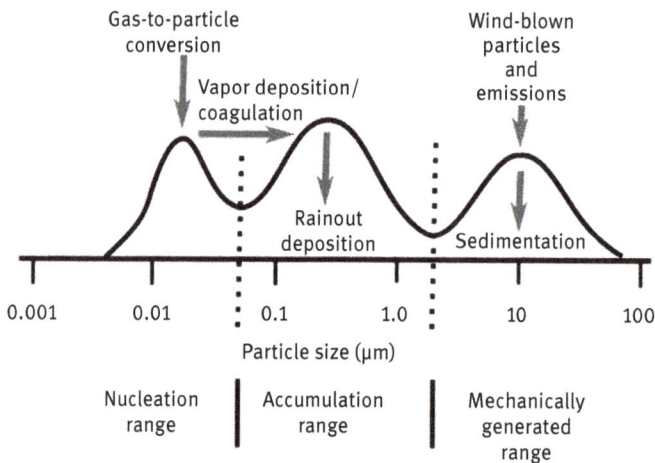

Figure 6.20: Size distributions for airborne dust.

PAHs will be found within the smaller size fraction and in most cases twinned with soot particles.

Dust from a production process, a dust separator or from an exhaust might become subjected to a direct particle separation by sieving as long as it is in a dry condition. Soil needs to be dispersed within a liquid before separation. This of course presents a serious limitation, since during dispersion, many attached compounds will become dissolved.

The classical way for dust or soil/sediment size fractionation is sieving by application of differently sized sieves (Table 6.7).

Before sieving, a surfactant is added in order to assist separation among the aggregated material dispersed in water. Sometimes, ultrasonic agitation helps to disintegrate the sample. If the analyst is interested in the organic part of different size fractions, these surfactants must not interfere with the subsequent analytical determination. A soil sieve, known as a gradation test, determines the particle size distribution of a solvent-dispersed sample. To conduct this test, a series of mesh screen sieves is ordered as a cascade from largest mesh size (top) to finest mesh size (bottom).

The dry or dispersed sample is added to the top, and the stack of sieves is moved onto a shaker. The shaker then agitates the soil in the sieves for a specified amount of time.

The screen arrangement and the result of such sieving separations are shown in Figure 6.21.

After gradation, the individual fractions are removed from the screens and analyzed by the respective analytical method. The found analytical masses then form a size distribution, given exemplarily in Figure 6.22. Each fraction is represented as a vertical bar volume in a histogram plot. Accumulating the sample masses to 100%, total volume (mass) forms the typical cumulative size distribution. The 50% value is

Table 6.7: Commonly applied mesh sizes of gradation screens used for bottom-ash analysis.

Mesh size	Sieve size opening (mm)
20	0.853
40	0.422
50	0.297
60	0.251
70	0.211
80	0.178
100	0.152
200	0.075
325	0.044

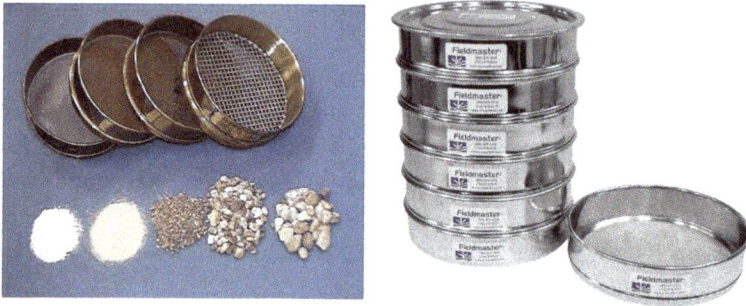

Figure 6.21: Series of stainless steel mesh screen sieves for soil or bottom-ash separation. (Left) The individual screens with retained particle fractions. (Right) The screen cascade.

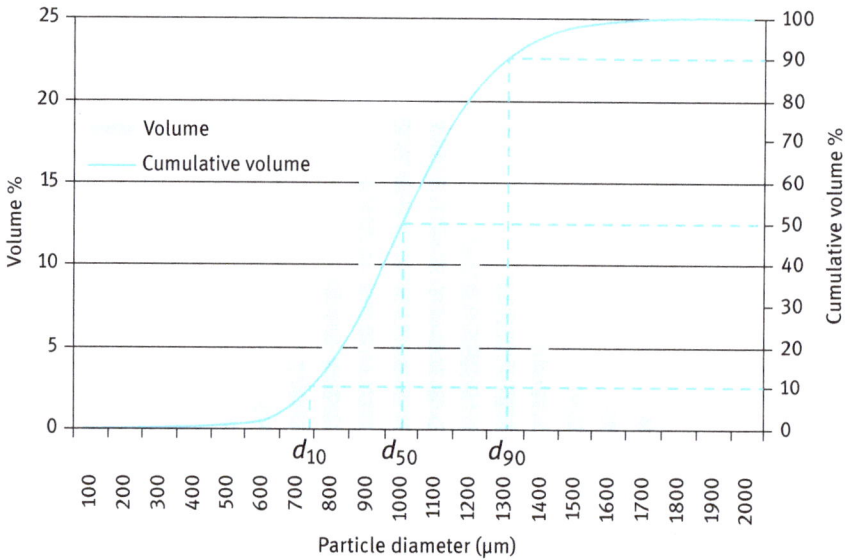

Figure 6.22: Cumulative size distribution of sieved soil or ash fractions.

the mean particle diameter d_{50}. The slope of the size distribution function between 10% value and 90% value reflects the dispersity grade.

Other techniques, like separation by sedimentation, flotation or ultracentrifugation, are applicable too. Ultracentrifugation is used to gain polymer size distributions since many decades.

In all cases, the liquid-state dispersion has to be adjusted to the subsequent analytical procedure. Ultrafine particles are very sensitive to changing pH conditions due to agglomeration or repulsion by particle charges.

6.4 Gaseous Samples and Aerosols

6.4.1 General Remarks

As has been pointed out for the subject of sampling trace gases and airborne particulate matter (see Chapter 4, Sections 4.3 and 4.7), enrichment is often combined to it. Since most air standards are related to 24-h sampling, the appropriate air volume taken is a concern. For aerosols with number concentrations up to 10^6 particles/mL, overloading of the sampling device may become a problem. To understand this, one has to be aware that particles in the suspended state move more or less independently of each other. Once deposited in a particle sampler, they are now intimately and closely accumulated, with the chance of particle–particle interactions.

Gas sampling seems less complex. Here, a breakthrough through the enrichment device, usually a chromatographic column or an absorber, is to be avoided. So, the optimization of ad- or absorption capacity of a gas sampler is of utmost importance. Since gases are in an equilibrium with the particulate matter, enrichment by separation always perturbs the existing equilibrium. So, it may happen that already deposited semi-volatile particulate matter starts to vaporize into the gas phase. Many organic compounds show such behavior. A prominent example is semi-volatile PAHs. During 24-h sampling, enormous losses of particle-bound PAHs have been observed. One way to correct such losses is parallel short-time sampling (shorter time intervals) with back extrapolation to retrieve the originally present amount. Also, adsorption of gas molecules at a receptor does not necessarily prevent from post-reactions.

6.4.2 Liquid Absorption of Trace Gases

The classic way to directly sample and enrich trace gases is the absorption within a reactive absorber solution in a gas washing bottle. The organic contaminant is bubbled through a liquid sorbent, whose capacity must exceed the expected analyte mass. Reaction kinetics also has to be fast. Often so-called impinger glass flasks are used for this (see Figure 6.23, left).

The usage of these impingers has some severe drawbacks. First, the contact time of the gas molecules within the gas bubbles is too short to reach the gas/liquid interface; hence, the collection rate is low. Second, with progressing sampling time, losses of the absorber solution by evaporation are unavoidable. Therefore, after sampling, adjustment to the initial absorber volume is needed. A much more efficient tool is the application of an absorber-filled glass bottle with a bottom glass frit (Figure 6.23, right). This enables the gas stream to separate into many small gas bubbles with much higher specific bubble surface. The observed reaction yield therefore becomes much better. Both cases suffer under the relatively large absorber volume, since most subsequent analytical techniques only need some microliters injection volume, or even much less. So, further concentration by volume reduction is usually applied.

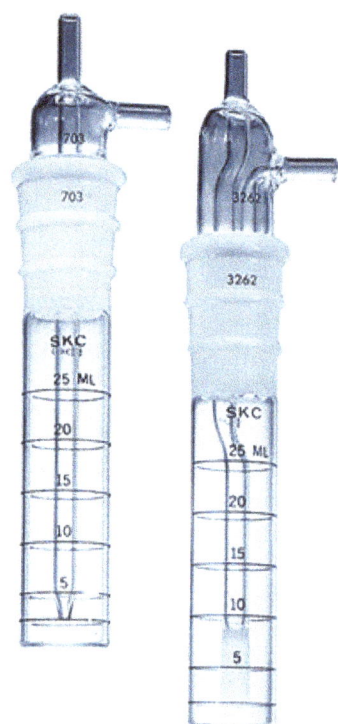

Figure 6.23: Samplers for gas sampling by impinging (left) or by gas dispersion with a glass frit (right).

6.4.3 Solid Adsorption of Trace Gases

The application of solid phases for gas sampling is very popular. The basic idea stems from adsorption chromatography and uses adsorber-filled short columns. Figure 6.24 depicts the structure of a sorption tube. The sorption capacity is determined by the specific surface area accessible and the adsorber properties.

Active charcoal or silica are not the only common adsorption materials; many other stationary phases are in use, much like in GC. Only several milligrams of a solid phase are necessary. After having pulled a prescheduled volume through the sorption bed, the adsorber is removed and extracted by an even more absorbing liquid (e.g., CS_2). Sometimes, direct thermal desorption and direct coupling with a gas chromatograph are possible. If desired, the desorbed stationary phase might be reused again for sampling. Important is the knowledge of the breakthrough volume for the requested analyte, which must not have exceeded during sampling. After sampling, the tube is removed from the sampling line, closed and stored cool under an inert atmosphere until analysis. Generally, the applicability of such sorption tubes has to be approved by test gas sampling under the expected humidity range, since water molecules often directly compete on the available sorption sites, hence reducing sampling capacity.

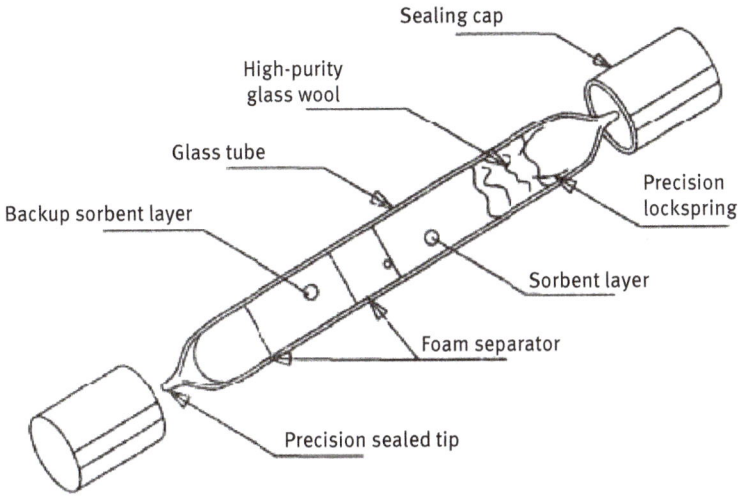

Figure 6.24: Schematic structure of a gas adsorption sampling tube.

Disadvantageous is the low-volume limitation. Only some hundred milliliters up to a few liters of gas volume can be probed. Otherwise, a breakthrough of analyte will occur. Therefore, such collection tubes are preferentially applied for occupational hygiene monitoring, where the concentrations of harmful substances are often in the ppmv range, and short-time sampling saves a lot of working time.

Solvent vapors or odorous substances are typical target substances for adsorptive gas sampling.

6.4.4 Filtration

Gas/particle separation and enrichment by membrane filters have a long-lasting tradition. Nowadays, membrane filters are made not only of glass or quartz fibers, but from a variety of polymers. The progress in clean air technology is connected with this development. Filtration can be successfully used for hydrocolloid separation too.

In filtration technology, two kinds of filter types are predominant: deep-bed and membrane types.

Deep-bed filters possess several cm thick filtration layers, in contrast to membranes, which are in the μm-range of thickness. Figure 6.25 shows both types of filters. Deep-bed filters are woven by statistically clustered long glass or quartz fibers (see Figure 6.25, left). For analytical applications, quartz fibers show a higher purity. *Membrane filters* are mainly made of polymer sheets, whose pores are either produced by stretch forming under thermal stress or by chemical etching after bombardment with alpha-tracks (Nuclepore™ filter; see Figure 6.25, middle). Different to them are

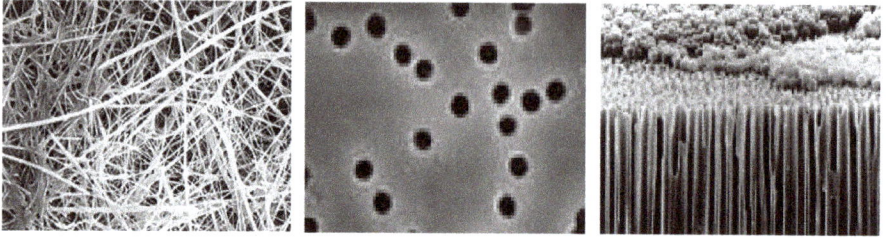

Figure 6.25: (Left) Deep-bed filter made of glass fibers, fiber diameter: 1 µm; (middle) Nuclepore™, trace-etched polycarbonate membrane filter, pore size 0.1 µm; (right) Anodisc™, inorganic membrane filter, pore size: 1 µm.

Anodisc™ membranes (see Figure 6.25, right), electrochemically produced highly porous aluminum oxide membranes. These filters are thermostable up to 450°C. Pore sizes can be as small as 20 nm. Even smaller pore sizes can be achieved with graphene nanosheets.

The parameters governing particle separation are
- face velocity
- filter thickness
- pore size
- fiber thickness
- porosity
- filter charge
- particle charge.

The technical goal is to remove any particle from a gas flow, regardless of its size. This is statistically impossible, but filter structures are available which approach this to a certain extent. The expense for this is an enormous pressure drop across the filter. But the existence of a so-called filter gap has to be accepted, where particle collection efficiency shows a minimum (typically in the range of 0.1–1 µm, depending on face velocity) (see Figure 6.26).

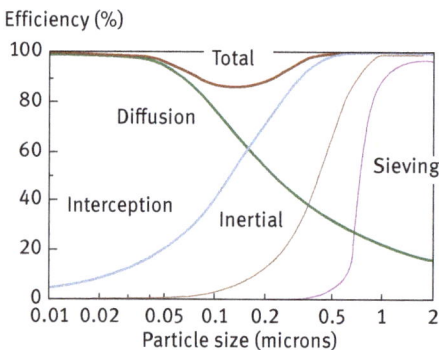

Figure 6.26: Particle removal efficiency of a filter as sum of different separation mechanisms.

Figure 6.27: Hollow fiber bundle for high-efficiency cleaning of water.

In water filtration nowadays, hollow fiber filters (tubular membrane filter bundles) allow a throughput of several liters per minute per module (e.g., see Figure 6.27).

More difficult is the removal of deposited material from such filter structures. In air filtration, membrane filters are most convenient, since after sampling, the filter is removed from the filter holder and subjected to an extraction step, for example, in a Soxhlet apparatus.

In all cases, such filters have to be conditioned prior to their use, by baking out or thorough rinsing with clean solvents. Problems may arise from adsorption of gaseous organic traces on deposited particles and the filter matrix itself. This causes losses when used as prefilter before enrichment or a direct-reading instrument. Instead of extraction with organic solvents, thermal desorption is also an option.

6.4.5 Diffusion Sampling (Denuder Sampling)

In order to avoid post-reactions by subsequently incoming reactive air constituents after collection of organic particulates on filter surfaces (e.g., ozone or NO_2 reacts with sampled PAHs), so-called diffusion separators ("denuders") have been developed during recent years.

Diffusion separators are tube bundles or single tubes attached to a subsequent particle filter. Its configuration is such that it lets submicron particles pass through the tubes but removes gaseous components of the probed aerosol by diffusion separation (see Figure 6.28).

This happens when an inner wall coating of such a denuder consists of a chemically reactive substance, which irreversibly binds gas molecules when hitting the wall by Brownian motion. Due to the simple geometry, particle losses become negligible

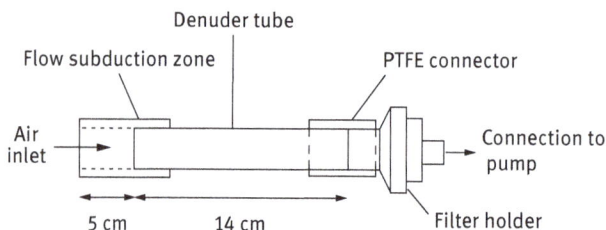

Figure 6.28: Diffusion separator ("denuder") for gas sampling with inlet and backup filter for particle sampling.

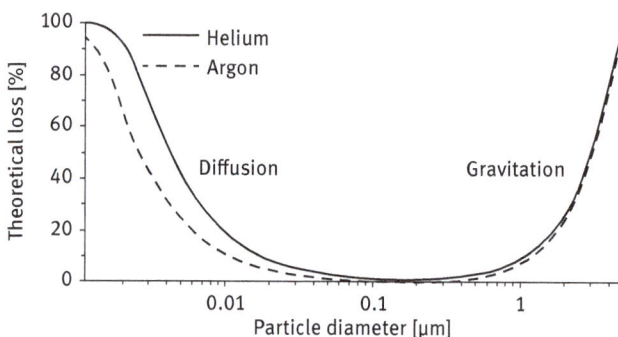

Figure 6.29: Theoretical particle losses within a denuder tube as a function of particle size and carrier gas.

(<5% particle mass) for typical environmental log-normal particle size distributions (mean particle diameter 50–200 nm; standard deviation around mean approx. ±90%). This is because particles in this size range possess diffusion coefficients several orders of magnitude lower than those of gas molecules. Figure 6.29 shows calculated particle losses for typical operational conditions of a horizontally arranged denuder tube. Also shown is the influence of carrier gas viscosity on particle deposition rate.

Developed first for sampling of inorganic gases from the atmosphere (NH_3, SO_2, HNO_3, H_2S, etc.), the applications now cover a wide range of organic gases too. After sampling, the denuder becomes removed from the aerosol conducting line. The deposited analyte inside is then extracted and analyzed. Published applications are nitrosamines, amines, nicotine, isocyanate, gaseous PAHs and many more. The success of denuder sampling is always linked with the finding of an effective sink material for discriminating gas molecules and particles.

To prevent oxidation of particle-bound PAHs already deposited on filter surfaces, ozone destruction by catalytically acting MnO_2 previous to particle collection filter has been reported.

6.4.6 Sampling by Condensation

To mimic nature is sometimes not the worst decision. Cloud, rain and fog droplets are very efficient collectors of trace gases and aerosol particles. Depending on the acting water supersaturation, aerosol particles become incorporated as the nucleus for a growing water droplet. In the dispersed state, such droplets are efficient scavengers for gas molecules due to their high specific surface area. The limits for such a collection method are given by the particle size of the aerosol particles, wettability of it and the accommodation behavior of gas molecules when bouncing the growing droplet surface.

A commercialized version is the particle-into-liquid sampler (PILS) (see Figure 6.30), which enlarges particles in supersaturated water vapor. The droplets grow up to a size easily to be collected by inertial impaction and then chemically analyzed offline or in real time. The liquid sample can be fractionated and analyzed by various analytical techniques. Small errors may happen due to volatilization losses and have to be corrected. Applications are the sampling and analysis of water-soluble carboxylic acids in air.

A combination of PILS with a TOF-MS for studying isoprene photooxidation products online is shown in Figure 6.31.

From the quartz impaction plate, a flow of HPLC water flushes the impacted liquid droplets to the HPLC pump responsible for injecting the sample into the ionization source of a "soft" ionization TOF-MS mass spectrometer. Also, the combination with capillary electrophoresis separation has been reported. Currently, the main application for it is the study of secondary aerosol formation, namely carboxylic acids and organic sulfates.

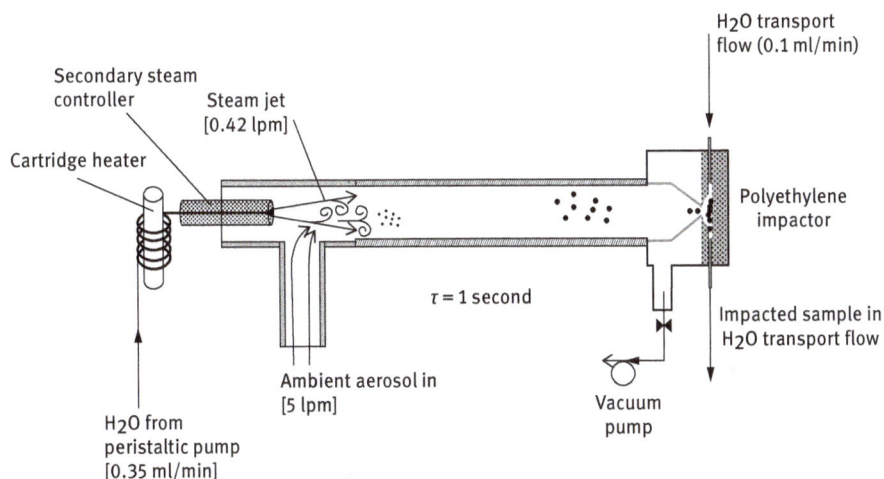

Figure 6.30: Particle-into-liquid sampler (PILS).

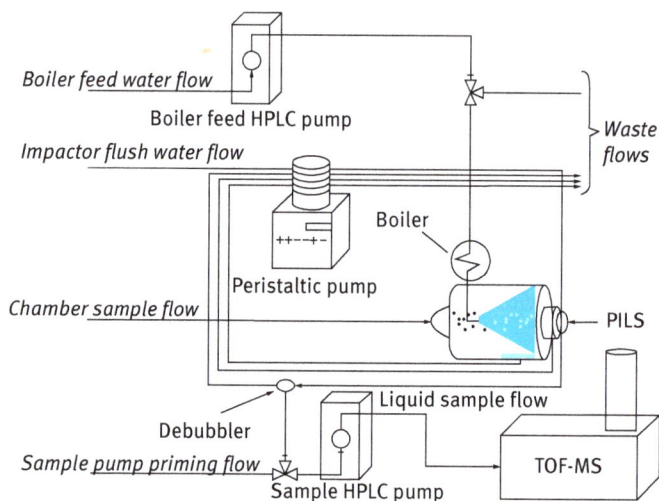

Figure 6.31: PILS coupled to time-of-flight mass spectrometry (PILS-TOF).

6.4.7 Size-Resolved Particle (Aerosol) Sampling

It is advantageous to know the size of aerosol particles together with their chemical composition. To achieve this, various techniques for hyphenating sampling and analysis are available in aerosol technology:
- impaction sampling with subsequent extraction of the collected material;
- diffusion separation of ultrafine particles by Brownian motion;
- sampling after electrical mobility particle separation;
- phoretic particle separation.

The most common combination is using an *impactor* for aerosol sampling. Figure 6.32 shows the principle of such a device.

The aerosol flow becomes accelerated when passing an orifice. Opposite, an impaction plate is placed. Only small particles can follow the stream lines surrounding this obstacle, whereas the larger ones will leave the flow lines and become impacted on the plate depending on flow rate and dimensions of orifice and distance to impaction plate. There are some peculiarities with impaction samples:
- losses by evaporation from the impaction spot (especially for semi-volatile compounds);
- reactions among impacted material may lead to chemical changes;
- reentrance of impacted material by particle-bouncing or ongoing erosion by jet force.

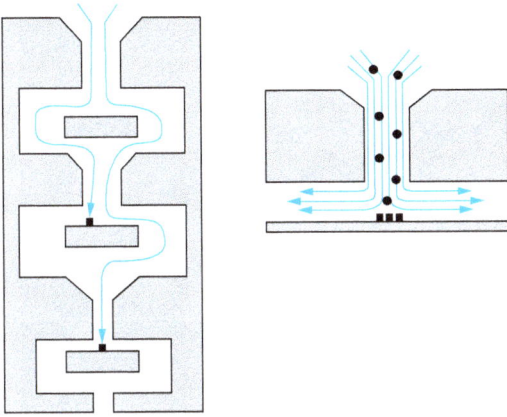

Figure 6.32: Scheme of a 3-stage impactor (left) and flow in a single-stage (right).

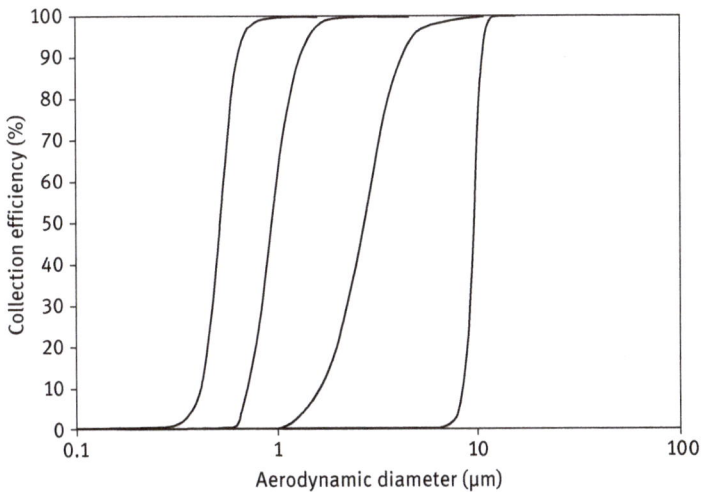

Figure 6.33: Separation characteristics of a four-stage impactor.

Multistage impactors, covering up to ten stages, allow size fractionated sampling in a size range of roughly 20 nm to 20 µm (aerodynamic particle size). Figure 6.33 shows the experimentally determined impactor separation function.

Diffusion separation can be achieved when only particles below 50 nm have to be characterized. So-called diffusion batteries (see Figure 6.34), which are in fact differently long denuder bundles, are operated in series.

The first stage of a diffusion battery contains only short tube length equivalents (e.g., bundled tube, screen, porous plate, parallel plates, etc.), thus enabling separation of only the ultrasmall particle fraction. With the increase in stage number, the

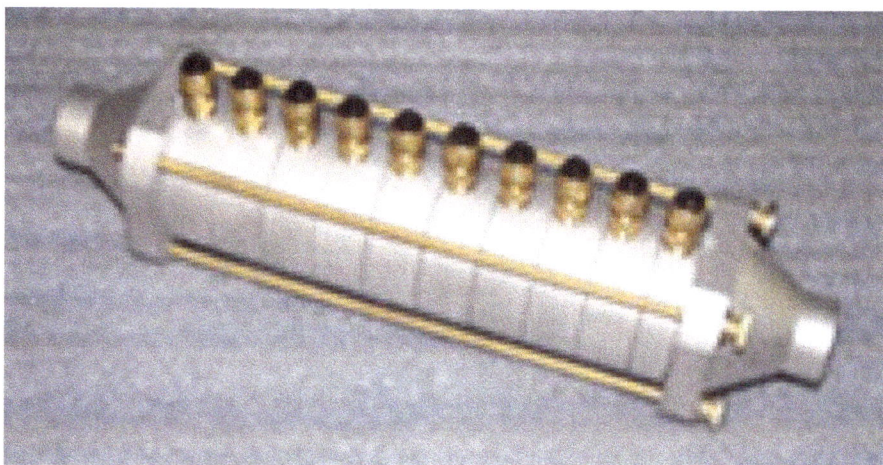

Figure 6.34: Ten-stage diffusion battery.

length of tube equivalents increases (up to kilometers); therefore, also larger particles become removed by only Brownian motion. From comparison, particle losses between different stages and knowledge of the particle transfer rates as a function of particle size one can assign the particle loss to a certain size fraction. Because of the exponential increase in diffusion coefficients with decreasing size, diffusion separation becomes powerful for particle sizes below 10 nm. Prerequisite is a stable aerosol source during measurement and nonexistence of larger particles. The latter is realized by an impactor as pre-collector.

Electrically charged particles can be separated according their size when unipolar charging is given. Otherwise, the electric mobility will yield same separation for double-charged particles but larger in size. This restriction sometimes makes it difficult, since with particles >300 nm, multicharging happens frequently (according to Boltzmann's distribution function). On the other hand, charging probability approaches only parts of percent when particles sizes are below 20 nm. Hence, the combined analytical determination must be extremely sensitive. Only a few applications for protein separation were known.

Phoretic particle sampling is possible, when an orthogonal force field becomes applied to an aerosol flow. Forces to be applied can be a temperature gradient between a hot and cold plate, light intensity from a laser beam, acoustic force, inertial forces within a spirally wound duct within a centrifuge and directed gas diffusion (diffusiophoresis). So far, only the Stoeber centrifuge has become commercialized. The aerosol particles become deposited on an inserted foil within the aerosol duct at locations, whose gravitational settling position has been predetermined by calibration. Figure 6.35 shows this device.

This sampling is applicable for particles with sizes between 0.1 and 6 μm.

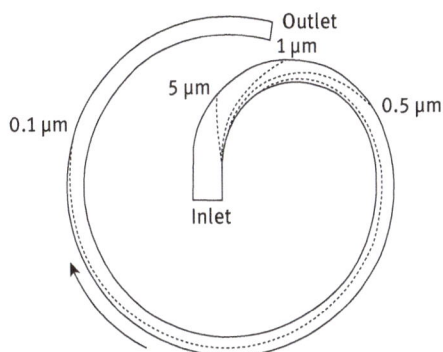

Figure 6.35: Locations for size-resolved aerosol deposition within a Stoeber aerosol centrifuge. Aerosol is introduced in the center axis of the running centrifuge.

Further Reading

Alpendurada MD. Solid-phase microextraction: a promising technique for sample preparation in environmental analysis. J Chromatogr A 2000;889(1-2):3–14.

Aulakh JS, Malik AK, Kaur V, Schmitt-Kopplin, P. A review on solid phase micro extraction-high performance liquid chromatography (SPME-HPLC) analysis of pesticides. Crit Rev Anal Chem 2005;35(1):71–85.

Bajgai Y, Hulugalle N, Kristiansen P, McHenry M. Developments in fractionation and measurement of soil organic carbon: a review. Open J Soil Sci 2013;3:356–60.

Barcelo D. Environmental analysis – techniques, applications and quality assurance. Amsterdam: Elsevier, 1996: 646.

Bjorklund E, Sporring S, Wiberg K, Haglund P, von Hillist C. New strategies for extraction and clean-up of persistent organic pollutants from food and feed samples using selective pressurized liquid extraction. TRAC Trends Anal Chem 2006;25(4):318–25.

Boeckelen A, Niessner R. Combination of micellar extraction of polycyclic aromatic hydrocarbons from aqueous media with detection by synchronous fluorescence. Fresenius' J Anal Chem 1993;346(4):435–40.

Bright DA, Richardson GM, Dodd M. Do current standards of practice in Canada measure what is relevant to human exposure at contaminated sites? I: a discussion of soil particle size and contaminant partitioning in soil. Human Ecol Risk Assess 2006;12:591–605.

Bruzzoniti MC, Checchini L, De Carlo RM, Orlandini S, Rivoira L, Del Bubba M. QuEChERS sample preparation for the determination of pesticides and other organic residues in environmental matrices: a critical review. Anal Bioanal Chem 2014;406(17):4089–116.

Capelo JL, Mota, AM. Ultrasonication for analytical chemistry. Curr Anal Chem 2005;1(2):193–201.

Chary NS, Fernandez-Alba, AR. Determination of volatile organic compounds in drinking and environmental waters. TRAC Trends Anal Chem 2012;32:60–75.

Cooper JJ. Particle-size measurements. Ceram Eng Sci Proc 1991;12:133–43.

Dai H, Xiao D, He H, Li H, Yuan D, Zhang C. Synthesis and analytical applications of molecularly imprinted polymers on the surface of carbon nanotubes: a review. Microchim Acta 2015;182(5-6):893–908.

Fenik J, Tankiewicz M, Biziuk M. Properties and determination of pesticides in fruits and vegetables. TRAC Trends Anal Chem 2011;30(6):814–26.

Froeschl B, Stangl G, Niessner R. Combination of micellar extraction and GC-ECD for the determination of polychlorinated biphenyls (PCBs) in water. Fresenius' J Anal Chem 1997;357(6):743–6.

Garcia, JI, Rodriguez-Gonzalez P. Isotope dilution mass spectrometry. London: Royal Society of Chemistry, 2013:453.

Garcia-Rodriguez D, Maria Carro-Diaz A, Antonia Lorenzo-Ferreira R. Supercritical fluid extraction of polyhalogenated pollutants from aquaculture and marine environmental samples: a review. J Sep Sci 2008;31(8):1333–45.

Hinze WL, Pramauro E. A critical-review of surfactant-mediated phase separations (cloud-point extractions) – theory and applications, Crit Rev Anal Chem 1993;24(2):133–77.

Katayama A, Bhula R, Burns GR, Carazo E, Felsot A, Hamilton D, Harris C, Kim YH, Kleter G, Koedel W, Linders J, Peijnenburg J, Sabljic A, Stephenson RG, Racke DK, Rubin B, Tanaka K, Unsworth J, Wauchope, RD. Bioavailability of xenobiotics in the soil environment. In Whitcare DM, editor. Reviews of environmental contamination and toxicology, vol 203. Heidelberg: Springer Verlag, 2010.

Latawiec AE, Reid, BJ. Sequential extraction of polycyclic aromatic hydrocarbons using subcritical water. Chemosphere 2010;78(8):1042–48.

LeDoux M. Analytical methods applied to the determination of pesticide residues in foods of animal origin. A review of the past two decades. J Chromatogr A 2011;1218(8):1021–36.

Moreda-Pineiro J, Moreda-Pineiro A. Recent advances in combining microextraction techniques for sample pre-treatment. TRAC Trends Anal Chem 2015;71:265–74.

Ribeiro C, Ribeiro AR, Maia AS, Goncalves, VM, Tiritan, ME. New trends in sample preparation techniques for environmental analysis. Crit Rev Anal Chem 2014;44(2):142–85.

Seidi S, Yamini Y. Analytical sonochemistry; developments, applications, and hyphenations of ultrasound in sample preparation and analytical techniques. Cent Eur J Chem 2012;10(4):938–76.

Souza Machado BA, Pereira CG, Nunes SB, Padilha FF, Umsza-Guez, MA. Supercritical fluid extraction using CO_2: main applications and future perspectives. Sep Sci Technol 2013;48(18):2741–60.

Stangl G, Niessner R. Micellar extraction – a new step for enrichment in the analysis of napropramide. Int J Environ Anal Chem 1995;58(1-4):15–22.

Tadeo JL, Ana Perez R, Albero B, Garcia-Valcarcel AI, Sanchez-Brunete C. Review of sample preparation techniques for the analysis of pesticide residues in soil. J AOAC Int 2012;95(5):1258–71.

Teo CC, Chong WPK, Ho, YS. Development and application of microwave-assisted extraction technique in biological sample preparation for small molecule analysis. Metabolomics 2013;9(5):1109–28.

Torres Padron ME, Afonso-Olivares C, Sosa-Ferrera Z, Juan Santana-Rodriguez J. Microextraction techniques coupled to liquid chromatography with mass spectrometry for the determination of organic micropollutants in environmental water samples. Molecules 2014;19(7):10320–49.

Vas G, Vekey K. Solid-phase microextraction: a powerful sample preparation tool prior to mass spectrometric analysis. J Mass Spectrom 2004;39(3):233–54.

Wang H, Adeleye AS, Huang Y, Li F, Keller AA. Heteroaggregation of nanoparticles with biocolloids and geocolloids. Adv Colloid Interface Sci 226(Part A):2015:24–36.

Wiilkowska A, Biziuk M. Determination of pesticide residues in food matrices using the QuEChERS methodology. Food Chem 2011;125(3):803–12.

Bibliography

[1]   Schwarzenbach RP, Egli T, Hofstetter TB, von Gunten U, Wehrli B. Global water pollution and human health. In Gadgil A, Liverman, DM, editors. Annual review of environment and resources, vol 35, 2010:109–36.

[2]   Koch DA, Clark K, Tessier, DM. Quantification of pyrethroids in environmental samples using NCI-GC-MS with stable isotope analogue standards. J Agric Food Chem 201361(10):2330–9.

[3]   Kistemann T, Hundhausen J, Herbst S, Classen T, Faerber H. Assessment of a groundwater contamination with vinyl chloride (VC) and precursor volatile organic compounds (VOC) by use of a geographical information system (GIS). Int J Hyg Environ Health 2008;211(3-4):308–17.

[4]   Daoussi R, Vessot S, Andrieu J, Monnier O. Sublimation kinetics and sublimation end-point times during freeze-drying of pharmaceutical active principle with organic co-solvent formulations. Chem Eng Res Des 2009;87(7A):899–907.

[5]   Caban M, Stepnowski P, Kwiatkowski M, Maszkowska J, Wagil M, Kumirska J. Comparison of the Usefulness of SPE Cartridges for the Determination of beta-Blockers and beta-Agonists (Basic Drugs) in Environmental Aqueous Samples. J Chem 2015.

[6]   Moser AC, Hage, DS. Immunoaffinity chromatography: an introduction to applications and recent developments. Bioanalysis 2010;2(4):769–90.

[7]   Kubo T, Hosoya K, Otsuka K. Molecularly imprinted adsorbents for selective separation and/or concentration of environmental pollutants. Anal Sci 2014;30(1):97–104.

[8]   Arthur CL, Pratt K, Motlagh S, Pawliszyn J, Belardi, RP. Environmental analysis of organic compounds in water using solid phase microextraction. HRC-J High Resolut Chromatogr 1992;15(11):741–44.

[9]   Vas G, Vekey K. Solid-phase microextraction: a powerful sample preparation tool prior to mass spectrometric analysis. J Mass Spectrom 2004;39(3):233–54.

[10]   Rezaee M, Assadi Y, Hosseinia, M-RM, Aghaee E, Ahmadi F, Berijani S. Determination of organic compounds in water using dispersive liquid-liquid microextraction. J Chromatogr A 2006;1116(1-2):1–9.

[11]   Fernandez P, Regenjo M, Bermejo AM, Fernandez AM, Lorenzo RA, Carro, AM. Analysis of drugs of abuse in human plasma by dispersive liquid-liquid microextraction and high-performance liquid chromatography. J Appl Toxicol 2015;35(4):418–25.

[12]   Berijani S, Assadi Y, Anbia M, Milani Hosseini M-R, Aghaee E. Dispersive liquid-liquid microextraction combined with gas chromatography-flame photometric detection – very simple, rapid and sensitive method for the determination of organophosphorus pesticides in water. J Chromatogr A 2006;1123(1):1–9.

[13]   Hou X, Zheng X, Zhang C, Ma X, Ling Q, Zhao L. Ultrasound-assisted dispersive liquid-liquid microextraction based on the solidification of a floating organic droplet followed by gas chromatography for the determination of eight pyrethroid pesticides in tea samples. J Chromatogr B-Anal Technol Biomed Life Sci 2014;969:123–7.

[14]   Saraji M, Boroujeni MK, Bidgoli, AAH. Comparison of dispersive liquid-liquid microextraction and hollow fiber liquid-liquid-liquid microextraction for the determination of fentanyl, alfentanil, and sufentanil in water and biological fluids by high-performance liquid chromatography. Anal Bioanal Chem 2011;400(7):2149–58.

[15]   Ketola RA, Virkki VT, Ojala M, Komppa V, Kotiaho T. Comparison of different methods for the determination of volatile organic compounds in water samples. Talanta 1997;44(3):373–82.

[16]   Kavcar P, Odabasi M, Kitis M, Inal F, Sofuoglu, SC. Occurrence, oral exposure and risk assessment of volatile organic compounds in drinking water for Izmir. Water Res 2006;40(17):3219–30.

[17] Belanger, JMR, Pare, JRJ, Turpin R, Schaefer J, Chuang, CW. Evaluation of microwave-assisted process technology for HAPSITE's headspace analysis of volatile organic compounds (VOCs). J Hazard Mater 2007;145(1-2):336–8.

[18] Starokozhev E, Sieg K, Fries E, Puettmann W. Investigation of partitioning mechanism for volatile organic compounds in a multiphase system. Chemosphere 2011;82(10):1482–8.

[19] Amaral, HIF, Berg M, Brennwald MS, Hofer M, Kipfer R. C-13/C-12 Analysis of ultra-trace amounts of volatile organic contaminants in groundwater by vacuum extraction. Environ Sci Technol 2010;44(3):1023–9.

[20] Liu H-W, Wu B-Z, Nian H-C, Chen H-J, Lo J-G, Chiu K-H. VOC amounts in ambient areas of a high-technology science park in Taiwan: their reciprocal correlations and impact on inhabitants. Environ Sci Poll Res 2012;19(2):303–12.

[21] Takei Y, Furuno M. Development of a Gas-liquid counter-current flow extraction and concentration/introduction system using spiraltube for determination of VOCs in aqueous samples by GC-MS. Bunseki Kagaku 2013;62(3):243–8.

[22] Perez, JFH, Sejeroe-Olsen B, Fernandez Alba AR, Schimmel H, Dabrio M. Accurate determination of selected pesticides in soya beans by liquid chromatography coupled to isotope dilution mass spectrometry. Talanta 2015;137:120–9.

[23] Cornelissen G, Rigterink H, ten Hulscher, DEM, Vrind BA, van Noort, PCM. A simple Tenax (R) extraction method to determine the availability of sediment-sorbed organic compounds. Environ Toxicol Chem 2001;20(4):706–11.

[24] Jonker, MTO, Koelmans, AA. Polyoxymethylene solid phase extraction as a partitioning method for hydrophobic organic chemicals in sediment and soot. Environ Sci Technol 2001;35(18):3742–8.

[25] Wang C, Zhu LZ, Zhang, CL. A new speciation scheme of soil polycyclic aromatic hydrocarbons for risk assessment. J Soils Sediments 2015;15(5):1139–49.

[26] Wang DH, Wang YM, Yin YM, Min S, Wang SY, Yu YL. Bioavailability-based estimation of phytotoxicity of imazaquin in soil to sorghum. Environ Sci Poll Res 2015;22(7):5437–43.

[27] Diez C, Traag WA, Zommer P, Marinero P, Atienza J. Comparison of an acetonitrile extraction/partitioning and "dispersive solid-phase extraction" method with classical multi-residue methods for the extraction of herbicide residues in barley samples. J Chromatogr A 2006;1131(1-2):11–23.

[28] Nguyen TD, Han EM, Seo MS, Kim SR, Yun MY, Lee DM, Lee G-H. A multi-residue method for the determination of 203 pesticides in rice paddies using gas chromatography/mass spectrometry. Anal Chim Acta 2008;619(1):67–74.

[29] Rashid A, Nawaz S, Barker H, Ahmad I, Ashraf M. Development of a simple extraction and clean-up procedure for determination of organochlorine pesticides in soil using gas chromatography-tandem mass spectrometry. J Chromatogr A 2010;1217(17):2933–9.

[30] Garcia Pinto C, Fernandez Laespada ME, Herrero Martin S, Casas Ferreira AM, Perez Pavon JL, Moreno Cordero B. Simplified QuEChERS approach for the extraction of chlorinated compounds from soil samples. Talanta 2010;81(1-2):385–91.

[31] Antonio Padilla-Sanchez J, Plaza-Bolanos P, Romero-Gonzalez R, Garrido-Frenich A, Martinez Vidal, JL. Application of a quick, easy, cheap, effective, rugged and safe-based method for the simultaneous extraction of chlorophenols, alkylphenols, nitrophenols and cresols in agricultural soils, analyzed by using gas chromatography-triple quadrupole-mass spectrometry/mass spectrometry. J Chromatogr A 2010;1217(36):5724–31.

[32] Lesueur C, Gartner M, Mentler A, Fuerhacker M. Comparison of four extraction methods for the analysis of 24 pesticides in soil samples with gas chromatography-mass spectrometry and liquid chromatography-ion trap-mass spectrometry. Talanta 2008;75(1):284–93.

[33] Rasche C, Fournes B, Dirks U, Speer K. Multi-residue pesticide analysis (gas chromatography-tandem mass spectrometry detection)-Improvement of the quick, easy, cheap, effective, rugged, and safe method for dried fruits and fat-rich cereals-benefit and limit of a standardized apple puree calibration (screening). J Chromatogr A 2015;1403:21–31.

[34] Lopes RP, Passos, EED, de Alkimim JF, Vargas EA, Augusti DV, Augusti R. Development and validation of a method for the determination of sulfonamides in animal feed by modified QuEChERS and LC-MS/MS analysis. Food Control 2012;28(1):192–8.

[35] Zhang LJ, Liu SW, Cui XY, Pan CP, Zhang AL, Chen F. A review of sample preparation methods for the pesticide residue analysis in foods. Cent Eur J Chem 2012;10(3):900–25.

[36] Prestes OD, Friggi CA, Adaime MB, Zanella R. QuEChERS – a modern sample preparation method for pesticide multiresidue determination in food by chromatographic methods coupled to mass spectrometry. Quim Nova 2009;32(6):1620–34.

[37] Pareja L, Fernandez-Alba AR, Cesio V, Heinzen H. Analytical methods for pesticide residues in rice. TRAC Trends Anal Chem 2011;30(2):270–91.

[38] Kim JH, Moon JK, Li QX, Cho, JY. One-step pressurized liquid extraction method for the analysis of polycyclic aromatic hydrocarbons. Anal Chim Acta 2003;498(1-2):55–60.

[39] Aguilar L, Williams ES, Brooks BW, Usenko S. Development and application of a novel method for high-throughput determination of PCDD/Fs and PVBs in sediments. Environ Toxicol Chem 2014;33(7):1529–36.

[40] Anitescu G, Tavlarides, LL. Supercritical extraction of contaminants from soils and sediments. J Supercrit Fluids 2006;38(2):167–80.

[41] Dean JR, Xiong, GH. Extraction of organic pollutants from environmental matrices: selection of extraction technique. TRAC Trends Anal Chem 2000;19(9):553–64.

[42] Harrison R, Bull I, Michaelides K. A method for the simultaneous extraction of seven pesticides from soil and sediment. Anal Methods 2013;5(8):2053–8.

[43] Beceiro-Gonzalez E, Gonzalez-Castro MJ, Muniategui-Lorenzo S, Lopez-Mahia P, Prada-Rodriguez D. Analytical methodology for the determination of organochlorine pesticides in vegetation. J AOAC Int 2012;95(5):1291–310.

[44] Guo L, Lee, HK. Microwave assisted extraction combined with solvent bar microextraction for one-step solvent-minimized extraction, cleanup and preconcentration of polycyclic aromatic hydrocarbons in soil samples. J Chromatogr A 2013;1286:9–15.

[45] Tadeo JL, Ana Perez R, Albero B, Garcia-Valcarcel AI, Sanchez-Brunete C. Review of sample preparation techniques for the analysis of pesticide residues in soil. J AOAC Int 2012;95(5):1258–71.

[46] Teo CC, Chong WPK, Ho, YS. Development and application of microwave-assisted extraction technique in biological sample preparation for small molecule analysis. Metabolomics 2013;9(5):1109–28.

[47] Bermejo P, Capelo JL, Mota A, Madrid Y, Camara C. Enzymatic digestion and ultrasonication: a powerful combination in analytical chemistry. TRAC Trends Anal Chem 2004;23(9):654–63.

[48] Wienhold BJ, Gish, TJ. Enzymatic pretreatment for extraction of starch encapsulated pesticides from soils. Weed Sci 1991;39(3):423–6.

[49] Nadeau RG, Howe RK, Burnett TJ, Lange, BD. Characterization of C-14 residues in the grain of rice plants grown in soil treated with phenyl-C-14-2-(diphenylmethoxy)acetic acid methyl-ester. J Agric Food Chem 1991;39(12):2285–9.

[50] Posyniak A, Zmudzki J, Semeniuk S. Effects of the matrix and sample preparation on the determination of fluoroquinolone residues in animal tissues. J Chromatogr A 2001;914(1-2):89–94.

[51] Yu C, Penn LD, Hollembaek J, Li WL, Cohen, LH. Enzymatic tissue digestion as an alternative sample preparation approach for quantitative analysis using liquid chromatography-tandem mass spectrometry. Anal Chem 2004;76(6):1761–7.

[52] Kaestner M, Nowak, KM, Miltner A, Trapp S, Schaeffer A. Classification and modelling of nonextractable residue (NER) formation of xenobiotics in soil – a synthesis. Crit Rev Environ Sci Technol 2014;44(19):2107–71.

[53] Jablonowski ND, Schaeffer A, Burauel P. Still present after all these years: persistence plus potential toxicity raise questions about the use of atrazine. Environ Sci Poll Res 2011;18(2):328–31.

[54] Vonberg D, Hofmann D, Vanderborght J, Lelickens A, Koeppchen S, Puetz T, Burauel P, Vereecken H. Atrazine soil core residue analysis from an agricultural field 21 years after its ban. J Environ Qual 2014;43(4):1450–9.

[55] Brenner KS. PU-foam-plug-technique and extractive co-distillation (Bleidner apparatus), versatile tools for stack emission sampling and sample preparation. Chemosphere 1986;15(9-12):1917–22.

[56] You IS, Bartha R. Evaluation of the Bleidner technique for analysis of soil-bound 3,4-dichloroaniline residues. J Agric Food Chem 1982;30(6):1143–7.

[57] Tekel J, Hatrik S. Pesticide residue analyses in plant material by chromatographic methods: Clean-up procedures and selective detectors. J Chromatogr A 1996;754(1-2):397–410.

[58] Ahmed FE. Analyses of pesticides and their metabolites in foods and drinks. TRAC Trends Anal Chem 2001;20(11):649–61.

7 Chromatography

7.1 General Remarks

After having performed all necessary steps before analysis (sampling, sample pretreat-ment, analyte enrichment and sample cleanup), we are now ready for the qualitative and quantitative determination.

In organic trace analysis, this still means, in most cases, the transfer of the sample with the mixed analytes into a chromatographic separation process. Identification is made by determination of retention behavior due to distribution processes within a separation column and comparison to analytes of known composition and quantity. Quantification and confirmation of the separated analytes is performed by a detector. The whole procedure is a noncontinuous one, as one has to await the termination of a separation process before the next analyte mixture may become injected into the chromatography system.

The term *"Chromatography"* stems from Greek language and means translated *"color writing."* More than 100 years ago, the first applications are found in separa-tion of leaf dye pigment extracts by adsorption chromatography (Mikhail Tsvet, 1903). Different color bands obtained on a calcium carbonate separation column were the results, and the term "chromatography" was born and thereon used. Later, driven by petroleum technology and nuclear chemistry, separation science rapidly expanded and is still the backbone for a multitude of chemical and life science technologies today.

Classically, its characteristics are:

- partition equilibria of analytes (adsorption or distribution) exist between two nonmiscible phases;
- one phase is stationary, the other one is mobile and in permanent contact with the stationary phase;
- separation happens due to subtle differences in partition coefficients for different analytes;
- when the analyte mixture gets injected into a steady flow of mobile phase at the beginning of partitioning, multiple repeated equilibrations lead to a spatial sep-aration *("introduced side by side"* → *"released one after the other"*), thus to a different retention time or retention volume on the column;
- depending on the presence of many analyte molecules within a sample, a high number of repeated equilibrations will be needed for a sufficient separation.

Nowadays, many approved versions of this methodology are already commercially available. The main difference is the applicability toward either polar analytes pos-sessing low vapor pressure or nonpolar analytes with often high vapor pressure.

DOI 10.1515/9783110441154-007

It has to be accepted that a separation process always stays incomplete. For statistical reasons, there will always be some different molecules found in each separated fraction. We can only approach a perfect separation but never reach it.

Today, the search for an ever-growing variety of molecules in whole functional living entities triggers the ongoing development of high-performance separation of organic substances, mainly in the trace concentration range. This is represented by the various "omics" areas: *metabolomics, proteomics, glycomics, lipidomics, etc.*

7.2 Chromatographic Separation

Chromatographic separation can be obtained when an analyte dissolved within a *mobile phase* is in contact with a *stationary phase* and exhibits different partitioning compared to a second analyte. This can be a one-time establishment of equilibrium (batchwise) or continuously repeated. Since differences in distribution coefficients for similar substances are very small, in organic trace analysis always a multitude of partitioning equilibria has to be applied in order to obtain sufficient differences in retention time. Applicable combinations for mobile and stationary phases are given in Table 7.1. It has to be noted that no separation can be achieved by pairing *solid–solid*, because no mass transfer from one phase to the other happens. Separation by *gas–gas* pairing would not be much successful due to Brownian motion, which prevents a good separation.

Gases or liquids as mobile phase are selected such that the affinity to the stationary phase compared to analyte molecules is negligible. The interaction of the analyte with the stationary phase compound has to be strong, but still reversible; hence, a high distribution coefficient can be expected.

As mentioned earlier, through repetition of a multitude of partitioning steps within sequentially arranged distribution sections along a column, we will obtain a spatial separation between at least two analytes with different distribution coefficients. This classical view of a transition from the movement of an analyte mixture through a serially arranged battery of vessels (liquid–liquid partitioning) to a connected arrangement, the "chromatographic column," is shown in Figure 7.1.

Table 7.1: Combinations for interacting mobile and stationary phases in chromatographic processes.

Mobile phase	Stationary phase	Application (example)
Gas	Liquid	Gas–liquid chromatography
Gas	Solid	Gas–solid chromatography
Liquid	Liquid	Liquid chromatography
Liquid	Solid	Adsorption chromatography; ion-exchange chromatography; size-exclusion chromatography; affinity chromatography

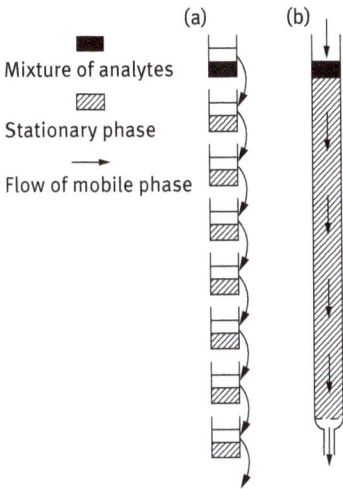

Mixture of analytes

Stationary phase

Flow of mobile phase

Figure 7.1: Transition from a serial arrangement of distributions in vessels to a chromatographic separation column. The separation process theoretically occurs by a multitude of partitioning steps (a) along the flow of the mobile phase through the columns containing the stationary phase (b).

The mobile phase carries the injected sample analyte mixture. After a multitude of repeated distributions and forward movement of the upper organic phase along the battery of vessels (here assumed for a liquid–liquid distribution), separation of the analytes will be achieved. Partitioning will lead to a Gaussian distribution of analyte concentrations at various locations of such a column after a certain time of interaction. The retention time or volume needed to leave the separation in an elution process is characterized by t_R or v_R. The most mobile analyte will leave the separation column first. For two differently mobile analytes, this is depicted in Figure 7.2.

The ideal separation would be represented by a non-overlap of the concentration distribution functions for two analytes applied to a distribution between the two phases (phase 1: mobile phase; phase 2: stationary phase). In practice, the criterion

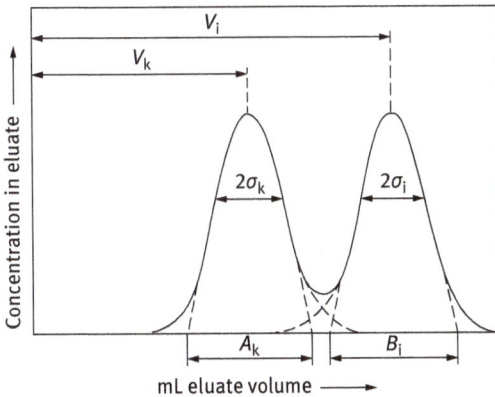

Figure 7.2: Separation of two analytes after a number of partitioning steps along a separation column. Plotted is concentration of analytes A and B as a function of retention volume.

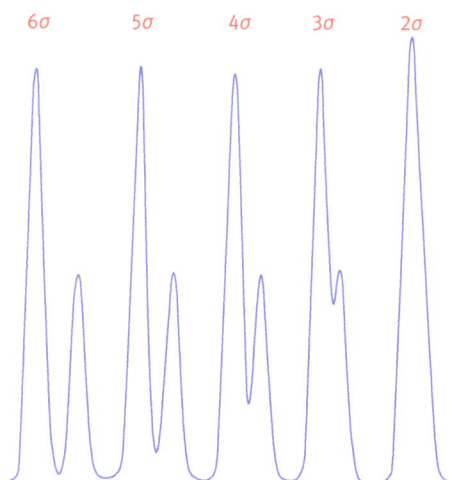

Figure 7.3: Resolution power expressed by peak width distances σ for two analytes.

for the separation of two peaks with the signal of the detector reaching (almost) the baseline (usually termed the resolution power) is defined as the ratio of the distance between the peak maxima to half the peak width (σ) at the points of inflection. The separation of a pair of analytes 2σ, 3σ, 4σ, 5σ and 6σ from each other is shown in Figure 7.3. The similarity of this convention to the definition of limit of detection (LOD) (see Chapter 2, Section 2.5) is obvious. It means a tolerable overlap of 1% of the respective analyte masses to each other.

For *baseline resolution*, the peak maxima should be separated by 6σ, but for most analyses (especially where peak height measurements are employed) a separation of 4 is usually sufficient.

In column chromatography, there are two common approaches: *elution* or *development on column*. The latter means that separation is continued and stopped when the faster moving analyte reaches the distal end of a separation column. Afterward, the column becomes sacrificed, and the different sections containing the separated analytes are then handled in another way. This approach is practiced in thin-layer chromatography (TLC) or paper chromatography, in which only a thin "slice of column" is used. This is depicted in Figure 7.4.

Elution means that the fastest analyte will leave the column prior to the next analyte arriving the exit. Usually, these eluates then become characterized in quality and concentration by an attached suitable detector system. Passing this, the analyte fractions can be sampled in a subsequent arrangement of collection vessels ("fraction sampling"; see Figure 7.5).

The continuous registration and readout of separated and eluted analytes passing a suitable detector represents the writing of a "chromatogram" (see Figure 7.2). This chromatogram contains all information that are necessary to evaluate the success of separation.

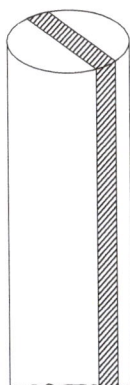

Figure 7.4: "Thin-layer chromatography, TLC", shown as a virtually excised layer from a separation column.

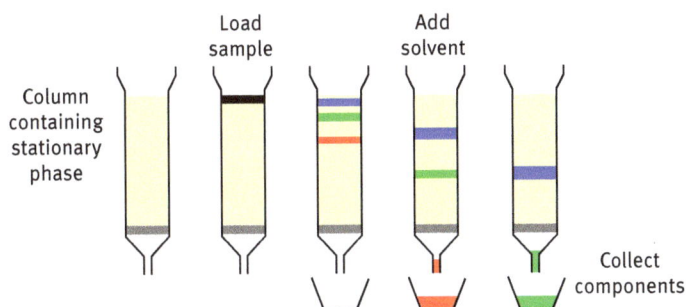

Figure 7.5: Sampling of fractionated analytes.

7.2.1 Adsorption Chromatography

Adsorption chromatography uses multiple equilibria establishment and is mainly performed when the separated analytes are separately quantified and evaluated on column (e.g., in the historical experiment executed by Tsvet; see Figure 7.6).

A different approach is the single establishment of a sorption equilibrium for solid-phase extraction (SPE) or affinity chromatography.

A prerequisite to adsorption chromatography in general is a linear range in the underlying adsorption isotherm. Only in such a case, a linear relationship between coverage of the adsorbent surface with concentration of adsorbing analytes can be expected. This means that there is an optimal concentration range for separation, which has to be defined. Exemplarily shown is this in Figure 7.7.

The stationary phases in use are selected according to their adsorption strength (depending on surface chemistry of the adsorbent and its specific surface area) toward an analyte and a given mobile phase. There are many known examples of adsorbents

An EtOH extract of leaf pigments is applied to the column. EtOH is used to flush the pigments down the column.

A glass column filled with powdered limestone ($CaCO_3$)

A series of colored bands is seen to form, corresponding to the different pigments in the orginal plant extract. These bands were later determined to be chlorophylls, xanthophylls and carotenoids.

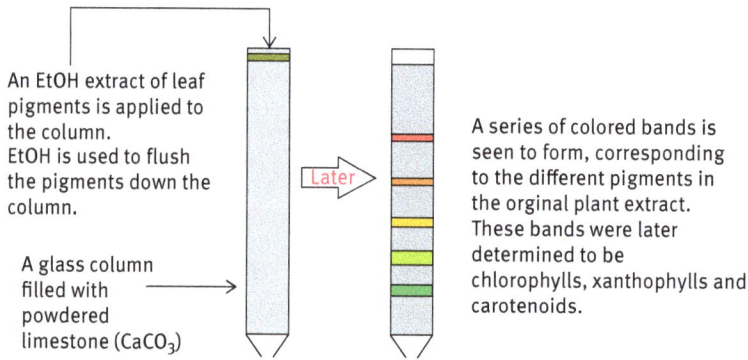

Figure 7.6: Adsorption chromatography with separation on column. Shown is the arrangement used by M. Tsvet.

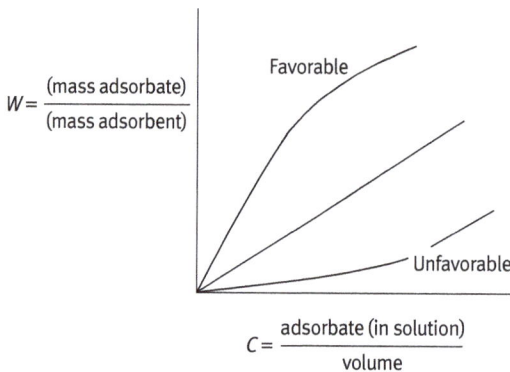

$$W = \frac{\text{(mass adsorbate)}}{\text{(mass adsorbent)}}$$

Favorable

Unfavorable

$$C = \frac{\text{adsorbate (in solution)}}{\text{volume}}$$

Figure 7.7: Adsorption isotherms and its meaning for chromatographic separation.

published, for example, Zechmeisters' series of adsorbents (descending adsorption power, for one selected analyte and mobile phase):

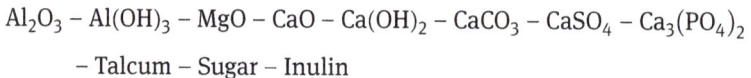

$$Al_2O_3 - Al(OH)_3 - MgO - CaO - Ca(OH)_2 - CaCO_3 - CaSO_4 - Ca_3(PO_4)_2$$

$$- \text{Talcum} - \text{Sugar} - \text{Inulin}$$

Of course, the variation of surface chemistry, for example of cellulose or polysaccharide, gives further possibilities for optimization. Also a variety of polymers have been developed in recent years. Strong and selective adsorption characteristics are achieved with natural or artificially produced affinity receptors (e.g., antibodies, molecularly imprinted polymers, solid-phase materials) [1, 2]. These are used for batch collection of analytes. This means that adsorbent and dissolved analytes are brought into contact until equilibrium is reached. Then, the adsorbent with the enriched analytes is removed and subjected to further analysis.

Eluents (the mobile phase) are selected in a similar fashion. They may be ordered according to their sorption power for a selected stationary phase and analyte. As an example, the eluotropic series of eluents with increasing polarity according to Trappe is shown:

Petroleum ether – CCl_4 – Trichloroethene – Benzene – CH_2Cl_2 – $CHCl_3$ – Diethylether

– Ethylacetate – Ethanol – Methanol – Water – Pyridine

Depending on the nature of the sorbent/solid interaction with the analyte, the dielectric properties of the solvent also play a role.

The usage of natural receptors, like antibodies, is of special interest in sample enrichment (see Chapter 6). Antibodies are glycoproteins, raised in mammals after having been in contact with antigens, and possess distinct properties to fix these antigens. The binding constant can be as high as 10^{15} mol^{-1}. Bound into short columns by tentacle chemistry or trapped inside a porous sol–gel matrix, they can be repeatedly used for analyte enrichment. Examples for their usage are toxin enrichment from milk, or pesticides from water.

A less pronounced distribution in application is found for so-called molecularly imprinted polymers (MIPs). MIPs (see Chapter 6) are chemically produced polymers, which show binding constants in the range of <10^6 mol^{-1}. Like antibodies, they find applications in enrichment of analytes from liquid matrices (food and drinking water).

Finally, refined polymer phases (e.g., restricted access polymer spheres) are also in use as adsorbing phases (100–1,000 mg adsorbent material) within small cartridges of 3–5 mL volume (SPME). A larger volume of liquid sample is sucked through it. Afterward, the adsorbed analytes become desorbed by a small addition of a strongly adsorbed compound forming a displacement reaction. These SPME cartridges are frequently used for analytical separation of polar or nonpolar contaminants from water.

Despite the expenses for refined receptors used in affinity chromatography, the possibility of using one affinity cartridge up to 100 times makes them very attractive for sample cleanup.

7.2.2 Ion-Exchange Chromatography

This technique was developed in the last century when separation technologies were needed for isotope separation and enrichment. Afterward, biotechnology, environmental chemistry and clinical diagnostics forced the development of ion-exchange chromatography (IC). In the 1970s, especially high-pressure instrumentation came up for the first time and enabled high-performance separations within short time. Nowadays, even flash IC has been published (separation of several ions within some seconds).

This separation principle is based on attraction or repulsion of charged species, following Coulomb's law.

Many organic molecules possess functional groups, which can be charged by adjusting the pH value when dissolved in aqueous phase. Especially carboxylic acids, sulfonic acids and amines or amino acids are successfully separated by IC. Large molecules, such as proteins and nucleotides carrying such groups, can be charged and separated too. Since these analytes are often present in an aqueous solution, they can be directly introduced to IC, whereas this is not possible, for example, for gas chromatography (GC) or liquid chromatography. Depending on nature of ion exchange applied, one can divide the applications into *cation-exchange chromatography* or *anion-exchange chromatography*.

The exchange equilibria for IC are

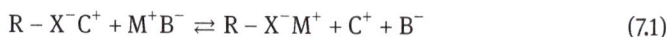

$$R - X^- C^+ + M^+ B^- \rightleftharpoons R - X^- M^+ + C^+ + B^- \tag{7.1}$$

for cation-exchange chromatography, or

$$R - X^+ A^- + M^+ B^- \rightleftharpoons R - X^+ B^- + M^+ + A^- \tag{7.2}$$

for anion-exchange chromatography.

R–X is the ion-exchanging backbone (e.g., for anion exchange ion-exchange resin containing positively charged groups, such as diethyl-aminoethyl groups) and is applied as stationary phases. The resins are packed as small spheres into a tubular column. For mobile phase, weak salt or buffer solutions of appropriate pH and ion strength are prepared.

The typical instrumentation needed for IC is shown in Figure 7.8. The sample is as usual injected (e.g., 100 μL) into the flowing mobile phase. The mobile phase (e.g., a $NaHCO_3/Na_2CO_3$ solution for anion separation) is pumped from a reservoir, degassed and filtered, to the separation column. After the injector, the sample enters first a so-called guard column (filled with a size-excluding gel-like phase), removing particulate

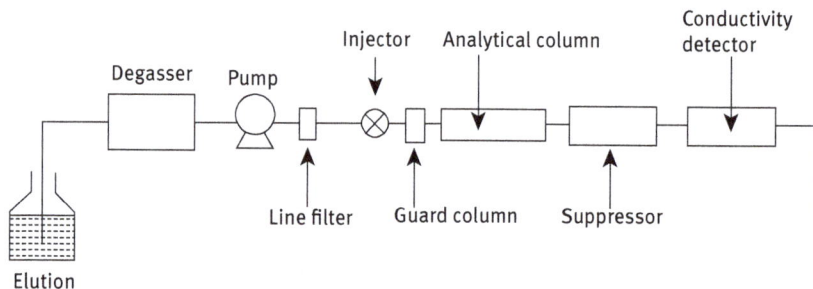

Figure 7.8: Schematic flow diagram for an ion chromatography system.

matter of undefined origin such as humic substances, which otherwise would block the analytical separation column over time. The separation happens within the analytical column, consisting of a weak anion exchanger when anions have to be separated. Depending on charge and molecular dimension for the differently mobile analytes, the residence time is different. Afterward, a so-called suppressor column follows filled with a cation exchanger. This is necessary as after the analytical column, the separated anions are still co-housed with the Na^+-containing eluent.

So, the separated trace amounts of anions would not be recorded well by a conductivity detector. Therefore, a suppressor column converts all Na^+ ions into H^+, forming the corresponding less dissociated hydrogencarbonate or carbonic acid, which then does no longer respond in a conductivity detector. But the (e.g., carboxylic) anions are transformed now into stronger dissociating acids, which are easily indicated because of their higher equivalent conductivity.

Two applications, one for aliphatic carboxylic acids, and the other one for aliphatic sulfonic acids, are shown in Figure 7.9.

7.2.3 Size-Exclusion Chromatography

Size-exclusion chromatography (SEC) or gel permeation chromatography (GPC) does not depend on the distribution of analytes among two immiscible phases. Rather, it is based on the differences in hydrodynamic size of analytes dissolved or dispersed within a mobile phase. In this case, the "stationary phase" represents a porous gel. Such an open pore gel, usually agarose for water as mobile phase, or polyacrylamide gel suited for hydrophobic carrier solvents, acts as a transient recipient for the very small molecules, when passing the separation column like in adsorption chromatography (see Figure 7.10).

Small molecules, due to their much higher hydrodynamic Brownian motion, will enter the very small pore volumes of the swollen gel matrix, whereas large analyte molecules will travel along the region of fastest flow lines. Therefore, large analytes leave the column first, followed by the smaller ones.

Typical applications are found in polymer chemistry and biotechnology (e.g., for protein purification). The mobile phase may consist of organic solvents.

As standards for calibration purposes, spherical polymer particles (e.g., polystyrene lattices) with known diameters as determined by transmission electron microscopy are frequently applied. From their retention volume (or time), the particle sizes, expressed as equivalent sizes to the hydrodynamic diameters of the used standard spheres, can be retrieved. Figure 7.11 exemplarily shows such calibration and the retention behavior of three differently sized polymers. As can be seen from this graph, the largest particle fraction is leaving the separation column first.

SEC/GPC is also advantageous for pre-separation of analytes from colloidal matrices (fermenter solutes, soil extracts, tissue digestion, etc.).

**Determination of aliphatic carboxylic acids
using the IonPac® NS1 column**

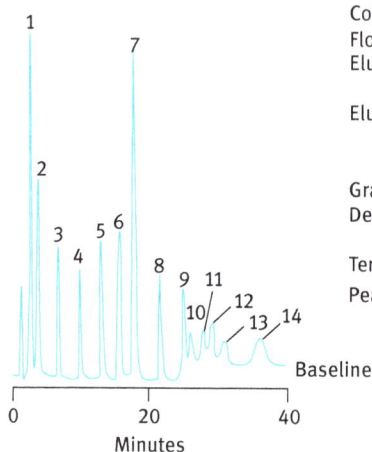

Column:	IonPac NS1 (10 mm)
Flow rate:	1.0 mL/min
Eluent 1:	24% acetonitrile,6% MeOH, 0.03 mM HCl
Eluent 2:	60% acetonitrile,24% MeOH, 0.05 mM HCl, 5 mM octanesulfonic acid
Gradient:	0–100% E2 in 20 min.
Detection:	Suppressed conductivity, AMMS´-ICE suppressor
Temp:	42 ˚C

Peaks:

1. Butyric	8. Dodecanoic
2. Pentanoic	9. Tetradecanoic
3. Hexanoic	10. Linolenic
4. Heptanoic	11. Linoleic
5. Octanoic	12. Hexadecanoic
6. Nonanoic	13. Oleic
7. Decanoic	14. Octadecanoic

**Separation of aliphatic sulfonic acids using
tetrabutylammonium ion as the ion-pair reagent**

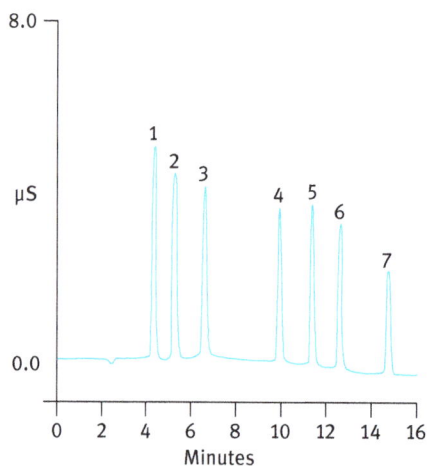

Column:	IonPac® NS1 (10 mm)
Eluent:	2 mM tetrabutylammonium hydroxide, 24% to 48% acetonitrile in 10 min
Flow rate:	1.0 mL/min
Inj. volume:	50 µL
Detection:	Suppressed conductivity, ASRS®-ULTRA, AutoSuppression®coupled with chemical regenerant mode
Regenerant:	10mN sulfuric acid

Peaks:

1. Methanesulfonic acid	5.0 mg/L
2. 1-Propanesulfonic aicd	8.6
3. 1-Butanesulfonic acid	8.7
4. 1-Hexanesulfonic acid	8.8
5. 1-Heptanesulfonic acid	8.9
6. 1-Octanesulfonic acid	8.9
7. 1-Decanesulfonic acid	9.1

Figure 7.9: Separation of aliphatic carboxylic acids (top) and of aliphatic sulfonic acids (bottom) by IC with subsequent ion suppression and conductivity detection.

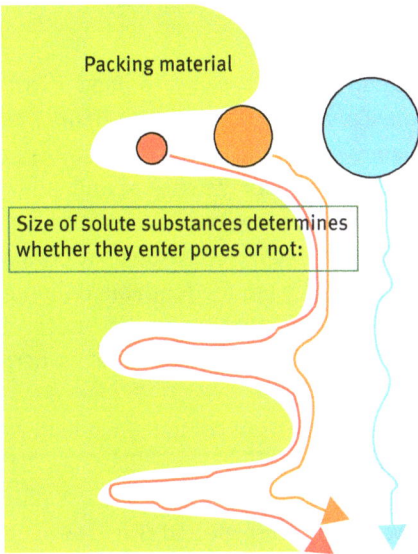

Packing material

Size of solute substances determines whether they enter pores or not:

Figure 7.10: Principle of size-exclusion separation due to Brownian motion of particles within liquid.

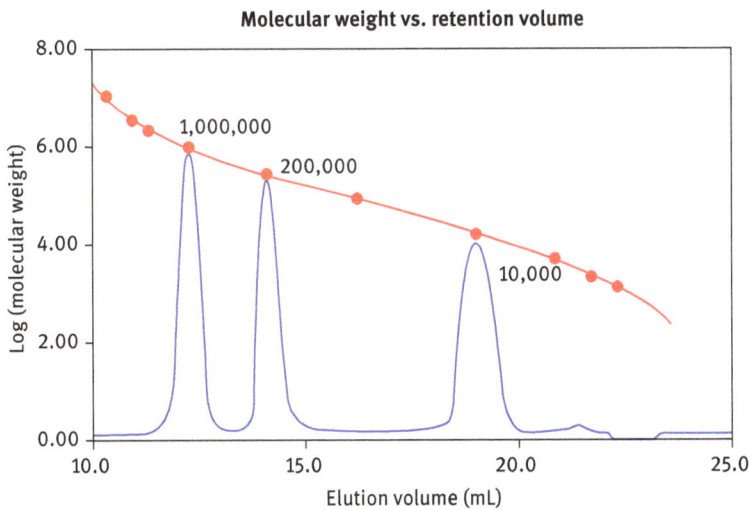

Molecular weight vs. retention volume

1,000,000

200,000

10,000

Log (molecular weight)

Elution volume (mL)

Figure 7.11: Calibration line for differently sized monodisperse particles and SEC separation of a mixture of three differently sized polymers.

7.3 Basics and Working Principles

Since there are many reviews and detailed application reports on a multitude of separation tasks available, we can concentrate on the most important issues in chromatography, and this might be the description of the resolving power for a given chromatographic process and the parameters to be used for its optimization.

7.3.1 Parameters to Achieve Best Resolving Power

As mentioned in Section 7.2, the optimal separation process based on a linear distribution isotherm leads to a Gaussian distribution of analyte molecules (see Figure 7.12, bottom left) along a separation column in the course of the continuous development of such a process, after experiencing many equilibrations. In the case of only a few equilibration steps and nonlinear distribution isotherms, for example, within a short column, we will observe a skewed peak shape, represented by a Poisson distribution (see Figure 7.12, bottom middle and right). We also mentioned the fundamental non-achievement of complete separation, as there will be always some molecules of different analytes still present at a column section or within an eluted fraction where we don't expect them.

Due to flow conditions prevailing in the usual separation columns, we have to anticipate a laminar flow profile, with the consequence of a broadened Gaussian distribution of separated analyte molecules across a volume element. Only for a plug flow by principle, an indefinitely small analyte peak could be expected. The task of an analyst therefore is the optimization in order to get maximal separation.

Figure 7.13 shows the ideal chromatographic peak shape.

From what was said before, not only a symmetrical peak shape is wanted. Rather, the peak shape should become narrow, because then the separated analyte is part of a

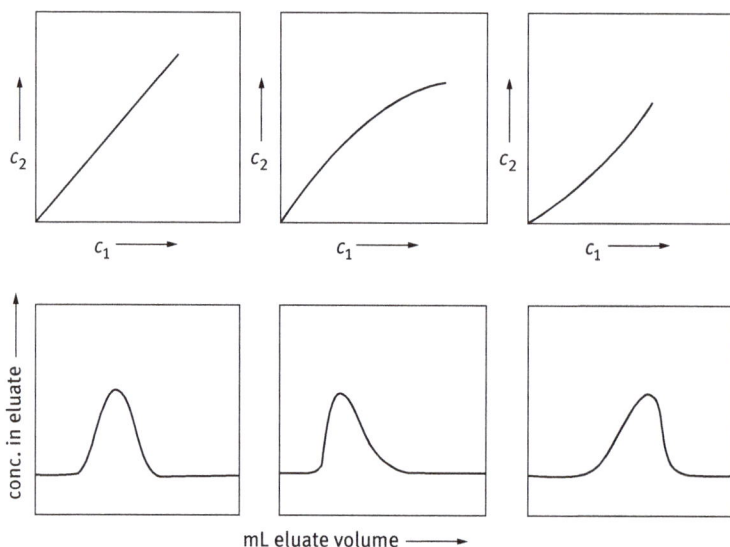

Figure 7.12: Impact of distribution equilibria (top: linear isotherm on the left, nonlinear isotherms middle and right) established between two phases on peak shape of distributed analyte during the separation process on column (bottom: Gaussian distribution, middle and right Poisson distribution).

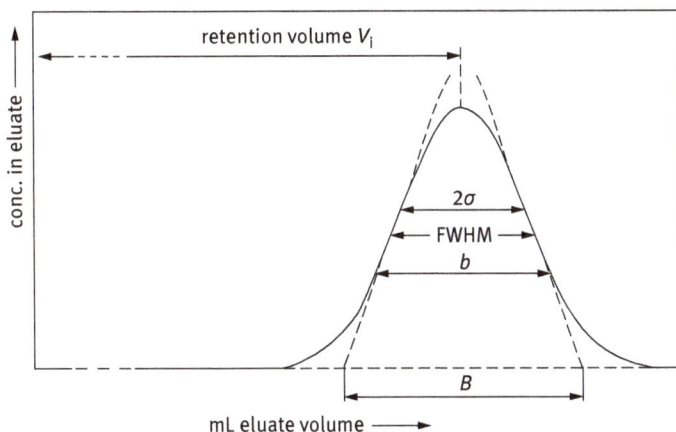

Figure 7.13: The ideal peak shape for best distribution and its characterization. FWHM stands for the full width at half maximum.

small eluate volume or accumulated within a short section of the separation column. This is represented by the full width half maximum (FWHM).

In chromatography, there are several ways to obtain narrow but still symmetric peak shapes. A common mean is the variation of the mobile phase or the character of the stationary phase. This was exemplarily discussed already in Section 7.2.1 for adsorption chromatography. Figure 7.14 depicts the effect of changing the polarity of the mobile phase in liquid chromatography. By principle, this holds for GC too.

Why is the narrowing of peaks so important? There are two reasons: First, broad peaks mean distribution of analytes within a larger eluate volume or a larger column section, hence high dilution of the analyte. Second, in the case of only a few analytes to be present within a sample, this may not cause much trouble. But typical samples,

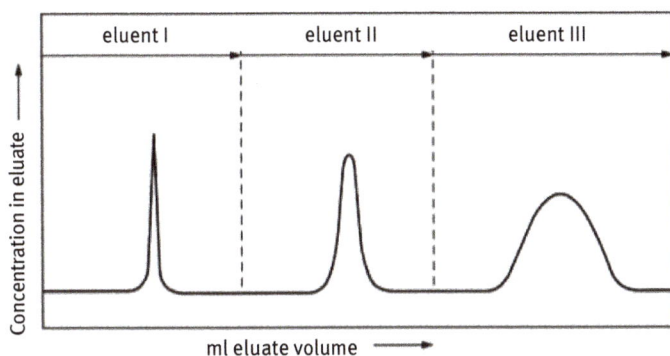

Figure 7.14: Effect of changing the mobile phase on FWHM during separation.

for example, from combustion or a petroleum fraction, consist of thousands of different compounds. In most cases, we can only roughly estimate the composition due to insufficient resolution power of the applied analytical techniques. So, there is a tremendous need to achieve as many as possible equilibrations within a given separation column. Or, derived from distillation technology, we like to have a high number of equivalents to a theoretical plate (height equivalent of theoretical plate, HETP). One HETP per column would mean a hypothetical zone or stage in which two phases, such as the liquid and vapor phases of a substance in a distillation process, establish an equilibrium with each other. It is clear that the more HETPs could be assigned to a chromatography column, the better the resolving power will be. This is shown in Figure 7.15.

The numbers of HETP can be derived from mathematical analysis of the shape of a Gaussian distribution (peak see Figure 7.12) and its retention time t_R from a chromatogram:

$$N = 8\ln(2) \cdot \left(\frac{t_R}{W_{1/2}}\right)^2 \tag{7.3}$$

where N is the HETP number, t_R is the retention time of chromatographic peak within a chromatogram and $W_{1/2}$ is the FWHM.

It is interesting to know the extent of HETP numbers for the different varieties of chromatography. GC is currently by far the best reaching HETP numbers per column, having a length of up to 100 m, of >100,000. An even refined multiple or fractional distillation column is in the range of 5–100. Capillary electrophoresis as a non-chromatographic separation technique achieves HETP numbers in the millions.

Martin and Synge (Nobel Prize 1952) were the first to apply such formalistic description for optimizing chromatographic processes. The Van Deemter equation (1956)

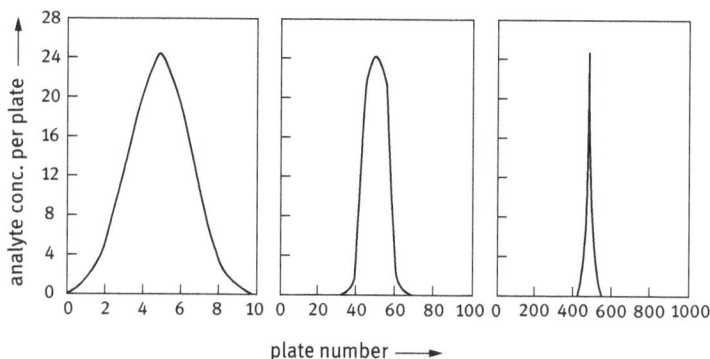

Figure 7.15: Distribution of an analyte across a separation column (consisting of many equilibration plate) with increasing HETP values.

summarizes this in the following way (eq. 7.4), representing the influence of physical, kinetic and thermodynamic properties of a separation process:

$$\mathrm{HETP} = A + \frac{B}{u} + (C_s + C_m) \cdot u \tag{7.4}$$

where HETP is the height equivalent to a theoretical plate; A is the Eddy diffusion parameter, related to channeling through a nonideal packing [m]; B is the diffusion coefficient of the eluting analyte molecules in the longitudinal direction, resulting in dispersion [m^2/s]; C is the resistance to mass-transfer coefficient of the analyte between mobile and stationary phase [s]; u is the linear velocity of mobile phase [m/s].

The graphical form, given in Figure 7.16, shows a minimum in HETP value as a function of mobile phase velocity. This means, at this minimum, which has to be found experimentally by the analyst, the best resolution power is achieved, since the highest number of equilibria for a given length of column is obtained. Or, in other words, too low mobile phase velocity will lead to peak broadening by lateral diffusion, and in contrast to this, too rapid movement of analytes through a separation column will not allow sufficient mass transfer for complete equilibration; hence, both are increasing HETP values, and thus decreasing the resolution power.

The van Deemter relation holds for most applications of chromatography. Of course, as with all thermodynamically or kinetically controlled processes, temperature could also be applied for optimizing a separation. But the strong influence of temperature on Brownian motion of the analyte molecules and viscosity of the carrier

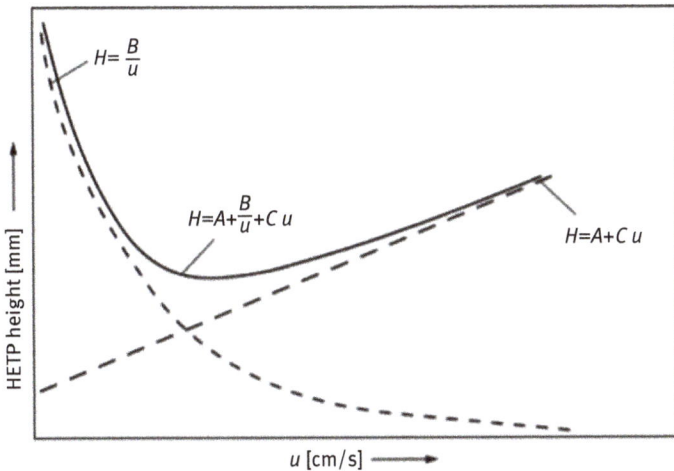

Figure 7.16: Resolution power in a chromatographic separation process expressed by HETP height value (abbreviated as H in this figure) as a function of mobile phase velocity u. For other parameters, see eq. (7.4).

medium restricts this. Nevertheless, temperature control of the separation column is always beneficial.

7.3.2 General Recommendations

The specific instrumentation needed for the different varieties in chromatography will be discussed later within the respective chapters in detail.

But some general rules may be given:

– A highly resolving detector (e.g., mass spectrometer [3]) is not always the first choice, and other detectors may be used [4, 5]. Pre-separation is the magic word! Even the best MS will surrender once the ionizer is blocked by unnecessarily deposited solids from concomitant material.
– There is never only one possible choice of universal stationary or mobile phase. Estimation of the polarity, vapor pressure or molecular size of an analyte makes decision about selection of adequate separation phase easier.
– It is reasonable to consider a change of state for an analyte, if its natural physical or chemical state prohibits easy separation or sensitive and selective detection. This might be addressed by chemical derivatization, ion-pair formation, encapsulation (within micelles or dendrimers), pyrolysis, backbone hydrogenation, etc.
– Overloading a separation column means to waste separation capacity. Take only the minimum sample volume appropriate for your problem.
– Providing a gradient during a separation run, either by temperature or mixing of mobile phases, is increasing resolving power a lot.
– Subsequent change of stationary phase after a first separation run often reveals new analytes not resolved in preceding separation run

$$(\rightarrow \text{ multidimensional chromatography}).$$

– In solid–liquid (gas) chromatography, particle size for stationary phase should be monodisperse, but not too small (ca. 1 μm). Microflow pattern or pressure drop across the column disturbs the resolving power.
– Purity of selected phases is of crucial importance.
– Stationary phase means that it should stay stationary. Otherwise, its lifetime becomes very short and a permanently shifting background is the result.
– Identification only by retention time t_R and comparison to standards is often used (so-called cochromatography) but is juridically questionable. Always, for more reliable identification, a second or third identification method should be applied.
– Selection of the appropriate detector saves money and prevents from frustrations. Not always the most expensive one is the best.
– Don't trust data bases in any case. Create your own data base.
– If available, use stable-isotope-labeled analytes for calibration.
– Check your analytical separation scheme by usage of certified reference materials.

7.4 High-Performance Liquid Chromatography

7.4.1 General Information

High-performance liquid chromatography (HPLC) is a well-established separation technique, providing a large variety of chromatographic conditions. The basic principle works by pumping a liquid, the mobile phase, consisting of one or more solvents, through (often first a pre-column for pre-purification and) a chromatography column. The column contains a matrix, the stationary phase, at which the analytes, which are injected before the column, are separated by different mechanisms at rather high pressure to increase the velocity of the eluent. The eluent is then passed through a detector system before it enters a waste container or can be collected for further analytical characterization (Figure 7.17). The detector signal plotted as a function of the solvent flow or of the time is the result of the chromatographic separation, that is, the chromatogram (Figure 7.18).

The possibility to vary each part of the HPLC system – the mobile phase, the stationary phase, the detector and the experimental conditions such as the pressure and the temperature – renders the system extremely flexible in terms of analyzing analytes of different physicochemical properties. HPLC separation can be performed for qualitative, quantitative and preparative purposes, the latter by using larger scale chromatographic columns.

7.4.2 Mobile Phase (HPLC)

One, two or more solvents can be used as the eluent in HPLC analysis. The choice of the solvents depends on the properties of the analytes and their respective distribution

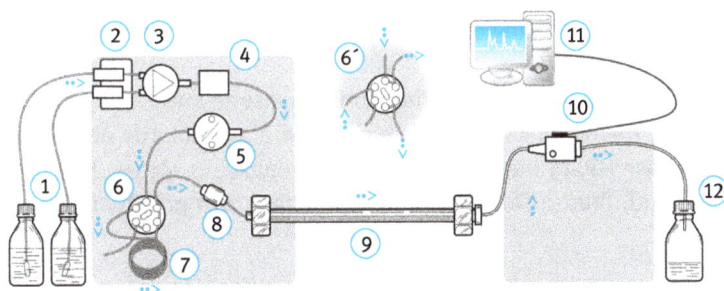

Figure 7.17: Scheme of a typical HPLC separation system: (1) solvents, (2) solvent degasser, (3) gradient valve, (4) mixing vessel for delivery of the eluent, (5) high-pressure pump, (6) switching valve in "inject position," (6') switching valve in "load position," (7) sample injection loop, (8) pre-column (guard column to remove impurities and particles), (9) analytical column, (10) detector, (11) data acquisition and analysis (computer), (12) waste (or the eluate is pumped to a fraction collector). ©: Wikimedia Commons.

(a)

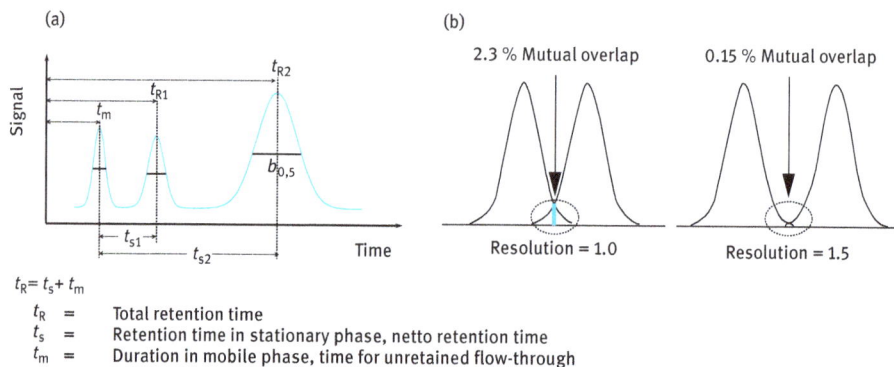

(b)

2.3 % Mutual overlap   0.15 % Mutual overlap

Resolution = 1.0   Resolution = 1.5

$t_R = t_s + t_m$

t_R	=	Total retention time
t_s	=	Retention time in stationary phase, netto retention time
t_m	=	Duration in mobile phase, time for unretained flow-through

Figure 7.18: (a) A schematic HPLC chromatogram. The capacity factor k can be determined by $k = (t_R - t_m)/t_m$ and allows a better comparison of analyte retention times due to normalization. (b) The resolution R of peaks, that is, how well they are separated, can be calculated by $R = 1.18 (t_{R2} - t_{R1})/(b_{0.5,\text{peak1}} + b_{0.5,\text{peak2}})$. Baseline separation of peaks is achieved with $R \geq 1.5$.

coefficients between the mobile and the stationary phase and on the properties of the matrix in the column, that is, whether it is a polar or a nonpolar matrix. Using polar matrices like silica gels, a nonpolar solvent like cyclohexane is appropriate; nonpolar matrices require the use of more polar solvents like acetonitrile, often in combination with water.

7.4.2.1 Polarity of HPLC Solvents
The polarity of some HPLC solvents together with other properties is given in Table 7.2.

In general terms, the polarity of solvents can be ranked according to chemical classes (Table 7.3).

It has to be taken into account that not all solvents are miscible with each other under all mixture conditions: Of course, immiscibility is a problem for liquid chromatography because of potential clogging of the system (see Table 7.4 for the miscibility of HPLC solvents).

In normal-phase HPLC, the polarity of the mobile phase is lower than that of a stationary phase and, therefore, sample components with low polarity are eluted quickly. In reversed-phase HPLC, the polarity of a mobile phase is higher than that of a stationary phase and, thus, a sample component with a high polarity is eluted quickly (see Figure 7.19).

7.4.2.2 HPLC Solvent Selection
The solvent selection for a given separation problem can start by matching the polarity of the solvent to that of the sample. The separation can then be refined: If the sample appears at the solvent front, then, the solvent has a too strong elution to allow the adsorbent to retard the sample. So it is proposed to choose a solvent with a lower elution strength, for instance in reversed-phase chromatography a more polar solvent (e.g., by

Table 7.2: Properties of typical HPLC solvents.

Solvent	Viscosity (cP, 20 °C)	Boiling point (°C)	UV cutoff (nm)	Snyder polarity index
Acetone	0.36	56.29	330	5.1
Acetonitrile	0.38 (15 °C)	81.60	190	5.8
n-Butyl alcohol	2.98	117.50	215	3.9
Chloroform	0.57	61.15	245	4.1
Cyclohexane	1.00	80.72	200	0.2
o-Dichlorobenzene	1.32 (25 °C)	180.48	295	2.7
Dichloromethane	0.44	39.75	233	3.1
1,4-Dioxane	1.37	101.32	215	4.8
Ethyl acetate	0.45	77.11	256	4.4
Ethyl alcohol	1.10	78.32	210	5.2
Ethyl ether	0.24	34.55	215	2.8
Heptane	0.42	98.43	200	0.1
Hexane	0.31	68.70	195	0.1
Isopropyl alcohol	2.40	82.26	205	3.9
Methanol	0.59	64.70	205	5.1
Pentane	0.23	36.07	190	0.0
Tetrahydrofuran	0.55	66.00	212	4.0
Toluene	0.59	110.62	284	2.4
Water	1.00	100.00	190	10.2

The UV cutoff suggests the lowest wavelength for use of an ultraviolet–visible (UV–vis) absorption unit as detector. The Snyder polarity index ranks the polarity of solvents based on their adsorption energy of alumina but is similar also for other matrices (low numbers for nonpolar, high number for polar solvents).

Table 7.3: Polarity of chemical classes.

Relative polarity	Chemical class	Compounds	Typical solvents
NONPOLAR	R–H	Alkanes	Petroleum, hexane
	Ar–H	Aromatics	Toluene, benzene
	R–O–R	Ethers	Diethyl ether
	R–Hal	Alkyl halides	Tetrachloroethane
	R–COOR	Esters	Ethyl acetate
	R–CO–R	Aldehydes, Ketones	Acetone
	R–NH$_2$	Amines	Triethylamine
	R–OH	Alcohols	Methanol, ethanol
	R–CONHR	Amides	Dimethylformamide
	R–COOH	Carboxylic acids	Ethanoic acid
POLAR	H–OH	Water	Water

Table 7.4: Typical HPLC solvents, miscible (m; two solvents can be mixed together in all proportions without forming two separate phases) or immiscible (i) with each other.

	Acetone	Acetonitrile	Cyclohexane	Dichloromethane	Ethylalcohol	Ethylether	Heptane	Hexane	Isopropanol	Methanol	Water
Acetone											
Acetonitrile	m										
Cyclohexane	m	i									
Dichloromethane	m	m	m								
Ethylalcohol	m	m	m	m							
Ethylether	m	m	m	m	m						
Heptane	m	i	m	m	m	m					
Hexane	m	i	m	m	m	m	m				
Isopropanol	m	m	m	m	m	m	m	m			
Methanol	m	m	i	m	m	m	i	i	m		
Water	m	m	i	m	m	m	i	i	i	m	

Normal phase   **Reversed phase**

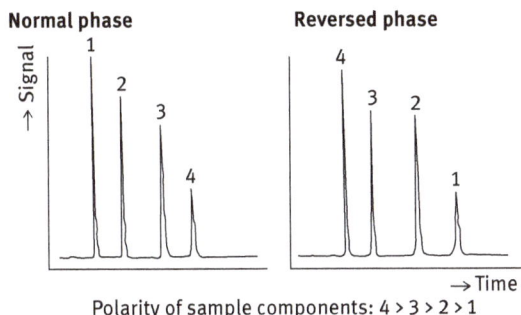

Polarity of sample components: 4 > 3 > 2 > 1

Figure 7.19: Elution of analytes with varying polarity from normal-phase and reversed-phase HPLC column matrices.

increasing the amount of water added to acetonitrile). Conversely, if the sample does not elute from the column in a reasonable time, choose a solvent with stronger elution strength (e.g., on reversed-phase columns by increasing the lipophilicity, for instance by reducing the amount of water added to acetonitrile).

7.4.2.3 Isocratic and Gradient Elution

If the mobile-phase composition during separation of analytes remains constant, we call this an isocratic elution (a Greek term meaning that the elution strength remains constant). The advantage of isocratic conditions is that the column matrix is equilibrated all the time and does not experience fast chemical changes. However, nowadays, often HPLC runs are based on a gradient of the mobile phase during elution, that is, the solvent strength is increased with time. For example, in reversed-phase chromatography, the mobile-phase composition at the start of the run is polar (for instance high amounts of water added to acetonitrile) and the percentage of the organic solvent (such as methanol or acetonitrile) is increased with time, thereby raising the elution strength. In normal-phase chromatography, the initial mobile-phase composition is usually unpolar such as hexane and then a more polar organic solvent is added, such as chloroform, tetrahydrofurane (THF), ethanol or isopropanol. Also, the ionic strength or the pH may be changed during chromatography by adding a suitable second solvent. Remember that some solvents such as methanol and acetonitrile do not mix with for instance hexane (Table 7.4). An example of a gradient elution is shown in Figure 7.20, but curved gradient types, convex or concave, are also common. The advantage of gradient runs is that the separation of the analyte mixture can be significantly improved and shortened compared to isocratic runs.

Solvents should be degassed in order to prevent bubble formation upon mixing under pressure; nowadays, degassing stations are included in commercial HPLC systems.

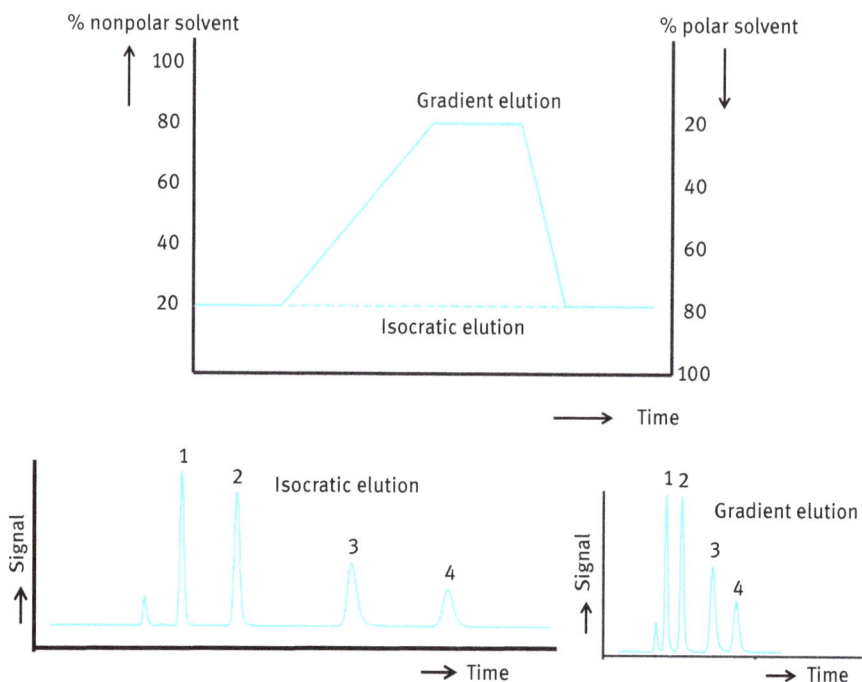

Figure 7.20: Example for a gradient elution under reversed-phase conditions (top) and resulting chromatograms of the same mixture of four analytes (bottom).

7.4.2.4 HPLC Pumping Systems

The solvents are introduced into the HPLC system by suitable pumps applying high pressures. Such pressure is necessary to overcome the resistance of the narrow stainless steel capillaries in which the solvents reach the column which is filled with small-sized matrix particles (3–10 μm). No solvent flow could be observed without such high pressure. Two types of systems are available: (1) Constant pressure pumps are driven by gas pressure that moves a single or dual piston to provide solvent flow. One disadvantage of this type of system is potentially a change in flow rate that can occur, for instance, if the viscosity of the solvent changes by mixing with another solvent (in gradient elution). Flow changes may lead to changes in retention times of analytes that might hamper the identification of compounds if based on cochromatography with reference substances. (2) Constant flow pumps deliver reproducible elution volumes and peak areas despite viscosity changes of the solvent or column blockage below a certain pressure limit. Most common for this type are dual piston pumps with one pump delivering the solvent, while the other at the same time is refilling. The flow profiles of the two pumps overlap so that any pulsation of pressure is minimized.

Mixing of two solvents can be performed by use of two high-pressure pumps, each connected to a mixing chamber in which the two flows are mixed for delivery to the injector and the column. Thus, mixing is accomplished on the high-pressure side of the pumps. The total flow rate equals the sum of the individual flow outputs of both pumps and is electronically controlled to obtain the desired ratio of flows. Such systems provide excellent mixing accuracy. In low-pressure systems, a single pump controls the overall flow rate and mixing occurs in a mixing chamber with proportioning valves that are controlled to open individually for a certain period of time. Mixing therefore occurs prior to the pump, that is, at the low-pressure side of the pump.

Pumps vary in pressure capacity, but their performance is measured on their ability to yield a consistent and reproducible flow rate. Pressure may reach as high as 8,700 psi (60 MPa). Even higher pressures are needed if particle sizes in the columns below 2 µm are used (ultra-high-performance liquid chromatography): Pressures can reach 17,400 psi (120 MPa).

7.4.2.5 Sample Injection (HPLC)

Samples to be injected into a HPLC system should be free of particles. That can be achieved by filtering or, if the filter retains part of the analyte, by centrifugation. Ideally, samples should be applied in a small volume to optimize the resolving power.

Samples can be injected manually with syringes through a septum on the chromatographic column, eventually through a frit, or into a sample loop with a defined volume (typically few µL–100 µL) by use of a six-port valve injection system (Figure 7.21). Besides the advantages of high-pressure resistance and good reproducibility of injection volumes in the sample loop delivery, a disadvantage is the loss of parts of the sample when filling the loop. Nowadays, injections can be programmed using automated systems (autosampler) in which a pressure-driven syringe withdraws the desired volume from one of a large set of sample vials (up to several hundred) and injecting the samples for instance overnight.

7.4.3 Stationary-Phase Materials

Column packings must withstand the high operating pressures and are available in different forms.

The most commonly used stationary phases in HPLC are reversed phase, normal phase and ion-exchange matrices similar to those described in SPE (see Table 6.1). Also, SEC and affinity chromatography will be described in the following.

7.4.3.1 Normal-Phase Matrices

Normal-phase chromatography started with silica packings and the name reflects that this was the first type. Silica has a high affinity for water and since chromatography

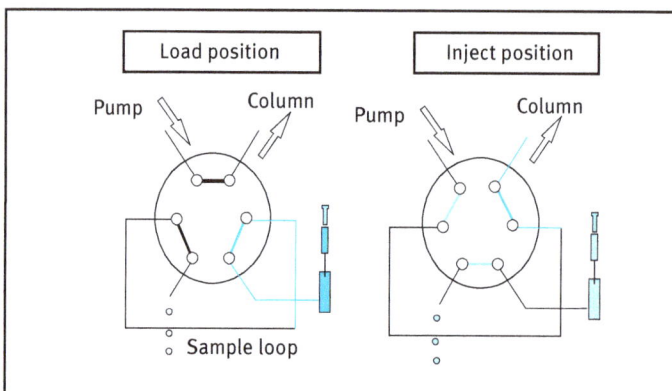

Figure 7.21: Six-port valve system for injection of samples on a HPLC column. In the load position, the sample is injected into the sample loop; switching to the inject position directs the flow to the chromatography column.

characteristics change after hydration, the reproducibility of analysis is not as easy as with other packings. Normal phase presents a polar surface and solvents used are nonpolar and nonaqueous; to increase the elution strength, the polarity of the eluent has to be increased. Also alumina can be used as matrix in normal-phase chromatography.

Hydrophilic interaction liquid chromatography (HILIC) is a variant of normal-phase liquid chromatography in which polar chromatographic surfaces are used, such as bonded phases with diols, amino or amide groups. Solvents for HILIC contain small amounts of water forming a layer on the matrix surface and creating a liquid–liquid extraction system besides hydrogen bonding and electrostatic interactions.

7.4.3.2 Reversed-Phase Matrices

In reversed-phase HPLC (RP-HPLC), a nonpolar stationary phase is used together with a moderately polar mobile phase. One common stationary phase is silica that has been silylated by reaction with RMe_2SiCl, with R representing a straight chain alkyl group containing for instance 18 carbon atoms (C_{18}). Nonpolar analytes will elute later than polar ones on such columns. The more polar the mobile phase, for example, by mixing a nonpolar solvent with some water, the longer the retention times of the analytes on RP columns. The most popular RP matrix for HPLC columns is C_{18}-bonded silica, followed by C_8-bonded silica and phenyl-bonded silica. Also cyano-bonded silica columns can be used in a reversed-phase mode depending on analyte and mobile-phase conditions. Surface functionalization of silica can be followed by reaction with short-chain organosilanes to cover remaining silanol groups by so-called end-capping.

The retention mechanism is based on the principle of hydrophobic interactions: The binding of the analyte to the stationary phase is proportional to the contact surface area of the nonpolar moieties of the analytes and depends on the surface tension of the eluent. Structural properties of the analyte molecule play an important role in its retention characteristics. In general, an analyte with a larger hydrophobic surface area (nonpolar atomic bonds, such as C–C, C–H, S–S) is retained longer on RP columns. On the other hand, more polar analytes containing for instance hydroxy, amino or carboxylate groups are less strongly retained because they are more soluble in the more polar mobile phase. Branched chain compounds elute earlier than corresponding linear compounds because the overall surface area is decreased.

The mobile phase pH may change the hydrophobic interactions of the analyte. Therefore, buffers are often used in the eluent to control the pH, to neutralize charges on the silica surface of the stationary phase and of potential analyte charges.

RP-HPLC can be used to estimate the log K_{OW} of chemical substances, that is, the partitioning of a substance between water and n-octanol, as a measure of its lipophilicity as important property determining the environmental fate of organic pollutants, such as adsorption, absorption, membrane transport, bioavailability and metabolism, as well as toxicity. This can be achieved simply by measuring the retention times of structurally related molecules with a known K_{OW} and using the regression line to determine the K_{OW} of an unknown compound by its retention time. Han et al. [6] established a correlation between the log k and log K_W of organic molecules: k is calculated by the retention times of model substances, t_R, normalized to the retention time t_0 of a chemical that has no affinity on RP columns [$k = (t_R - t_0)/t_0$] such as sodium nitrate. K_W represents the extrapolated retention factor of the chemicals when using 100% water as eluent in RP-HPLC. First, log K_W versus log k regressions of the test compounds were investigated, validated and compared with each other for silica-based C_8 and C_{18} stationary phases, respectively, using polychlorinated biphenyls (PCBs), and benzene derivatives for which reliable experimental K_{OW} values were known. For C_8 stationary phase, the following relation was established: Log $K_{OW} = 1.32 \log K_W - 0.25$, and for C_{18} stationary phase, log $K_{OW} = 1.13 \log K_W + 0.04$.

7.4.3.3 Ion-Exchange Matrices

The retention in ion-exchange chromatography is based on the ionic (Coulomb) interaction of analyte ions and charges of the stationary phase. Analyte ions of the same charge as the charged sites on the stationary phase are eluted without interaction, whereas analyte ions are retained if they carry the opposite charge. The strength of the interaction is determined by the number and location of the charges on the analyte and on the functional groups involved. After binding to the matrix, the analyte is eluted by changing the solvent conditions, such as increasing the salt concentration of the eluent or changing the pH of the solvent.

If the mobile phase contains a buffer, the pH must be between the isoelectric point (pI) or acid dissociation constant (pK_a) of the charged molecule and the pK_a of the charged group on the matrix. For example, in cation-exchange chromatography, using a functional group on the solid support with a pK_a of 1.2, an analyte with a pI of 8 may be investigated in a mobile phase buffer of pH 6. In anion-exchange chromatography, an analyte with a pI of 7 may be run in a mobile phase buffer at pH 8 if the pK_a of the solid support is 10. A decrease in pH reduces the retention time of an analyte in cation exchange because more hydrogen cations compete for the binding sites, whereas an increase in pH reduces the retention time in anion exchange.

Types of ion exchangers include polystyrene resins, sepharose, cellulose and dextran gels or porous silica modified with ionic groups (see Section 7.2.2 and Table 7.5).

7.4.3.4 Size-Exclusion (SEC) Matrices

Size-exclusion HPLC is used for separating large molecules such as proteins, polysaccharides or other polymers from lower molecular weight molecules by trapping the latter in the pores of a suitable matrix (principles see Section 7.2.3). Aims of SEC are purification of environmental samples, desalting, concentration of samples and the determination of relative molecular masses. Typical SEC matrices are composed of polyacrylamide, agarose or styrene divinylbenzene or surface-modified silica and elution is most often performed in an isocratic mode. SEC flow rates are usually lower than those when using other separation matrices, typically in the range of 0.5 mL/min for columns with 4.6 mm inner diameter.

Trace priority pollutants analyses rely on highly sensitive techniques such as GC/MS or LC/MS. Such analyses are significantly better to perform if complex, high molecular weight interferences such as lipids, pigments, proteins or dissolved natural organic matter are first removed from the sample. Otherwise, the matrix effect may impede the target analysis and also the analytical columns become contaminated by such materials that often have much higher concentrations than the analyte of interest. Here, size-exclusion HPLC can be applied for purification as excellent tool

Table 7.5: Examples for cation- and anion-exchange matrices in HPLC.

Anion exchangers	Functional group
Diethylaminoethyl	$-O-CH_2-CH_2-N^+H(CH_2CH_3)_2$
Quaternary aminoethyl	$-O-CH_2-CH_2-N^+(C_2H_5)_2-CH_2-CHOH-CH_3$
Quaternary ammonium	$-O-CH_2-CHOH-CH_2-O-CH_2-CHOH-CH_2-N^+(CH_3)_3$

Cation exchangers	
Carboxymethyl	$-O-CH_2-COO^-$
Sulfopropyl	$-O-CH_2-CHOH-CH_2-O-CH_2-CH_2-CH_2SO_3^-$
Methyl sulfonate	$-O-CH_2-CHOH-CH_2-O-CH_2-CHOH-CH_2SO_3^-$

to efficiently isolate broad classes of organic low molecular weight molecules from complex matrices.

Other examples for use of HPLC and SEC are the determination of apparent molecular weights of naturally dissolved organic carbon, DOC [7], or the monitoring of DOC removal from drinking water after ion exchange chromatography [8]. For this type of analysis, an organic carbon detector can be used after UV-persulfate oxidation of the DOC to CO_2 which is quantified by a mass spectrometer.

7.4.3.5 Affinity Matrices

The principles of high-performance liquid affinity chromatography (HPLAC) have been described in Chapter 6, Section 6.2.3, when discussing SPE. In HPLAC, a chromatographic column contains an immobilized biologically related agent (e.g., a protein or receptor) as the stationary phase. These immobilized affinity ligands can then be used to study the binding of an analyte to the column. The separation of the analyte from other components in a sample is based on the specific and reversible interactions like the binding of an antibody with an analyte. If the interaction is strong with binding constants $>10^5-10^6$ M^{-1}, an elution buffer and a change in the pH, temperature or mobile-phase composition are usually required to elute the analyte from the column. In the case of lower binding constants, the target analyte may be eluted under isocratic conditions.

Originally, for affinity chromatography with larger glass columns containing the affinity matrix inexpensive supports and non-rigid materials was used, such as agarose gels or other carbohydrate-based materials. For HPLAC, the matrix needs to have a sufficient mechanical stability and still good binding efficiency because of the higher pressure and flow rates of the mobile phase. Materials to meet these requirements are modified silica or glass, aza-lactone beads, and hydroxylated polystyrene media and other organic polymers. Also so-called monolithic matrices, inorganic, organic and hybrids, can be used for HPLAC. Such supports can also be used in HPLC microcolumns with small inner diameters of as low as 0.5 mm and short length (5–15 cm) [1]. A summary of recent developments in HPLC stationary phases was given by Chester) [9].

Different types of experiments can be performed with HPLAC: zonal elution profiles, in which a small plug of analyte is injected under isocratic conditions. The retention time and also its response on changing the elution conditions (pH, temperature, ionic strength, polarity or adding a competing second analyte) will provide data on the interaction type of the analyte with the binding ligands. Figure 7.22 shows a competition study with zonal elution showing a decrease in the retention of the analyte with increasing concentration of a competitive analyte, indicating that direct competition or a negative allosteric effect between these two compounds occurred.

In frontal affinity chromatography, the analyte is continuously applied to a column, and the amount of the eluting analyte from the end of the column is monitored. As the target binds to the immobilized affinity ligand, the column is becoming saturated, and the amount of the eluting analyte increases with time resulting in a

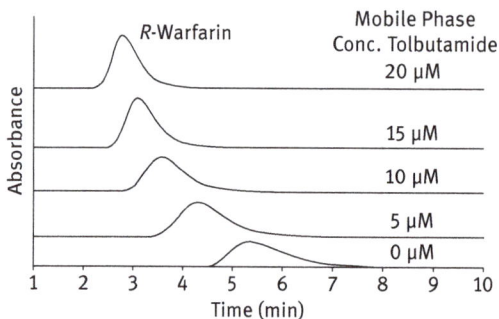

Figure 7.22: Competition zonal elution with R-warfarin as a site-specific probe onto an HPLC affinity column with immobilized human serum albumin in the presence of various concentrations of tolbutamide in the mobile phase [1].

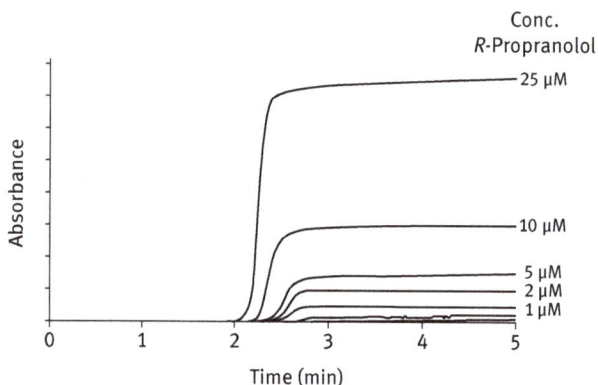

Figure 7.23: Breakthrough curves obtained in frontal HPLAC by application of various solutions R-propranolol to a column containing immobilized high-density lipoprotein as binding ligand [1].

breakthrough curve from which the association equilibrium constant between the analyte and the ligand and total moles of active binding sites can be determined (Figure 7.23).

Other examples of HPLAC applications are aflatoxin analyses in cheese and butter by use of a AFLAKING immunoaffinity column. The recovery of spiked aflatoxin was higher than using a Florisil column and analysis resulted in a low limit of quantitation (LOQ) (0.1 µg/kg) [10]. Others analyzed aflatoxins in rice by a similar method [11].

7.4.4 Columns

HPLC columns are usually made of stainless steel and are highly pressure resistant up to about 8,000 psi (55 MPa). The dimensions of analytical columns, that is, the inner diameter (0.1–10 mm) and length (5–25 cm), and the size of the matrix particles (see above) will affect the efficiency of separation and the speed of analysis. Short column lengths (3–5 cm) allow short-run times and low backpressure, whereas longer columns (25–30 cm) will perform better in terms of resolution of separated analytes but with longer analysis times, more solvents and increased costs.

Depending on the chromatographic application, analytical, semipreparative and preparative columns can be chosen (Table 7.6).

Larger columns (inner diameter above 10 mm) are used to purify high amounts of substances because of their large loading capacity. Analytical scale columns (often inner diameter 4.6 mm) are mostly used in analytical laboratories, though smaller columns are more and more favored. Narrow-bore (microbore) columns (0.5–1 mm) are used if higher sensitivity is needed; also, the amount of solvent used for elution is lower than in larger columns. Capillary columns (inner diameter below 0.3 mm) are often used with highly sensitive detector systems such as mass spectrometry (MS). They are usually made from fused silica capillaries, rather than the stainless steel tubing that larger columns employ.

When moving from analytical (anal) to preparative (prep) columns, the proper flow rates (Flow) can be calculated by referring to the inner diameter of the columns (D) according to

$$
\text{Flow}_{prep} = \text{Flow}_{anal} \times \frac{D^2_{prep}}{D^2_{anal}} \tag{7.5}
$$

A flow rate of 0.1 mL/min on a 4.6-mm column will correspond to about 5 mL/min on a 10-mm (50 mm) preparative column or 120 mL/min for a 50-mm column.

Capillary HPLC uses smaller column internal diameters (ID) than traditional HPLC. Smaller ID columns, for fixed amounts of injected analyte, will produce taller peaks due to a lower dilution factor and, thus, better detection limits for concentration-sensitive detectors. For the same amount of material injected, the peak height is inversely proportional to the inner diameter of the column. Down-scaling the column inner diameter further, for example, from 4.6 mm of a traditional analytical column to below 100 μm in nano-HPLC, would result in a further gain in sensitivity. Nano-HPLC columns are predominantly packed in fused-silica capillaries with

Table 7.6: Column dimensions and flow rates in HPLC systems.

Type	Column inner diameter (mm)	Flow rate (mL/min)
Preparative HPLC	10–50	5–100
Traditional ("analytical") HPLC	1–10 (often 4.6)	≥ 1
Microbore HPLC	0.5–1	0.1–0.5
Capillary HPLC	0.1–0.5	0.001–0.025
Nano-HPLC	<0.1	<0.001

internal frits to keep the stationary phase. Stationary-phase particle sizes for nano-HPLC columns are similar to traditional columns. Nano columns typically require flow rates of less than 0.001 mL/min. All other system components should be downscaled as well, that is, the tubing connections, injector and the interface to the detector; also, special pumps have to be used being able to produce such low flow rates.

7.4.5 HPLC Detectors

An electronic signal from the detector is used to quantify the analytes eluting from the chromatographic column. The detector should provide a linear response related to a certain concentration range of the solute, a wide linear dynamic range and a high signal-to-noise ratio.

The *sensitivity* corresponds to the detector output as a function of the analyte concentration. When using calibration solutions of an analyte – in a calibration diagram signal versus concentration – the sensitivity corresponds to the slope of the calibration curve in its linear range. The LOD is given by the sensitivity and the background signal B (noise) and its standard deviation of the detector and is often defined as the amount of the analyte giving a signal $\bar{x}_{bl} + 3s_{bl}$ (see Chapter 2), that is, the detector output without analyte during the same period of time as the analyte signal. The LOQ similarly corresponds to ten times the background signal.

The *dynamic range* of a detector means the change of the electronic output in relation to the analyte concentration in-between the LOD and the highest concentration at which a signal change is observed. For quantification, the linear relation of analyte concentration versus detector signal should be used, which is not given over the whole dynamic range because it flattens off at the higher range of the detector output.

The detector cell should be small enough to minimize backdiffusion and peak broadening but still providing enough analyte amounts to get a significant signal output. For instance, for UV absorption detectors, this is accomplished by special constructions of the flow through cell in HPLC, that is, the Z cell (see below).

Different types of detectors are commercially available, each with certain properties, working principles and selectivities: For instance, fluorescence and electrochemical detectors are selective for analytes, whereas UV absorption detectors are nonselective and universal (for all absorbing analytes at certain wavelengths). The most common types are discussed in the following.

7.4.5.1 UV–Vis Absorption Detectors

The UV–vis detector measures the absorbance of monochromatic light at certain wavelengths in the UV or visible wavelength range (typically between 190 and 800 nm) against a reference beam and relates the analyte absorbance to its concentration when passing a flow cell behind the chromatographic column.

Analytes suitable for UV detection, that is, containing unsaturated bonds, aromatic groups or functional groups containing heteroatoms, absorb the radiation by elevation of electrons in outer electron shells to higher electronic levels, π and σ nonbonding orbitals. These nonbonding orbitals contain a wide distribution of lower (vibrational and rotational) energy levels that lead to a distribution of absorbance energies: therefore, UV absorption spectra are relatively broadly spread over a certain wavelength range.

The concentration of the analyte can be determined from Lambert-Beer's law (equation 7.6):

$$A = \varepsilon \cdot c \cdot d \tag{7.6}$$

with ε being the molar extinction coefficient (M^{-1} cm^{-1}), c the concentration (mol/L), A is the absorbance, and d the path length of the light through the flow cell (cm). If the molar extinction coefficient of the analyte is unknown, standard solutions of known concentrations can be used to determine it and to establish a calibration curve (concentration vs. absorbance). Analytes with no or too low absorption properties can obviously not be analyzed by this detector type. Lambert–Beer's law is only valid for monochromatic light because the molar extinction depends on the wavelength. Therefore, for HPLC detectors narrow bandwidths of about 10 nm are typical. The problems of straylight because of defects in the diffraction gratings affecting the measured absorbances will not be discussed here.

The absorbance A equals the negative decadic logarithm of the ratio intensity of light emerging from the sample cell (I) divided by the intensity of light from the light source (I_0) entering the sample cell (the ratio I/I_0 is called transmittance). An absorbance of 1.0 means that 90% of the irradiating photons are attenuated by the solution in the detector cell and only 10% reach the detector; if A equals 2 or 5, 99% or 99.999% of the photons are extinguished and only very small amounts of light reach the detector. Reliable measurements below 0.001% of the light intensity reaching any UV–vis detector are not viable for commercial instruments and other detection types must be used like photoacoustic detectors in such situations [12]. To be on the safe side, absorbance much above 1 should be avoided with commercial instrumentation. Not only the analytes will absorb UV–vis light but also the solvent absorption properties have to be taken into account: For each HPLC solvent used, the wavelength cutoff has to be considered below which no absorbance measurements should be performed (see Table 7.2) because the solvent becomes opaque below the cutoff. It has also to be noted that the absorbance of the solvent will change in gradient elution.

The cell should be as long as possible because the absorbance depends on the optical path length d (eq. 7.6). On the other hand, the volume should be small, preferably one-half of the maximum allowed injection volume, to avoid band broadening. In HPLC, the cell construction represents a compromise with typical dimensions of 10-mm length and 8-μL volume in a double L- or Z-shape design (Figure 7.24).

Imagine an analyte (molecular weight 200 D) having a molar extinction coefficient ε of 5,000 1/(M cm) at a certain wavelength. For an HPLC cell with d = 1 cm and

Double-L-cell Z-cell

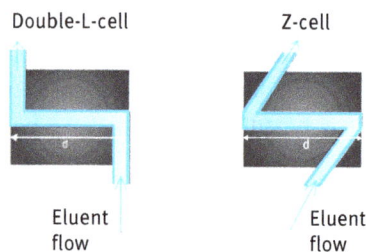

Eluent flow Eluent flow

Figure 7.24: Eluent flow through HPLC detector cell types.

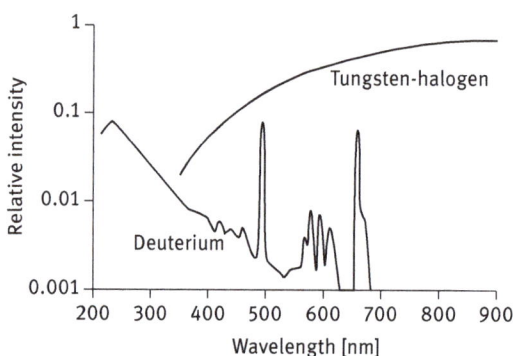

Figure 7.25: Spectra of deuterium and tungsten lamps (simplified). The relative positions of the spectral lines depend on the lamp wattage.

$V = 10\,\mu L$, the lowest measurable absorbance of about $10\,\mu AU$ (AU: absorbance unit) will correspond to a concentration of 2×10^{-9} mol/L and 4×10^{-12} g/10 μL (4 pg/10 μL), respectively. Assuming that the peak volume is ten times the cell column, this corresponds to an absolute mass detection limit of 40 pg. Often, the molar extinction of analytes is however lower than the assumed value.

Polychromatic light from a deuterium (UV) or tungsten (visible) lamp (Figure 7.25) is focused onto a monochromator which transmits a narrow band of selected wavelengths, via the use of a grating mounted on an electromechanical turntable, to the exit slit and ultimately through the flow cell. Analyte absorbance is measured using a beam splitter to compare light intensity reaching a sample photodiode with a reference photodiode, which is off-axis from the flow cell. This type of detector is called a variable-wavelength detector (Figure 7.26).

In diode-array detection (DAD), radiation from the lamp is collimated through the flow cell followed by a mechanically controlled or fixed width slit. Radiation is dispersed via a holographic grating into individual wavelengths of light that are detected using a photodiode array. Each photodiode receives a different narrow wavelength band and a complete spectrum may be obtained for any point within the chromatogram (Figure 7.27), or a chromatogram may be deconvoluted to show only those signals because of a chosen single or narrow range of wavelengths. By comparison with the

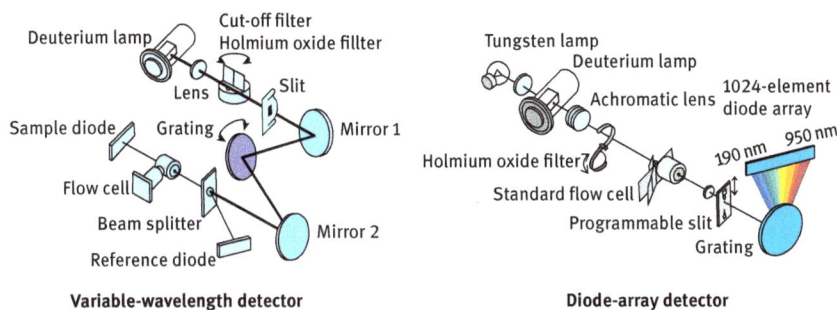

Variable-wavelength detector

Diode-array detector

Figure 7.26: Schemes of UV–vis absorption detectors for use in HPLC systems: left variable-wavelength detector, right diode-array detector (Agilent).

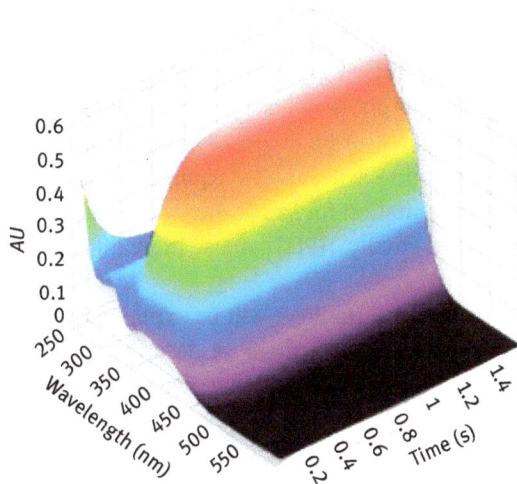

Figure 7.27: Diode-array spectrum of an analyte eluting from an HPLC column.

absorption spectra of the analytes, targeted analysis, component identification and also the simultaneous quantitative analysis of signals at several discrete wavelengths may be achieved.

In HPLC, the rapidly available DAD spectra at certain points during peak evolution in a chromatogram can be used to check whether the peak contains one or more coeluting components, if they have different absorption spectra.

The correct wavelength at which to measure the absorbance of the analyte is important to reach the best sensitivity. This can be achieved by studying the UV spectra of standards of the analytes of interest or by examining its spectrum within the chromatogram by DAD. The sample solvent, the pH of the solution and the temperature may change the position and intensity of absorption bands of molecules.

7.4.5.2 MS Detector

Several mass analyzers – single quadrupole, triple quadrupole, ion trap, time of flight (TOF) and quadrupole-TOF (Q-TOF) – can be used in LC/MS applications: These types will be described in Chapter 9.

LC–MS applications provide a very good sensitivity in analyzing a broad variety of analytes combining their separation, quantification and identification. Focusing on specific mass/charge (m/z) ratios, analytes can even be analyzed in the presence of other components such as in extracts from environmental media. However, precleaning of such extracts by the methods described earlier (Chapter 6) is always beneficial even for such a highly specific technique as LC–MS.

When standard analytical columns (inner diameter often 4.6 mm) are used for HPLC, the eluent has to be split for introducing into the mass spectrometer, often in a ratio of 10:1. The split of eluent allows to simultaneously use an additional detector such as UV–vis or fluorescence. Unlike in GC/MS, the eluent of HPLC is liquid; therefore, the analyte needs to be transformed to the gas phase for ionization and subsequent MS analysis occurring under vacuum. A typical HPLC flow is 1 ml/min which, when converted from the liquid to the gas phase, corresponds to about 1 L/min, way too much for an MS which can tolerate flows of up to about 1 ml/min of gas. In addition, LC operates at near ambient temperature, whereas MS requires elevated temperatures. There is literally no mass range limitation for samples analyzed by HPLC; there are only limitations for an MS analyzer. Finally, HPLC can use inorganic buffers but MS prefers volatile buffers.

How the eluent is transferred to the mass spectrometer will be explained in Section 7.4.8.1 and Chapter 9 (MS). Several interfaces will there be described, that is, electrospray ionization and atmospheric pressure ionization as the two most common interfaces used for HPLC/MS.

7.4.5.3 Refractive Index Detector

Refractive index (RI) detectors in HPLC systems are used to analyze samples that lack strong chromophores in ultraviolet or visible regions and are nonfluorescent. The RI describes the propagation of light passing different materials: The deflection at the interface between two media is

$$n_1 \cdot \sin \theta_1 = n_2 \cdot \sin \theta_2 \qquad (7.7)$$

with n_1 and n_2 the refractive indices of the two media and θ_1 and θ_2 the angles of incoming and deflected light with respect to the perpendicular line to the interface.

The RI detector is a universal detector (all analytes deflect light to a certain extent) but with limited sensitivity, about two to three orders of magnitude less than a UV–vis detector. In HPLC applications, the detector measures the ability of an analyte to deflect light of the eluent in a flow-through cell relative to a reference cell (Figure 7.28). Both the analyte and the solvent have their own RI. The deflection of light

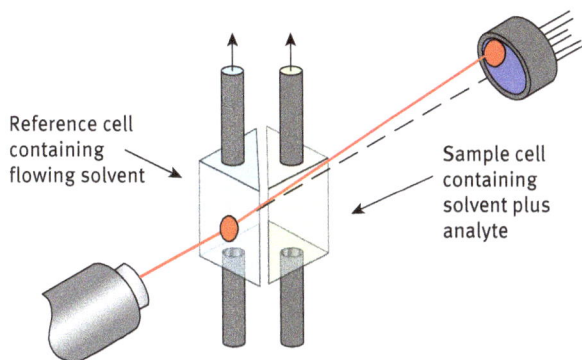

Figure 7.28: Scheme of a (two-cell) refractive index detector for use in HPLC.

is proportional to the concentration of the analyte (as always only in a certain range). Since the RI of the solvent is strongly changing in gradient elution, RI detectors can't be used under such conditions. Also, temperature, pressure and flow rate have effects on the RI, which therefore should be as constant as possible.

7.4.5.4 Fluorescence Detector in HPLC

Fluorescence is a kind of photoluminescence in which (fluorescent) chemicals excited by light at a certain wavelength emit light at a longer wavelength, that is, a red shift is observed. If a photon of the exciting light source is absorbed by a fluorescent molecule, valence electrons are elevated from its electronic ground state to a higher energetic vacant state. Then, the electron relaxes rapidly (10^{-11}–10^{-14} s) into the lowest energy excited state and subsequently will fall into the electronic ground state thereby emitting a photon, that is, fluorescence.

Fluorescence detectors offer a much higher sensitivity and selectivity than for instance UV–vis detectors and allow to quantify analytes at trace concentration levels. Only analytes who have fluorescent properties are "seen" by the detector. Nonfluorescent analytes can, however, be derivatized with fluorescent functional groups (see Section 7.4.7).

Unlike in UV–vis absorption measurement, where the transmittance of the incoming light is somewhat lowered if an absorbing and scattering analyte is present in the flow cell (which produces the signal of the detector), here the detector uses the emitted light from the analyte for delivering a signal. If the concentration of the analyte is zero, the photodiode output of the fluorescence signal is also zero. Also the signal is proportional to the intensity of the excitation light and can therefore be improved by using strong light sources such as lasers.

The fluorescence intensity I_f is a function of the incoming emission light with intensity I_0, the quantum yield Φ (the probability of an absorbed photon to produce a photon of emitted light), the molar extinction coefficient ε (at a certain emission

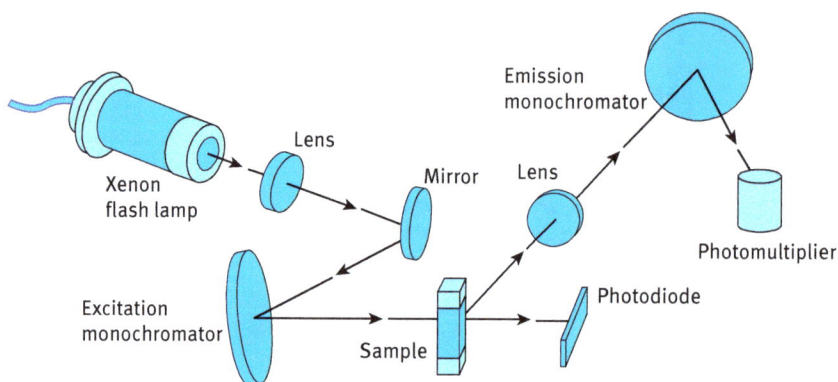

Figure 7.29: Scheme of a fluorescence detector for use in HPLC systems.

wavelength λ), the concentration of the fluorescent analyte and the light path length d of the flow cell:

$$I_f \propto I_0 \times F \times \varepsilon_{\lambda\,excitation} \times c \times d \tag{7.8}$$

The fluorescence light is measured at $90°$ (or off-axis) with respect to the emission light (see Figure 7.29) and, thus, allows very sensitive detection by separating it from the excitation light.

7.4.5.5 Evaporative Light-Scattering Detector
The evaporative light-scattering detector (ELSD) is a universal one that can be used for almost every component, also for those that can't be detected with absorption or fluorescent detectors, such as carbohydrates. The column effluent is passed through a nozzle and is mixed with nitrogen gas, forming aerosol droplets. The size of the droplets can be adjusted by changing the gas flow. These droplets are then passed through a heated tube to evaporate the mobile phase leaving solid particles suspended in the gas phase. These particles pass to a cell where a laser light beam is scattered by the particles. If the particle size is in the order of the wavelength of the light, Lorenz–Mie scattering occurs; if they are substantially smaller, Raleigh scattering will occur, and if they are much larger, the interaction is best described by reflection and refraction. The scattering can be detected generating an electric signal that is mass proportional.

The ELSD consists of three parts: a nebulizer, a heated drift tube and a light-scattering cell (Figure 7.30).

The ELSD is compatible with HPLC gradient elution because the eluent is removed before a signal is produced. As only requirement, the analytes must be less volatile than the eluent. The detector can be used with solvents that could not be used with UV–vis detectors because of their strong absorbance, such as acetone. Unlike the detectors presented so far, the ELSD is a destructive detector and the sample cannot

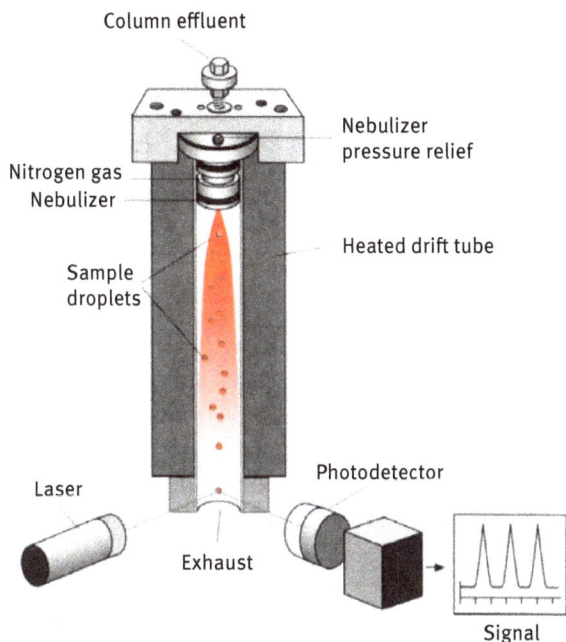

Figure 7.30: Scheme of an evaporative light-scattering detector (ELSD).

be investigated further. Although being less sensitive than other detector types, the universal applicability is of great advantage for a variety of HPLC separations.

7.4.5.6 Conductivity Detector

This detector measures the conductivity of the total mobile phase in the flow cell as a bulk property (see also chapter 7.2.2). Charged analytes, inorganic ions or ionizable organic chemicals such as amines or amino acids can be measured if their conductivity differs from that of the eluent. In low conductivity solvents, the separation of charged analytes may become a problem. On the other hand, in buffered eluents with better separation properties, the basal conductivity is high, and differences in conductivity in presence of the analyte are difficult to detect. In this case, the charges of the buffering components have to be compensated by use of an ion suppressor between the chromatography column and the detector. The ion suppressor will reduce the background conductivity and relatively increase the conductivity of the analyte. The principle is explained by a simple analysis of the chloride anion (NaCl) in a $NaHCO_3$ containing eluent. The ion suppressor column contains a strong cation exchanger in its protonated form ($R\text{-}SO_3^-H^+$). The following exchange reactions will occur:

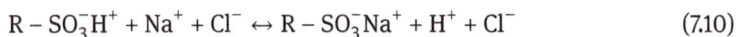

$$R - SO_3^-H^+ + Na^+ + HCO_3^- \leftrightarrow R - SO_3^-Na^+ + H_2O + CO_2 \tag{7.9}$$

$$R - SO_3^-H^+ + Na^+ + Cl^- \leftrightarrow R - SO_3^-Na^+ + H^+ + Cl^- \tag{7.10}$$

The bicarbonate of the eluent becomes neutralized (eq. 7.9) so that the conductivity of the eluent is significantly lowered. The analyte chloride is not affected but its counter ion Na^+ is exchanged for a proton (eq. 7.10). The conductivity detector will analyze the presence of both, the analyte (Cl^-) and the counter ion (Na^+); so, overall the sensitivity is significantly increased.

The electrical resistance R of the column effluent in the flow cell is given by

$$R = \frac{\rho \times d}{a} \qquad (7.11)$$

with ρ the electrical resistivity, d the path length of the cell and a the cross-sectional area of the cell. The resistivity of the volume of the mobile phase in the cell is dominated by the concentration of charged, ionic species present. Unlike for UV–vis detectors, in which the cell path length is directly proportional to the path length d, in conductivity detectors the concentration is proportional to the ratio d/a, that is, as long as this ratio is kept constant, the cell sizes can vary and become a component in miniaturized systems.

The conductivity detector consists of two electrodes situated in a suitable flow cell (Figure 7.31). In contact cells, the electrodes, made of stainless steel, platinum or gold, are directly positioned inside the eluent. In the contactless version, the electrodes are separated from the effluent by insulators and are capacity coupled to the sample.

In the electric circuit, the two electrodes are arranged to form the impedance component in one arm of a Wheatstone Bridge. When ions move into the sensor cell, the electrical impedance between the electrodes changes and the signal from the bridge is fed to a suitable electronic circuit. The output from the amplifier is either digitized for computer data treatment, or the output is passed directly to a potentiometric recorder. The detector actually measures the electrical resistance between the electrodes which

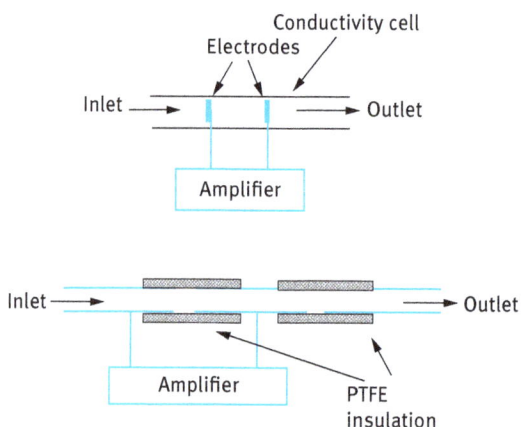

Figure 7.31: Scheme of a conductivity detector (top: contact cell; bottom: contactless cell).

by suitable nonlinear amplification can provide an output that is linearly related to solute concentration. An alternating current voltage is used across the electrodes to measure the cell impedance to avoid electrode polarization.

7.4.5.7 Electrochemical Detector

This detector type, employed mainly in reverse-phase or ion-exchange separations with their relatively polar eluents, uses the electrochemical properties of analytes for their determination. Examples for appropriate analytes are aromatic amines, chinones, phenols, indols, thiols and nitro derivatives. An inert electrolyte is sometimes added to enhance conductivity of the eluent. Electrochemical detectors are especially for some analytes the most sensitive choice and can detect in favorable cases femtomolar (10^{-15} M) concentrations; furthermore, they are selective because only oxidizable and reducible analytes can be detected.

Such detectors are constructed in a similar way as contact conductivity detectors with the electrodes in direct contact with the eluent. The principle is based on the measurement of electron flow at the electrode surface upon oxidation or reduction of an analyte. Upon oxidation, free electrons from the analyte are released to the counter electrode. In the case of reduction, electrons are provided from the counter electrode to the analyte. In coulometric detection, the charge generated is proportional to the amount of analyte that is transformed. Alternatively, in amperometric detection, the analyte moving with the eluent along the electrodes will produce a current that depends on the analyte's concentration.

Solid electrodes are made from carbon or metals (gold, platinum, nickel and copper) and are used for oxidative detection; mercury on the other hand is used for reductive applications. Three electrodes are needed, a working electrode (where oxidations and reductions occur), an auxiliary electrode and a reference electrode which compensate for changes on the conductivity of the eluent (Figure 7.32).

Especially sensitive analyses have been performed by HPLC using electrochemical detectors in the area of neurotransmitters like dopamine or serotonin [13].

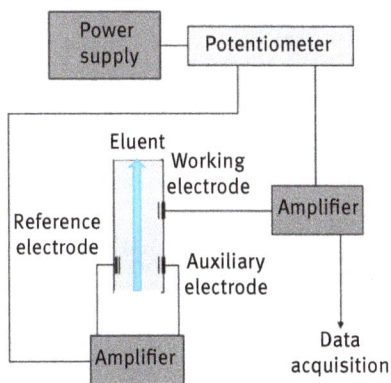

Figure 7.32: Scheme of an electrochemical detector.

A comparative summary of some analytical specifications of all presented detector types is given in Table 7.7.

7.4.6 Separation Parameters in HPLC

7.4.6.1 Resolution

Separation parameters describe the efficiency of HPLC to separate a mixture of several components during elution. The void volume of a column is the space outside of the column's internal packing material filled with the eluent. It is measured from a chromatogram as the first peak (with the retention time t_m) detected of a substance having no interaction with the matrix, often the so-called peak solvent in which the sample was injected. The void volume is used as a correction factor.

The efficiency in separation describes the ratio of the analytes peak area relative to the width of the peak at the baseline. Narrow peaks mean that an analyte is efficiently removed from a mixture indicating a high efficiency that depends on the column used and the HPLC method. The efficiency depends on the number of theoretical plates (see Section 7.3.1).

Retention factor (kappa prime) measures how long a component of the mixture stuck to the column, measured by the area under the curve of its peak in a chromatogram (since HPLC chromatograms are a function of time). Each chromatogram peak will have its own retention factor (e.g., kappa1 for the retention factor of the first peak). This factor may be corrected for by the void volume of the column.

Separation factor (alpha) is a relative comparison of how well two neighboring components of the mixture were separated (i.e., two neighboring bands on a chromatogram). This factor is defined in terms of a ratio of the retention factors of a pair of neighboring chromatogram peaks and may also be corrected for by the void volume of the column. The greater the separation factor value is over 1.0, the better

Table 7.7: Analytical specifications of HPLC detectors.

Detector	Specificity	LOD [ng]	Linear range
UV-vis absorption	No	0.1–1	10^5
Mass spectrometry	Yes	0.1–1	10^2–10^3
Refractive index	Yes	100–1,000	10^2
Fluorescence	Yes	0.001–0.01	10^3
Evap. light scattering	No	1	10^2
Conductivity	No	0.5–1	10^2–10^3
Electrochemical	Yes	0.001–1	10^4–10^5

LOD depends on analyte properties, for example, the molar extinction coefficient for UV–vis absorption detectors.

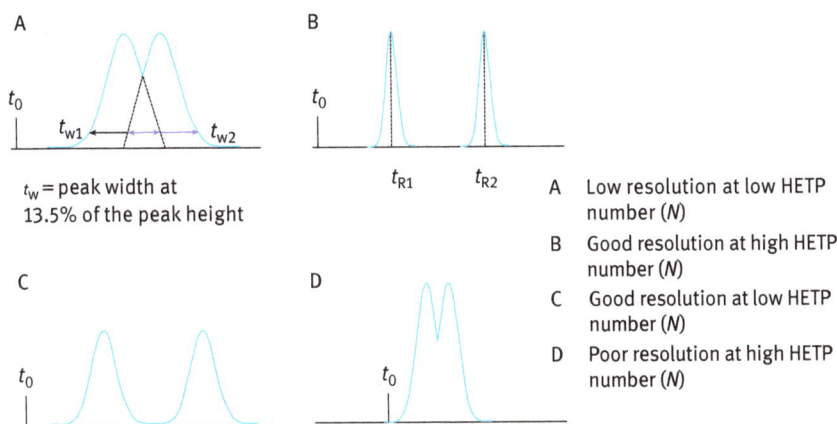

A

t_0

t_{w1} ← → t_{w2}

t_w = peak width at
13.5% of the peak height

B

t_0

t_{R1} t_{R2}

C

t_0

D

t_0

A   Low resolution at low HETP
    number (N)
B   Good resolution at high HETP
    number (N)
C   Good resolution at low HETP
    number (N)
D   Poor resolution at high HETP
    number (N)

Baseline separation at $R \geq 1$

Figure 7.33: Resolution of HPLC signals.

the separation, until about 2.0 beyond which an HPLC method is probably not needed for separation.

The resolution (R) relates these three factors (eq. (7.12); Figure 7.33):

$$R = 2 \times \frac{t_{R2} - t_{R1}}{t_{w1} + t_{w2}} \tag{7.12}$$

We can estimate what effective HETP number (N_{eff}) when choosing a HPLC column is necessary for a given separation problem (eq. (7.13), Table 7.8):

$$N_{eff} = \left(\frac{4\,Rr}{r-1}\right)^2 \quad \text{where the relative retention } r = \frac{t_{R2} - t_m}{t_{R1} - t_m} \tag{7.13}$$

We see from Table 7.8 that the more close two signals in a chromatogram are and the higher the resolution to be achieved, the higher is the number of effective theoretical plates.

Table 7.8: Estimation of the effective HETP numbers (rounded).

r	1.01	1.05	1.10	1.50
Necessary HETP numbers N_{eff} ($R = 1.0$)	163,000	7,100	1,900	140
Necessary HETP numbers N_{eff} ($R = 1.5$)	367,000	16,000	4,400	320

7.4.6.2 HPLC Column Dimensions

The internal diameter of an HPLC column is an important parameter that influences the detection sensitivity and also determines how much of an analyte can be loaded on the column. A decrease in the internal diameter of a column will increase the back pressure. Larger (preparative) columns have a lower sensitivity and resolving power compared to the smaller analytical columns. The efficiency of separation is also proportional to the column length: The longer a HPLC column, the better the efficiency (doubling column length increases the resolution by a factor of about 1.4).

7.4.6.3 Particle Size and Pore Size

The particle size has a great influence on the separation efficiency: Smaller sizes increase the efficiency of the column, that is, reduce the plate height (H); also, the flow velocity at which the minimum (optimum) plate height is achieved is increased (see Figure 7.34). However, the back pressure for smaller particles becomes higher.

Also, the pore size is a parameter influencing the efficiency of separation. The effect of different particle and pore sizes is shown in Figure 7.35.

Due to the temperature dependence of the partitioning processes and the viscosity of the mobile phase, columns should be temperature controlled. Typically, retention times of analytes will decrease about 1–2% with each 1 °C increase (see Figure 7.36).

The decrease in viscosity of the mobile phase with increasing temperature will result in lower pressures in the HPLC columns and better diffusion in the chromatographic system, resulting in narrower peaks. Simultaneously, shorter retention times will generate narrower peaks and, thus, lower detection limits.

7.4.7 Derivatization Methods (Pre-Column, Post-Column) in HPLC

Many reasons exist why analytes are derivatized in HPLC applications. Derivatization of analytes can improve the separation from other analytes or components by HPLC,

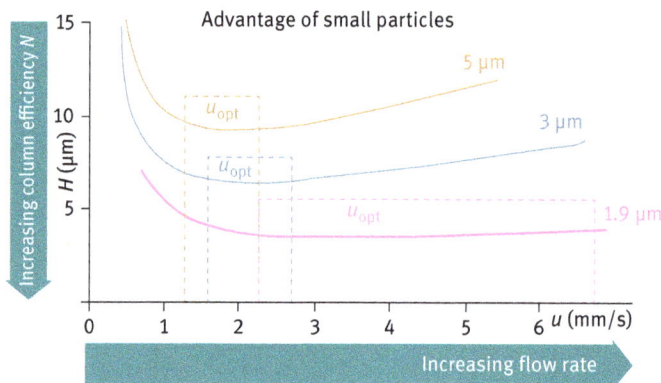

Figure 7.34: Van Deemter relations (plate height H vs. solvent flow u) for various silica particle sizes in normal-phase chromatography. u_{opt} describes the optimal flow rates (at the lowest H).

(a)

(b)

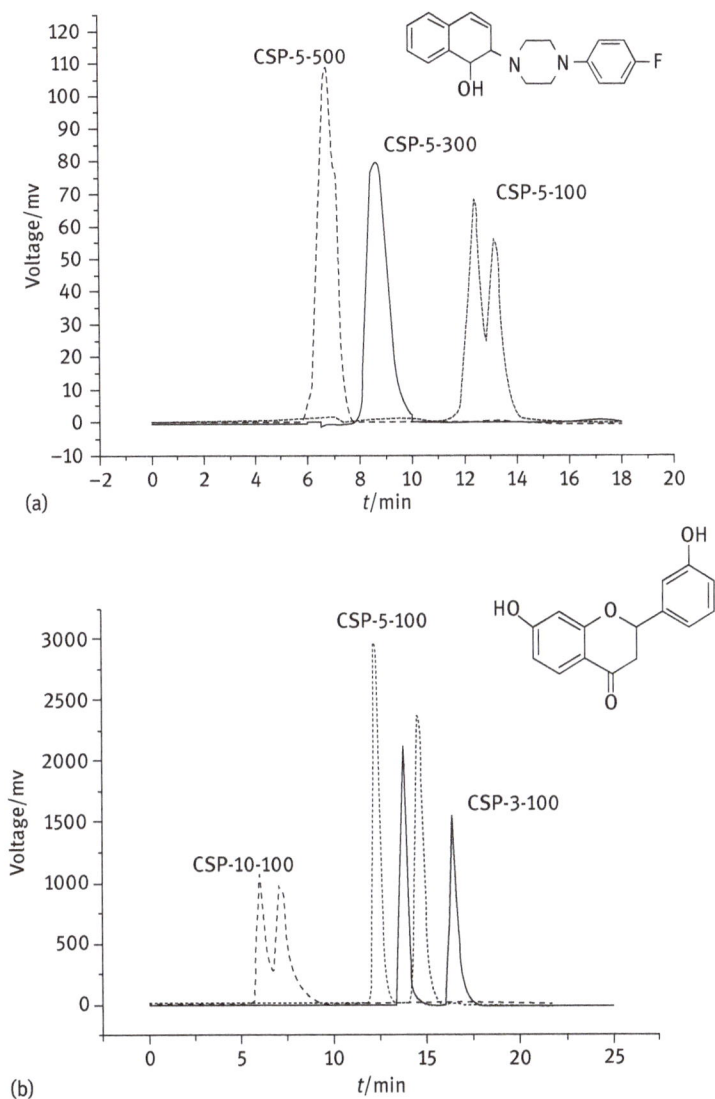

Figure 7.35: Amino-functionalized silica gel containing an immobilized ligand (CSP) with different pore (100, 300 and 500 Å) and particle sizes (3, 5 and 10 µm) were compared in terms of separation of racemates. (a): effect of pore sizes, normal-phase HPLC; (b): effect of particle sizes, reversed-phase HPLC. The retention factor and resolution of racemates generally decreased with increasing pore size; the column prepared with the smallest (3 µm) silica gel particle size gave the best column performance and enantioselectivity in both the normal- and the reversed-phase modes (Qin et al. (2015), Jornal of Separation Science 33, 2582–2589).

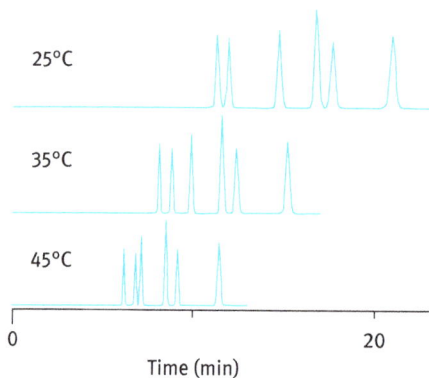

Figure 7.36: Effect of temperature on the elution of analytes in HPLC analysis.

for example, in reversed-phase runs after the introduction of a nonpolar group or in ion-exchange methods by a charged functional group. Introduction of a permanent charge in a molecule, for example, a quaternary ammonium ion, can increase the ionization efficiency in electrospray ionization MS. In addition, derivatization with a defined structural element may enable the prediction of specific fragmentation reactions in MS/MS applications. Another aim of derivatization is to transform analytes which are not visible in common HPLC detectors, such as UV absorbance and fluorescence, to derivatives that can be detected because a chromophoric or fluorophoric group was introduced.

Modification of an analyte before chromatographic separation is called pre-column derivatization. If the derivatization procedure is carried out after separation of components (i.e., after the column), we call it post-column derivatization.

Both methods have their advantages over the other: The pre-column derivatization takes place outside the chromatography system; thus, it uses simpler equipment than the post-column derivatization, has no restriction on the reaction conditions (reaction time, temperature, number of reagents), separation of the derivatization reagent from the derivatization product is possible (no interference in detecting the analyte) and (in some cases) the separation conditions after derivatization are improved. On the other hand, post-column derivatization occurs online, which also has advantages over the pre-column method: The reaction can be automated, and due to the short time on the column, unstable derivatization products are less problematic. However, the reaction reagent must not interfere with the detection of the derivatization product (for instance, for a UV detector, the absorption maxima should be different), and the reaction time should be as short as possible (to avoid band broadening).

Some examples of derivatization reagents are summarized in Table 7.9 (UV-vis detection) and Table 7.10 (fluorescence detection). Many more reagents, also for other detector types, are available and cataloged by chemical companies.

A lot of literature is available on chemical derivatization methods for HPLC detection (see General Literature). A recent review of (post-column) derivatization reactions and the optimization of HPLC performance was published by Jones et al. [14].

Table 7.9: Derivatization techniques for HPLC (for UV–vis detection).

Target analytes	Derivatization reagent
Amines and amino acids	*Acyl chlorides* (e.g., benzoyl chloride)
	Arylsulfonyl chlorides (e.g., *p*-TSCl, BSCl, DABSCl)
	Nitrobenzenes (e.g., FDNB, picrylsulfonic acid)
	Isocyanates and isothiocyanates (e.g., PIC, DABITC)
Carboxylic acids	Phenacyl bromide, methylphthalimide, *p*-NBDI
Hydroxy compounds	*Acyl chlorides* (e.g., benzoyl chloride, *p*-nitrobenzoyl chloride, PIC)
Carbonyl compounds	2,4-DNPH, PMP, *p*-nitrobenzylhydroxylamine

p-TSCl, toluenesulfonyl chloride; BSCl, benzenesulfonyl chloride; DABSCl, dimethylaminoazobenzenesulfonyl chloride; FDNB, 1-fluoro-2,4-dinitrobenzene; PIC, phenyl isocyanate; DABITC, 4-*N,N'*-dimethylaminoazobenzene-4'-isothiocyanate; *p*-NBDI, *p*-nitrobenzyl-*N,N'*-diisopropylisourea; 2,4-DNPH, 2,4-dinitrophenylhydrazine; PMP, 3-methyl-1-phenyl-2-pyrazoline-5-one.

Table 7.10: Derivatization techniques for HPLC analyses (for fluorescence detection).

Target analytes	Derivatization reagent
Amines, amino acids, peptides	*Sulfonyl chlorides*: Dns-Cl
	Halogenonitrobenzofurazans: NBD-Cl
	Isocyanates/Isothiocyanates: fluorescein isothiocyanate
	Fluorescamine: MDPF
Carboxylic acids	*p*-Bromophenacylbromide, 5-bromomethylfluorescein
Hydroxy groups	Diacetyldihydrofluorescein
Nitro-polyaromatic hydrocarbons	Online reduction with Pd/Pt yielding the corresponding amine
Carbonyl compounds	Dansylhydrazine, semicarbazide, 2-aminopyridine, DPE

Dns-Cl, 5-dimethylaminonaphthalene-1-sulfonyl chloride; NBD-Cl, 4-chloro-7-nitrobenzofurazan; MDPF, 2-methoxy-2,4-diphenyl-3(2*H*)-furanone; DPE, 1,2-diphenylethylenediamine.

7.4.8 Hyphenated Techniques (HPLC–MS, HPLC-NMR)

7.4.8.1 HPLC–MS

As described in 7.4.5.2, the liquid eluent of HPLC separations and the need for the gaseous state for MS analysis with its high vacuum conditions need special interfaces to combine both methodologies. The challenge is to get rid of the solvent, to maintain the adequate vacuum level in the mass spectrometer and to generate the gas-phase ions. Some methods will be described in Chapter 9 (MS) that enable the combination of HPLC and MS, that is, electron impact ionization and atmospheric pressure ionization (Figure 7.37).

7.4.8.2 HPLC-NMR

The combination of HPLC and NMR provides another effective tool for structural elucidation of analytes. Although NMR besides MS has been well known as a most

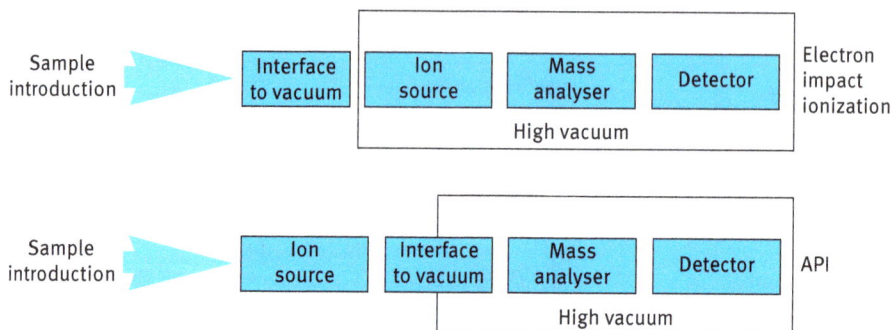

Figure 7.37: Most common interfaces for HPLC–MS applications: electron impact ionization and atmospheric pressure ionization (API).

important technique for structural elucidation since decades, it was generally considered to be too insensitive for the analysis of organic compounds in environmental samples. Compared to HPLC–MS, this technique is much less sensitive but new methods in HPLC-NMR were developed to increase the sensitivity significantly (see later). It is beyond the scope of this book to explain the basics of NMR spectroscopy; excellent text books are available (e.g., see Further Reading). Only the basic principle is briefly summarized. The technique is based on the interaction of nuclei of certain isotopes – especially well suited are ^{1}H, ^{13}C or ^{19}F – with a strong external magnetic field (up to about 20 T). Such nuclei can be imagined as spinning on their axes providing them with a magnetic momentum. According to quantum mechanics, a nucleus with the spin I will have 2I + 1 possible orientations; for instance, a nucleus with spin 1/2 as for ^{1}H nuclei will have two possible orientations. In the absence of an external magnetic field, these orientations are of equal energy. If an external magnetic field is applied, then the energy levels split, one representing the alignment of the spinning nuclei along, the other opposed to the external magnetic field: The higher the external magnetic field, the larger is the energy difference of the two energy levels. The alignment of the spinning nucleus along the external magnetic field corresponds to the energetic ground state that opposed to it represents the excited energy level.

At the ground level, there are slightly more spinning nuclei at room temperature (according to the Boltzmann distribution) than in the excited state. Like in other spectroscopic techniques, irradiation by an electromagnetic field, for NMR this is in the radiowave range, can excite the nuclei in the ground state to the higher energetic level, if the low-energy difference between the two energetic states is matched by the energy of the electromagnetic irradiation: In this case, the radiowave energy is absorbed by the nuclei, and we observe a resonance signal. The excited nuclei relax by basically two mechanisms in order to return to the ground state: spin–lattice relaxation (the transfer of energy from the excited nuclear spin system to the sample "lattice") and spin–spin relaxation (a neighboring nucleus with the same spinning frequency in the lower energy level will be excited, whereas the excited nucleus relaxes to the lower

energy state, that is, the average lifetime of the nucleus in the excited state will decrease. Now, most importantly for the interpretation of NMR spectra, the resonance
frequency of a certain nucleus depends on its chemical environment: for instance, a
^{13}C atom (^{13}C occurs naturally at ca. 1% abundance) surrounded by electron donat-
ing groups (e.g., amino, ether or alkyl) has a different chemical shift (location of the
NMR signal) than a ^{13}C atom in the vicinity of electron withdrawing (e.g., carbonyl
groups, halogens and ammonium) groups. This shift is due to the different shield-
ing of the nucleus against the external magnetic field because the electron density
around the nucleus changes in dependence of the neighboring functional groups.
In summary, the frequency of the resonance signals shifted by the chemical environment together with further spectroscopic information (like the energetic coupling
of a nucleus with other neighboring nuclei) comprise the basic information of NMR
investigations: The chemical structure of unknown analytes can be unambiguously
determined. For HPLC-NMR , ^{1}H- and ^{19}F-NMR spectroscopy is the best choice at the
present since they are superior to other suitable NMR isotopes in terms of sensitivity.

In the following, we will explain the combinatory use of both HPLC separation
and NMR analyses which represents that HPLC-NMR is a nondestructive method, that
is, the sample can be retrieved after analysis and then analyzed by other methods. The
sample is transferred into the NMR machine directly as a "peak" eluting from the HPLC
column, that is, it arrives with the peak concentration in the eluent. NMR spectra are
then acquired either on-flow (continuously during chromatographic separation) or in
stopped-flow mode (whereas a selected peak is retained in the NMR probe and the
chromatography is stopped). Each eluting peak can also be parked intermediately in
individual sample loops and introduced to NMR after HPLC separation. It is also possible to immobilize selected peaks after the chromatographic separation by SPE with
the advantage that the eluent can subsequently be removed by eluting the analyte
with deuterated solvents from the cartridges into either an NMR flow probe or into
NMR sample tubes. This method still increases the sensitivity of HPLC-NMR since a
significant enrichment by SPE is achieved: Analytes in the low-to-medium nanogram
range can be analyzed. The experimental setup is shown in Figure 7.38.

The conventional NMR tubes used in stand-alone systems are replaced by U-
shaped glass flow-through cells if combined with HPLC. The cell has inner diameters
of 1.7–4 mm and is inserted vertically into the magnet; the cell volume has about 30–
200 µL, that is, similar to the HPLC peak volume passing the cell in a laminar flow.
If the eluent peaks were measured directly by NMR, the signals of the solvent would
completely dominate the weak signals of the analyte. Two options to circumvent this
difficulty are available: First, the SPE technique described above with subsequent elution with a deuterated solvent and second by selective suppression (saturation) of the
solvent signals.

Alves et al. [15] investigated wastewater for the presence of nontargeted pollutants. By combination of HPLC-(UV/MS)-SPE-NMR, the authors detected nonspecific
contaminants in such complex matrix at micromolar and in some cases nanomolar

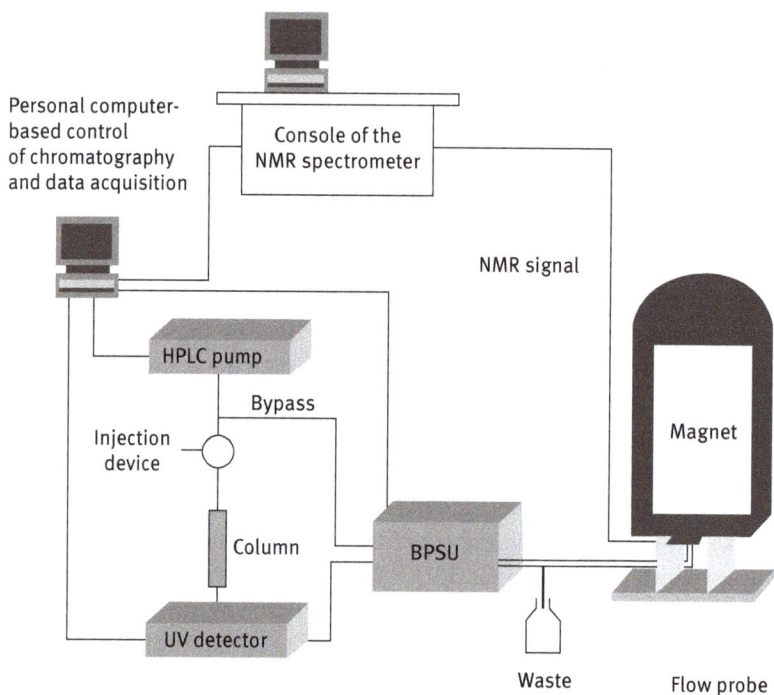

Figure 7.38: HPLC-NMR coupling for structural elucidation of unknown analytes. The sampling unit (BPSU) can contain various sample loops in which the peaks are retained or an automatic SPE unit from which the analytes can be eluted by deuterated solvents (Patel et al. (2010), Pharmaceutical Methods 1, 2–13).

concentrations and were able to elucidate the corresponding structures of the pollutant. Analyses by HPLC coupled to SPE-NMR and time-of-flight MS revealed residues of explosives, for instance trinitrotoluene and several degradation products in groundwater. The concentrations of the contaminants ranged between 0.1 and 48 µg/L [16].

7.5 Gas chromatography

7.5.1 General Remarks

GC belongs to the best performers trace analysis has. Based on an idea presented by Martin and Synge first already 1941, the publication of James and Martin (1952) is generally seen as the birth of GC. GC instrumentation is nowadays found in all analytical labs. Not only its performance in trace analysis makes it so attractive, rather the possibility for preparative separation is an additional advantage.

GC uses gas as the mobile phase, with the consequence of being suitable as only volatile gaseous compounds as analytes become separated. First applications were separations of volatile fatty acids, followed later by characterization of petroleum fractions. At present, volatile organic compounds (VOCs: e.g., benzene, toluene and xylenes), fragrances or odorous substances are typical analytes measured by GC. Stationary phases can be either liquid or solid. The range of applicability of GC can be extended considerably by derivatization of nonvolatile compounds in order to make them volatile. GC exists in many combinations with a variety of more or less selective and sensitive detection techniques. Especially, this option makes it suitable as a robust tool, for example, for process analysis. In recent years, multidimensional separation in GC, coupled with highly selective MS detection, became the most powerful instrumentation to treat analytical tasks of highest complexity ("comprehensive analysis") [17]. This opens the way for tasks within the field of -*omics* (metabolomics, etc.) with its nearly indefinite wealth of analytes.

The typical setup for GC separation with its basic parts is depicted in Figure 7.39.

A gas supply provides an inert carrier gas as mobile phase. Within this carrier gas, a small volume of analyte mixture becomes injected into the injector block. Separation of the analyte between the mobile phase and the solid or liquid stationary phase happens within a column usually placed within a thermostat oven. In GC, only elution of separated analytes is wanted. The column-leaving substances are recorded and quantified by a detector. In preparative GC, the fractioned samples are collected afterward.

Already in the 1970s, attempts to miniaturize such systems became known. Photolithographic techniques helped to shrink down to microsystems ("µGC"). Not only the separation columns, but also the whole apparatus inclusive carrier gas supply possesses cm-to-mm dimensions. Examples of two miniaturized GC columns are seen in Figure 7.40.

In the following, the characteristics of GC with its individual parts will be presented and discussed.

7.5.2 Separation Columns and Stationary Phases (GC)

In the early days, *packed columns* with powder inside (brick flour or diatomaceous earth) with and without coating were used. A widely applied chromatography support is polysilicic acid (in the hydrated amorphous silica form) having a porous structure and containing varying amounts of metal oxides. The most commonly used type being

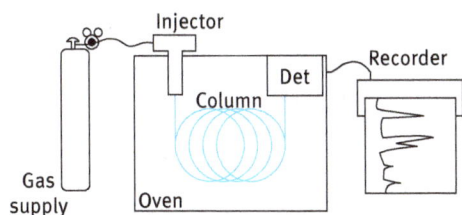

Figure 7.39: Basic scheme of a gas chromatography system.

Figure 7.40: In-silico miniaturized GC separation columns (www.bo.imm.cnr.it/site/book/export/html/355).

Chromosorb™. These columns are several mm in diameter and have a length of up to several meters. Also, alumina salts or molecular sieves find application.

Golay (1958) was the first to introduce *capillary columns* (open tubular columns) with diameters of some hundred μm, but lengths of up to 100 m. The production is similar to fiber-optical technology. In contrast to the first packed columns, these capillaries are characterized by plug flow instead the peak-broadening laminar flow. The disadvantage is the extreme low loading capacity of less than about 100 ng. This might be surpassed by usage of capillary bundles. Usually, for preparative GC separation, only packed columns are in use. Table 7.11 gives an overview on the different column types.

The *stationary phase* can be a solid support or a liquid. The latter consists of a compound of low volatility that is deposited as a multilayer on the inner surface of the capillary, or on a particulate support filled in. This enables a quick equilibration, since mass transfer to and from stationary phase has to pass only few nm layer thickness. Materials for such stationary phases range from high-boiling hydrocarbons, silicon oils, polymers up to ionic liquids. A variety of so-called "wax" polyethylene glycol phases are also frequently used. For optimal GC separation, phase selection is based on the general principle that "likes dissolves like," that is, a nonpolar column is best

Table 7.11: Properties of typical GC columns.

	Wall-coated capillary	Support-coated cap.	Packed column
Length [m]	10–150	10–50	1–5
Internal diameter [μm]	0.1–0.8	0.5–0.8	2–4
Liquid film thickness [μm]	0.1–0.5	0.8–2	10
Capacity per peak [ng]	<100	50–2,000	10,000
Resolving power in HETP number [−]	>200,000	<40,000	<1,000

for the analyses of nonpolar substances and polar columns will separate polar compounds well. For solving complex problems, hyphenation of differently polar phases
in a sequential order is possible.

A critical problem is bleeding of stationary-phase material from the support at
higher temperatures. Therefore, the vapor pressure of the stationary phase must be
low. The distribution of analytes among these phases is based on dispersive forces.
Separation of special molecular arrangements can be favored, for example, by chiral
phases for racemic mixtures, or liquid-crystal lamellar phases for coplanar congeners.

The theoretical plate heights (H, also named HETP) in GC are higher than those
in liquid chromatography applications such as HPLC ($H_{GC} > H_{HPLC}$). This is because
the diffusion coefficients of the analytes in the gas phase are much higher than in
liquids. Nevertheless, the resolution power of capillary GC is much better than that in
HPLC since the length of the capillary (L) can be up to about 100 m compared to the
cm lengths of HPLC columns, and thus, the number of theoretical plates N becomes
larger in GC ($N = L/H$). Thus, the separation efficiency of HPLC is significantly poorer
than that of GC.

The longer a separation column is, the better its resolving power will be, but at
the expenditure of pressure loss. It should be noted that doubling column length will
not double resolution (resolution increases only according to the square root of the
column length). A good compromise is a column length of 30–60 m.

A demonstration of *resolving power* of capillary columns versus packed columns
is seen in Figure 7.41.

Since there are so many possibilities to develop and apply new stationary phases
and column configurations, a standardized testing of them is important. Grob et al.
(1978) were the first to propose a test containing 12 compounds of different polarity,
acidity and vapor pressure (see Table 7.12) that allowed characterization and comparison of capillary columns in one thermo-programmed GC run: Assessment criteria
are the separation efficiency, adsorption to the stationary phase, film thickness and
retention times.

7.5.3 Mobile Phase

As mentioned above, the *mobile phase* in GC consists of an inert carrier gas like N_2,
Ar, He, CO_2 or H_2. Inert means, neither a reaction of the analytes with the stationary
phase nor intermolecular reactions of the different analytes during transport through
the columns. The selection of the mobile phase is determined by the attached detector
system. In the case of flame ionization, hydrogen is the first choice. Since not only the
temperature, but also the carrier gas velocity controls the distribution of the analyte
between mobile and stationary phases, a variable adjustment of gas flow with highest
precision is needed. Critical orifices or mass-flow controllers are used for this. The
viscosity also determines the flow pattern within the column. Helium is here preferred.

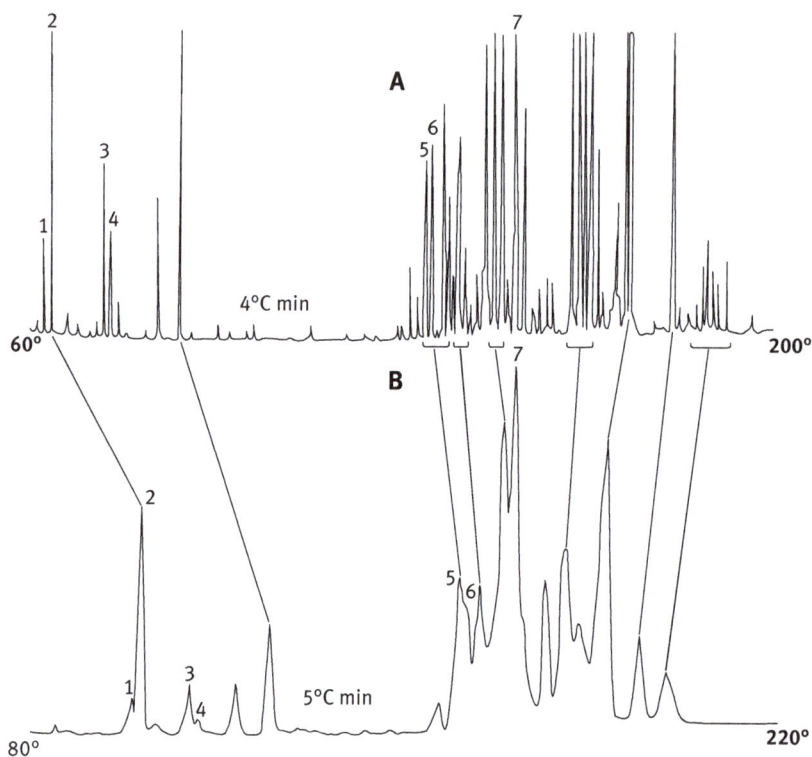

Figure 7.41: Comparison of resolving power of packed column (length: 4 m) with capillary column (length: 50 m) and same stationary phase applied on Calamus oil separation (Bartle and Myers 2002). Reproduced with permission from Elsevier.

Table 7.12: Grob's GC test mixture for GC separation column assessment.

Composition of the test mixture			
Component	Concentration (mg/L)	Component	Concentration (mg/L)
C_{12}-acid methyl ester (E_{12})	41.3	Nonanal (al)	40
C_{11}-acid methyl ester (E_{11})	41.9	2,3-Butanediol (D)	53
C_{10}-acid methyl ester (E_{10})	42.3	2,6-Dimethylaniline (A)	32
Decane (10)	28.3	2,6-Dimethylphenol (P)	32
Undecane (11)	28.7	Dicyclohexylamine (am)	31.3
1-Octanol (ol)	35.5	2-Ethylhexanoic acid (S)	38

Special attention has to be paid to the purity of the carrier gas. Usually, a purity of >99.999% is required. This still means that a permanent contamination of 10 ppmv has to be tolerated. For gaining higher purity, reactive catalytic *in line* traps can be installed in order to improve this. Remaining impurities of oxygen or water are dangerous to the stability of the stationary phase, as this may become oxidized or hydrolyzed at higher column temperature. Hydrocarbons as contaminants will raise the detector background to a certain extent.

7.5.4 Isothermal versus Temperature-Programmed GC

For practical reasons, temperature T is the most convenient parameter to manipulate the separation efficiency. Once the stationary and mobile phases and the column length are decided, besides T, only the carrier gas velocity u could be adjusted to improve separation. But adjustment of u is technically not so easy to achieve as for T. Also, u will become enhanced in parallel with increased T by viscosity dependency.

There are two ways to make use of T: the *isothermal* or *temperature-programmed* GC (TPGC). Both situations are depicted in Figure 7.42.

The effect of this is demonstrated for separating C_6 to C_{21} hydrocarbons with a wide range of boiling points. In the case of isothermal separation (see Figure 7.43), the chromatogram only shows a good separation at the beginning, but at the end, the peaks show an uneven shape and a progressing time interval in between. The separation in total lasts more than 1.5 h.

In the case of a temperature-controlled run (ramping rate: 50–250 °C at 8 °C/min), not only separation for all hydrocarbons is fine, but also the whole chromatography run is finished after roughly one-third of time. This is shown in Figure 7.44.

Prerequisite for TPGC is non-bleeding of the stationary phase. Otherwise, an increasing background of the detector signal would be observed. Also, after reaching the highest temperature, cooling by air ventilation is needed to reduce the temperature within a reasonable time back to T at start of chromatography.

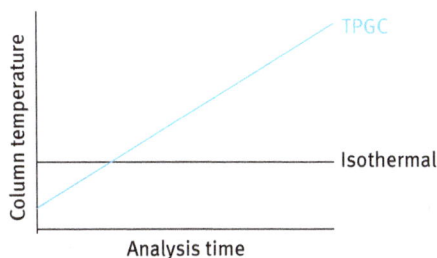

Figure 7.42: Isothermal heating of GC separation column versus temperature-programmed ramping.

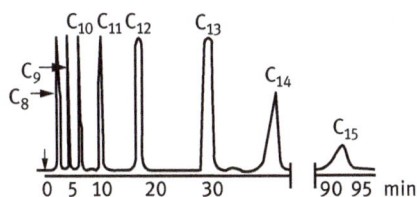

Figure 7.43: Isothermal separation (138 °C) of C_6 to C_{21} hydrocarbons by GC.

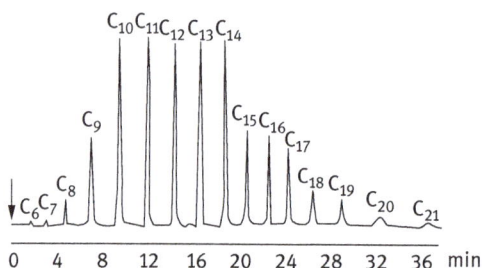

Figure 7.44: Effect of thermoprogrammed ramp heating during GC separation of C_6 to C_{21} hydrocarbons.

7.5.5 Injection Techniques for GC

There are several ways to introduce a sample into a GC system: direct, on-column, split and splitless injection. The sample size injected through a septum by a microsyringe ranges from tenths of microliters to several microliters. Capillary columns possess an extremely small capacity; so, the sample has to be split for them up to a factor of 500.

Direct injection means the transfer of an already gaseous sample as gained from headspace or purge and trap, or applications of solid-phase micro-extraction. In all cases, no solvent vapor should be present, in order to keep the applied gas volume small. Small inlet liners are used to avoid peak broadening.

On column injection describes the direct application of a liquid sample extract on the column head. This is shown in Figure 7.45.

The reproducibility for such on-column injection is not the best, since the location where the sample becomes deposited will not be always the same (due to bubble formation and early loss of low volatile ingredients). It is only applicable with temperature programming.

A *split/splitless injection* port is depicted in Figure 7.46.

Here, the extract or sample is vaporized and mixed with carrier gas in the inlet port. The carrier gas transfers the entire gas cloud into the separation column. Since the column is held at a lower temperature by 10–50 °C, the lowest boiling analytes condense on the column inlet. Then, either by closing or opening the split outlet, only a small part of the evolved gas cloud becomes transferred to the column. The bulk of vaporized substances is vented away. Sometimes, wool plugs are inserted for better gasification.

A general problem may pose the septum material by bleeding or by degradation of the polymer.

7.5.6 GC Separations at Extreme Temperatures

The usual temperature range for GC is about 30–300 °C. But some analytical tasks will ask for temperatures beyond or below.

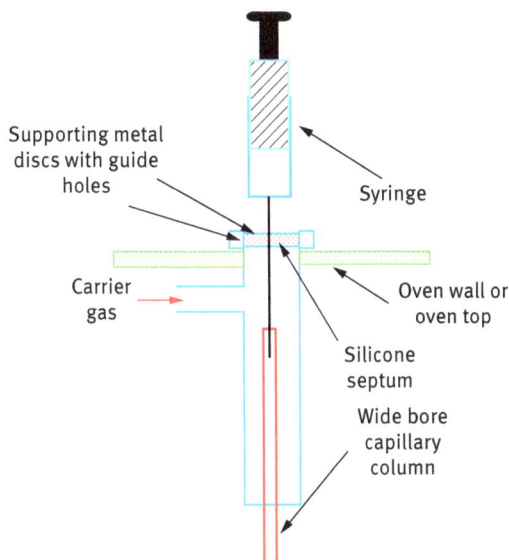

Supporting metal
discs with guide
holes

Syringe

Carrier
gas

Oven wall or
oven top

Silicone
septum

Wide bore
capillary
column

Figure 7.45: On-column injection for a large bore open tubular column.

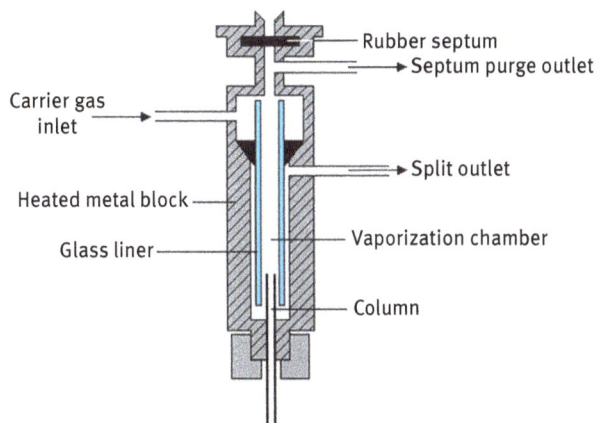

Rubber septum
Septum purge outlet

Carrier gas
inlet

Split outlet

Heated metal block

Glass liner

Vaporization chamber

Column

Figure 7.46: Split/splitless GC injector.

Cryogenic GC separation has been applied for various low-boiling compounds, such as organo-metallic (e.g., alkylmercury or organotin compounds) or odorous species (e.g., R-S-R′). For this, the separation column is inserted into a Dewar filled with liquid nitrogen ($-196\ °C$) or a cold atmosphere produced by admission of cold nitrogen gas. The challenge is to get rid of water vapor before entering the Carbopack™ (graphitized carbon black) column. Usually, the analytes are sampled at ambient temperature by selective sorption, followed by a transfer to the GC column. Currently, aside H_2/D_2 separation, increased biogas production asks for alkane/alkene/alkyne analysis within a crude matrix.

High-temperature GC (>300 °C) is needed for separations of stable semipolar or other low-boiling substances, for example, motor oil compounds, waxes, glycerides, steroids, hopanoids or lipids, in general. For such purpose, the selection of an appropriate stationary-phase material with very low vapor pressure is essential. Newer materials for this are polysiloxanes or imidazolium ionic liquids. Temperatures up to 400 °C have been reported for GC separation for such analytes.

7.5.7 Derivatization Techniques in GC

The application range for GC separation is limited by the availability of an analyte to serve as a gaseous compound needed for distribution and transport through the column. Today, more and more questions are linked with biopolymers or polar or semipolar compounds, especially from life sciences or natural product analysis, and the different fields of -*omics* (e.g., lipodomics, glyconomics, etc.). Created by bio-based chemical synthesis, we are commonly faced with a variety of similarly configured molecules extremely difficult to separate and quantify. GC seems ideal for tackling these tasks due to its enormous resolving power.

There is a second reason for changing the volatility of analytes prior GC separation: the thermal instability of the analytes. Many functional groups will act as leaving groups when heated up during a thermoprogrammed chromatography run for thermodynamic reasons, forming small compounds like CO_2 or others. So, the analyte is partially destroyed on-column.

Furthermore, less volatile substances will need a longer retention time with the risk of peak broadening. A short retention time is therefore wanted.

Finally, depending on the derivatization chemistry, the transformed analyte might become detectable at much lower concentrations, for example, by introducing easily detectable halogen atoms to the analyte molecule and subsequent analysis by an electron capture detector (ECD).

In rare cases, derivatization is performed to lower the volatility of an analyte. This is for small organic molecules that otherwise have to be separated under cryogenic conditions.

Over time, routes have been developed for improving the volatility of compounds. Aside from fragmenting a large molecule by pyrolysis (which will be discussed in Chapter 7.5.8), chemical derivatization by introducing a more hydrophobic character (and by this enhancing its volatility) seems more appropriate. This approach, without changing to much the chemical identity of an analyte molecule, is a general procedure used, either performed prior transfer to a GC separation (*extractive derivatization*), or by so-called *on-column derivatization*, where the derivatization reaction is started

Table 7.13: Formation of chemical derivatives from low-volatile substances.

Functional groups at analyte	Derivatization by	Target substances formed
R–COOH	Methylation (e.g., by CH_2N_2; BF_3/methanol; $(CH_3)_4OH$ salt, pentafluorobenzyl bromide)	R–COOCH$_3$
R–OH	Methylation (CH_3I/Ag_2O); silylation (trimethylsilyl donor e.g., HMDS, BSTFA); acylation $(R'-CO)_2O$	R–O–CH$_3$ R–O–Si–$(CH_3)_3$ R–O–CO–R$'$
R–NH$_2$	Acylation $(R'-CO)_2O$; silylation (trimethylsilyl donor e.g., HMDS, BSTFA)	R–NH–CO–R$'$ R–N–[Si–$(CH_3)_3$]$_2$
R–SH	Acylation $(R'-CO)_2O$; silylation (trimethylsilyl donor e.g., HMDS, BSTFA)	R–S–CO–R$'$ R–S–Si–$(CH_3)_2$
Chiral substances, for example, amines	Diastereomer formation of each of the enantiomers (e.g., (N-trifluoroacetyl-L-prolyl chloride)	Chiral separation and detection needed

HMDS, hexamethyldisilazane; BSTFA, N,O-bis(trimethylsilyl)trifluoroacetamide.

during injection onto the heated column, producing the volatile derivatives just at the starting point of separation. Table 7.13 shows only some possibilities.

7.5.8 Change of State During GC Separation (Pyrolysis, Carbon Skeleton Formation)

Modern high-performance GC techniques also allow the indirect analysis of otherwise difficult accessible samples as there are paints (in forensic analysis), wood samples, resins, fabrics, (bio)polymers or other high molecular weight compounds. This is achieved by pyrolytic cracking under inert atmosphere in front of the GC (Py-GC) separation column. Such an arrangement is shown in Figure 7.47.

There are two ways for a quick flash heating (up to 1,000 °C). Resistive heating of the deposited sample (on a Pt wire or in a quartz boat) creates high temperature, or the sample holder is made of a ferromagnetic alloy (Curie point pyrolysis) which receives the needed energy by an inductively coupled radiofrequency (RF) pulse from a transmitter nearby. Flash pyrolysis is achieved by dropping the sample cup into the hot furnace and sweeping the pyrolysate to the GC column for subsequent analysis. Often, before analyzing the pyrolysis products the gases evolved at lower temperature become analyzed first (evolved gas analysis) by MS. Afterward, the released gaseous fragments are separated and analyzed by GC/MS.

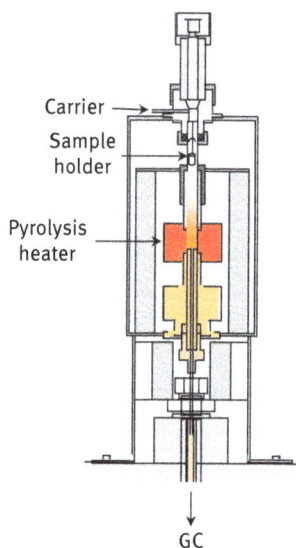

Figure 7.47: Pyrolysis stage for GC analysis.

Since only comparison ("fingerprinting") with known pyrolysis processes allows identification, a pyrolysis data bank created by treatment of known products is helpful.

Although the disadvantage of PY-GC clearly is the formation of a wide range of fragments, the *carbon skeleton technique* allows one to reduce complexity of chromatograms by, for example, dechlorination and hydrogenation. The analysis of chlorinated paraffins (CP) is an example for this. Such compounds are present to a high extent in flame retardants, plasticizers, additives, cutting fluids and sealing materials [18]. Produced by chlorination (up to 70% w/w) of C_{10} to C_{30} paraffins, an unmanageable amount of homologues and isomers are present in such products. Here, the catalytic reduction of the raw sample within a catalytic zone under hydrogen atmosphere, directly inserted within the GC injector block, drastically simplifies the analytical task (see Figure 7.48).

Similar successful attempts are published for other halogenated compounds like PCBs, polychlorinated naphthalenes or toxaphene pesticide mixtures.

7.5.9 Analyte Detection in GC (Electrical, Electrochemical, Optical)

7.5.9.1 General Requirements

Quantitative detection of the separated analytes is the final step in gas chromatographic determination.

Nowadays, combinations with MS detection represent a major pillar of quantification. Since there are various configurations possible, and MS for itself can be seen as an autonomous separation method applicable to organic trace compounds, it seems justified to treat them in own chapters (see Chapters 9 and 11). Such GC–MS

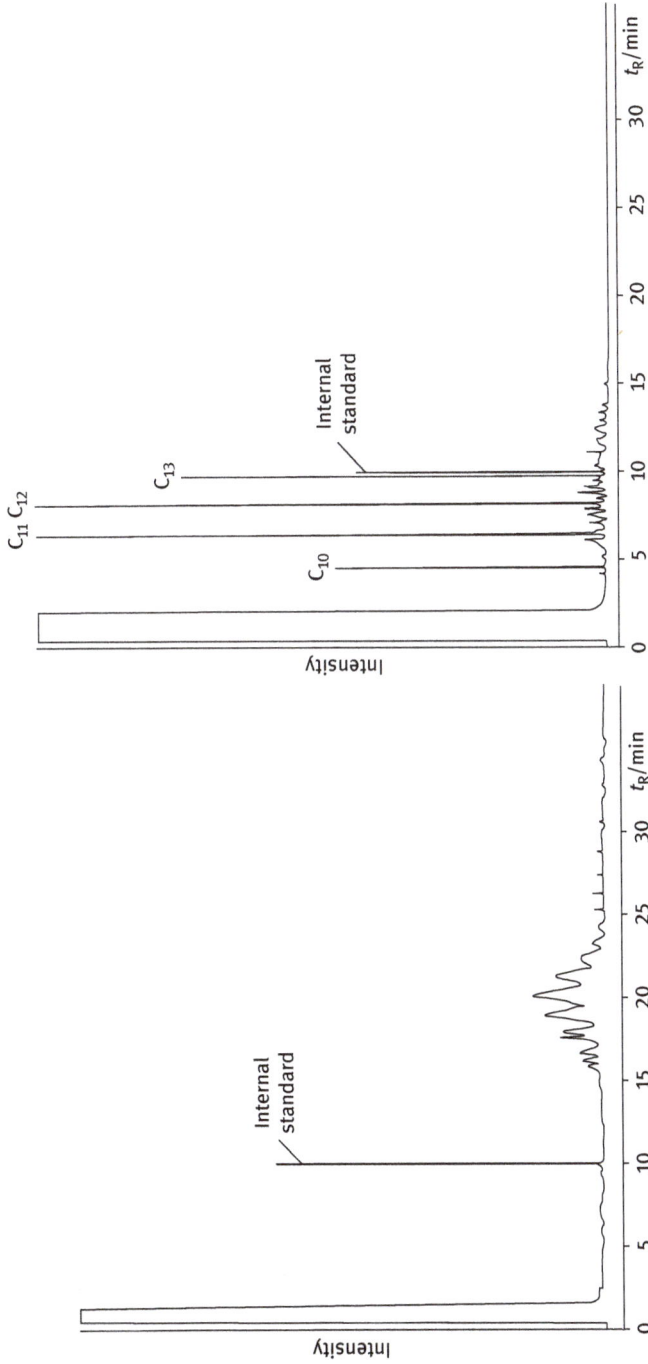

Figure 7.48: GC/FID analysis of chlorinated paraffin mixture CP 56 (left: before hydrogenation; right: after dechlorination and hydrogenation.

configurations exhibit tremendous performance, but for routine analysis or dedicated identification of known analytes, there exists a variety of approved detectors, which sometimes are even much better performing and are definitely less expensive than a MS system.

The requirements for a GC detector are in general the following:

- sensitivity of <1 ppmv to separated analytes within the gaseous eluate (within the peak volume the concentration of analyte is high for a short time);
- no back mixing during detection (small detector volume, if collection for preparative GC is wanted);
- linear concentration response function (at least a differentiable relationship and linearizable);
- wide dynamic concentration range (depending on resolving power a change of concentrations of up to six orders of magnitude is possible within ms);
- quick response time (in order to follow peak evolution);
- low background noise (due to carrier gas contamination or electronically);
- no shifting of baseline (due to a gradient or bleeding of septa or columns);
- insensibility to instability of carrier gas flow rate;
- simple and reliable operation (no engineer needed).

7.5.9.2 Most Common GC Detectors

To fulfill the above-mentioned requirements by a GC detector isn't easy. Over the last decades, some well-tried detector principles have developed, which will be presented in an overview first (see Table 7.14). Common to all named detector principles is a high sensitivity, but low chemical selectivity in response.

Modern chemical and bioanalytical sensor development will offer new possibilities for hyphenation to a GC separation. Especially, the miniaturization of GC systems will require new routes. It is a conspicuous fact that not many optical detectors for GC separation are known, in contrast to liquid chromatography, possibly because a

Table 7.14: Overview on common GC detectors.

Detector	Specificity	LOD [ng]	Linear range
Flame ionization (FID)	C	0.1	10^7
Alkali FID (AFID or NPD)	P, S, N, halogens	0.01 (P)	10^3
Electron capture (ECD)	Halogens, $R-NO_2$, $R-CN$	10^{-6}	10^5
Thermal conductivity (TCD)	None	5	10^5
Flame photometry (FPD)	P, S	0.01 (P)	none
Electrolytic (ELCD)	Halogens, S, N	10	10^3
Mass spectrometer	None	10^{-4}	10^6

NPD, nitrogen–phosphorus detector; ECD, electron capture detector; FID, flame ionization detector; AFID, alkali FID; TCD, thermal conductivity detector; FPD, flame photometry detector; ELCD, electrolytic conductivity detector.

photometric detection would need a long optical path length for getting sensitive re-cordings (according to Lambert–Beer's law). An emission-based optical sensing would need a very high excitation radiation intensity. Some rare examples, like photoacous-tic detection or microwave excitation, are published, but not extensively used. With the advent of cheap and strong UV and white-light LEDs, this will probably change.

In the following, the most common GC detectors become presented with its peculiar features.

Flame Ionization Detector (FID) The FID was developed in the 1950s of the last century. Its configuration is rather simple and shown in Figure 7.49.

A small permanently burning hydrogen flame is doped by the eluate from the sep-aration column. Therefore, often hydrogen is used as mobile phase. Within the flame, all organic compounds (except CO_2 or CO) are destroyed and form a variety of ionic fragments. These gas ions are measured by collection through a collector electrode within an electrostatic field. A sensitive electrometer measures the resulting electric current continuously. The ions are proportionally formed to the analyte concentration. The response to any carbon atom within a larger organic molecule is equal despite its molecular surroundings. Other elements contribute less.

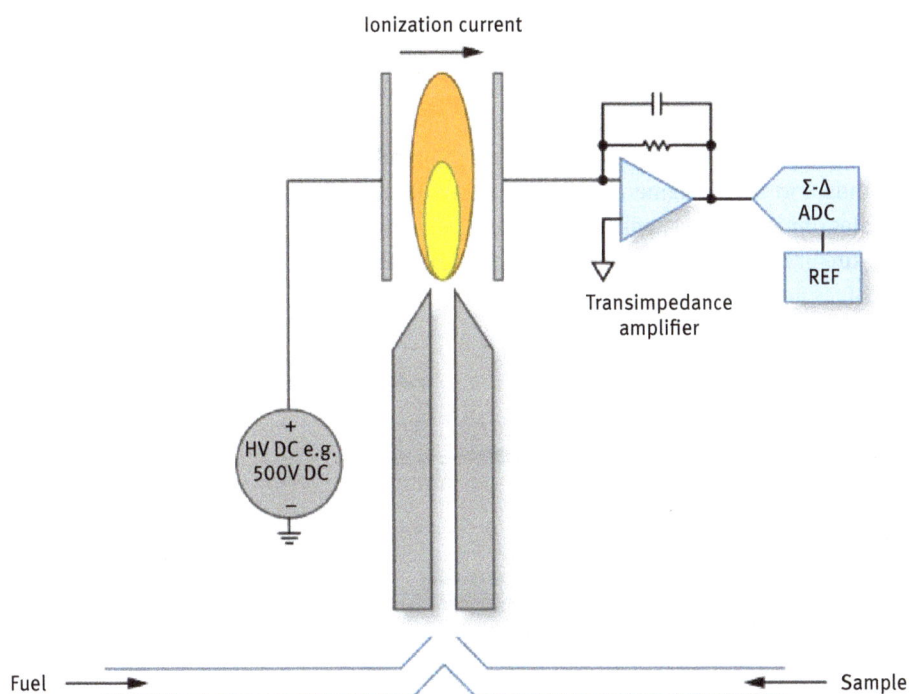

Figure 7.49: Basic working principle of FID.

Using a so-called *methanizer*, also oxidized carbon species ("greenhouse gases") are feasible without changing retention time after catalytic reduction to methane.

Beneficial is the simple and rugged construction, as well as the wide dynamic range. Disadvantageous is the permanent burning of a flame in a lab surrounding, and to a certain degree its nonselective response.

Alkali FID (AFID) or Nitrogen–Phosphorus Detector (NPD) Here, a rubidium or cesium silicate bead is hanging above the exit of the burner for an oxyhydrogen flame. It is externally heated such that it ignites catalytically the oxyhydrogen and forms a "cold" plasma at the bead surface with a high ion yield. It has been found that preferentially N- or P-containing chemicals are converted into ions in an enhanced mode. It is assumed that impinging N- or P-species (when hitting the surface) lower the work function of the silicate surface and thermoelectrons become released. Adversely rated is the short life time of the salt bead.

Electron Capture Detector (ECD) This detector has been invented by Lovelock first in 1948. After several improvements published 1958, it became the most powerful detector for halogenated organic compounds. Its working principle (see Figure 7.50) is based on the presence of a tiny amount of a β-emitting radionuclide, usually ^{63}Ni, inside the detector space. Other sources for β-radiation are given for comparison in Table 7.15.

The electrons emitted possess a kinetic energy in the keV range, thus ionizing all incoming analytes from the column. Therefore, nitrogen as carrier gas with a low attachment coefficient has to be applied. The high-energy electrons collide with the analyte molecules producing fragmentation and electron attachment. These compounds and their fragments, containing electronegative substituents like halogens, nitriles or nitro groups, will become preferentially combined with electrons ("electron capture"). Hence, they cause for a moment of presence a base line current drop. This becomes amplified and recorded.

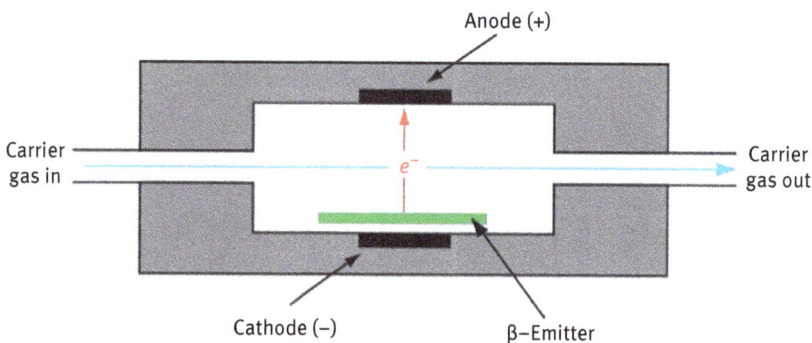

Figure 7.50: Schematic view of an electron capture detector (ECD).

Table 7.15: Properties of various β-emitter sources for ECD.

Properties	^{63}Ni foil	^{3}H Ti foil	^{3}H Sc foil	^{55}Fe on Ni foil
β-Particle energy [keV]	66	18	18	5.387–5.640
β-Particle range [mm]	10	2.5	2.5	0.5
Maximum activity [mCi/cm^2]	10	170		3
Upper temperature limit [$^\circ$C]	350	220	325	400
Maximum current [pA]	9 (15-mCi source)	30 (500-mCi source)		0,5 (5-mCi source)
Rate of electron production [sec^{-1}]	6×10^{10}	2×10^{11}		3×10^9
Noise level [pA]	1.5	3		0.1

Up to now, ECD has its merits for analyzing halogenated volatile substances, nitro-compounds like explosives [19] and organometallic compounds.

Disadvantageous is the need for a license to handle ECDs because of the radioactive components.

Thermal Conductivity Detector (TCD)　The thermal conductivity detector (TCD) can be seen as the only universal detector in GC. Its function is based on heat conductivity of gases. Since all gases, when impinging on the surface of a heated resistor, will change the temperature of it, and consequently the resistivity too, all kind of analytes can be detected. Proportionality to the gas concentration is given. As the heat conductivity is not so much different for different molecules, a rough concentration estimation is also possible.

Classically, it is achieved by a differential resistance measurement (e.g., by Wheatstone bridge circuit) across two resistance-heated filaments (see Figure 7.51).

One arm of the TCD is permanently flushed by a pure gas, for example, He. This constant flow creates a reference resistance in a Wheatstone bridge resistivity

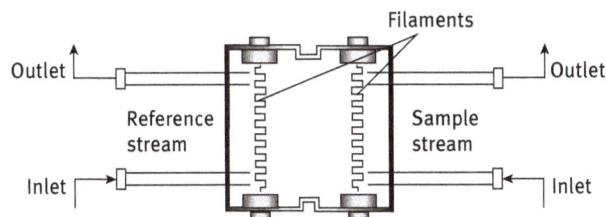

Figure 7.51: Cross section of a thermal conductivity GC detector.

measurement. The other arm receives the carrier gas from the GC column. In case a separated analyte crosses this part, the heat conductivity will change for a moment. Even slightest deviations in heat conductivity are realized by this configuration and become recorded.

The TCD is by far not as sensitive as an FID but applicable to all substances, robust and cheap. It is often used in process analysis. The TCD is also ideal for miniaturization purposes. Many attempts have been published about µTCD fabrication. It is one of the rare nondestructive detectors, hence suitable for preparative GC.

Flame Photometric Detector (FPD)   The flame photometric detector (FPD) has a unique history. It became developed in 1962 by the German Drägerwerk, Lübeck, for the detection of perfluorinated phosphorus esters, known as a part of chemical warfare agents (Tabun, Sarin, Soman).

It makes use of the emission of a chemiluminescence radiation after introduction into a hydrogen flame and spectroscopic observation of the reducing part of the flame. The simplified radical reaction scheme for excitation is

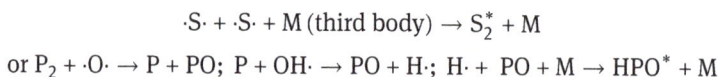

$$\cdot S \cdot + \cdot S \cdot + M \, (\text{third body}) \rightarrow S_2^* + M$$

$$\text{or } P_2 + \cdot O \cdot \rightarrow P + PO; \, P + OH \cdot \rightarrow PO + H \cdot; \, H \cdot + \, PO + M \rightarrow HPO^* + M$$

The observed luminescence is attributed to the excited fragments of HPO* and S_2*. A photomultiplier tube measures the characteristic chemiluminescence emission from these species (see Figures 7.52 and 7.53).

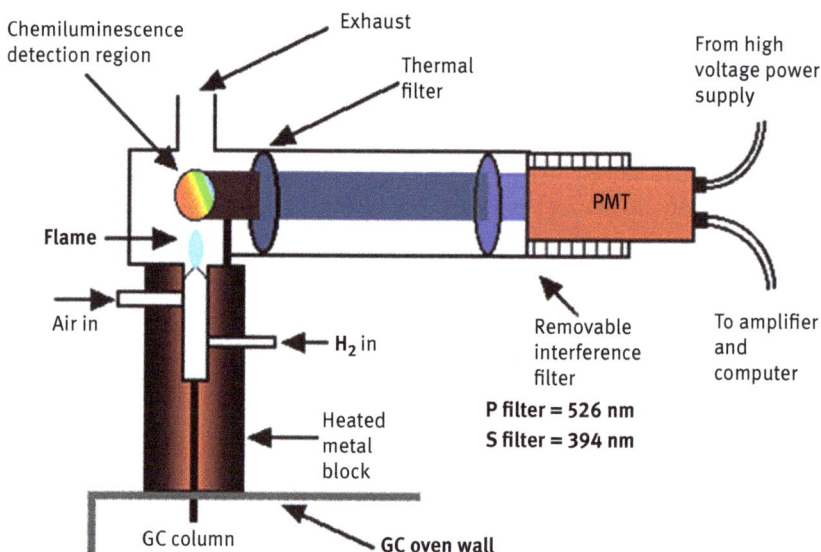

Figure 7.52: Scheme of an FPD.

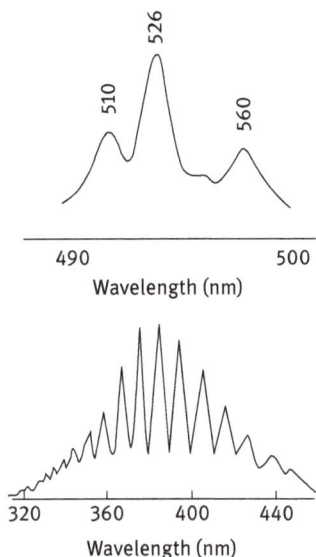

Figure 7.53: Luminescence of S_2^* and HPO^*.

The detector response to phosphorus is linear, whereas the response to sulfur depends on the square of the concentration. The FPD is excellent for sulfur organic compound detection [20]. The use of two parallelly arranged photomultipliers with respective optical filters allows the simultaneous view of light of 394 nm for sulfur measurement and 526 nm for phosphorus (see Figure 7.54).

Together with a P-sensitive channel identification of pesticide traces containing P and S is very easy.

Electrolytic Conductivity Detector (ELCD)   This detector (see Figure 7.55) makes use of the change of electrolyte conductivity within an aqueous solution, which is absorbing the combustion products.

When a halogen-, sulfur- or nitrogen-containing analyte leaves the separation column, it becomes catalytically converted into the equivalent mineral acid. Amines are converted into ammonia, thus increasing the pH. Due to the high conductivity, equivalents of H^+ or OH^--sensitive detection in the ppbv range are possible.

Disadvantageous is the need for a clean hydrogen gas source.

Thermionic Ionization Detector (TID)   This detector is similar to the FID or AFID/NPD by generating ionic fragments, but without flame. The GC effluent meets a very hot surface, where electron-drawing fragments from pyrolysis, especially from nitro groups (nitro-PAHs or explosives) or halogens (highly volatile halogenated hydrocarbons), are formed and mixed with the thermoelectrons emitted from the hot surface. This situation is shown in Figure 7.56.

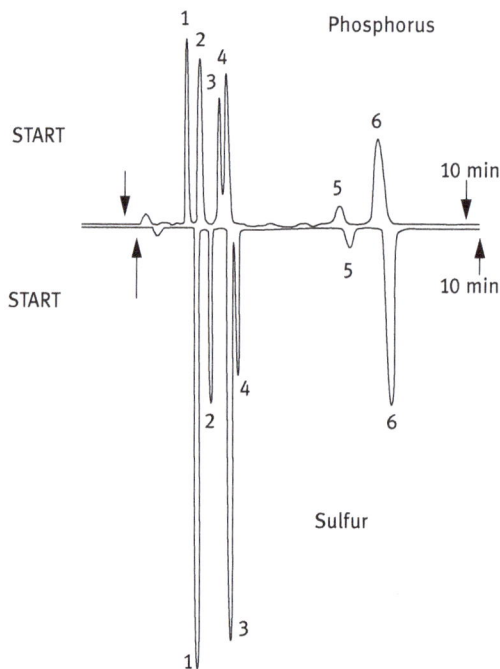

Figure 7.54: Simultaneous detection of phosphorus- and sulfur-containing compounds. 1 = disyston; 2 = methyl parathion; 3 = malathion; 4 = parathion; 5 = methyl trithion and 6 = ethion.

Photoionization Detector (PID) Selective photoionization of gaseous compounds by UV irradiation of the GC effluent is applied in a photoionization detector (see Figure 7.57).

When energy-rich photons encounter organic molecules and become absorbed, photoionization may take place. Depending on the so-called ionization potential (IP), the molecules will lose a photoelectron:

$$R + h\nu \rightarrow R^{+} + e^{-}$$

The released ions are measured by an integrated electrometer. Whether this happens requires not only that the photon energy exceeds IP. The recombination probability for photoelectrons and competition with cell-surface-released photoelectrons sometimes makes quantification difficult. In favorable cases, especially aromatic compounds due to their delocalization potential of formed cationic species can preferentially be detected in the pptv range. Benzene, as an example, has an IP of 9.24 eV. Available Pen-Ray™ UV lamps with a radiation edge of 9.5 eV would selectively ionize benzene, whereas C_{10} alkanes or alcohols (IP > 9.5 eV) present in the same sample would not become ionized.

It should be mentioned here that selective *multiphoton ionization* and coupling with MS as detector opens a way to extremely selective organic trace analysis of

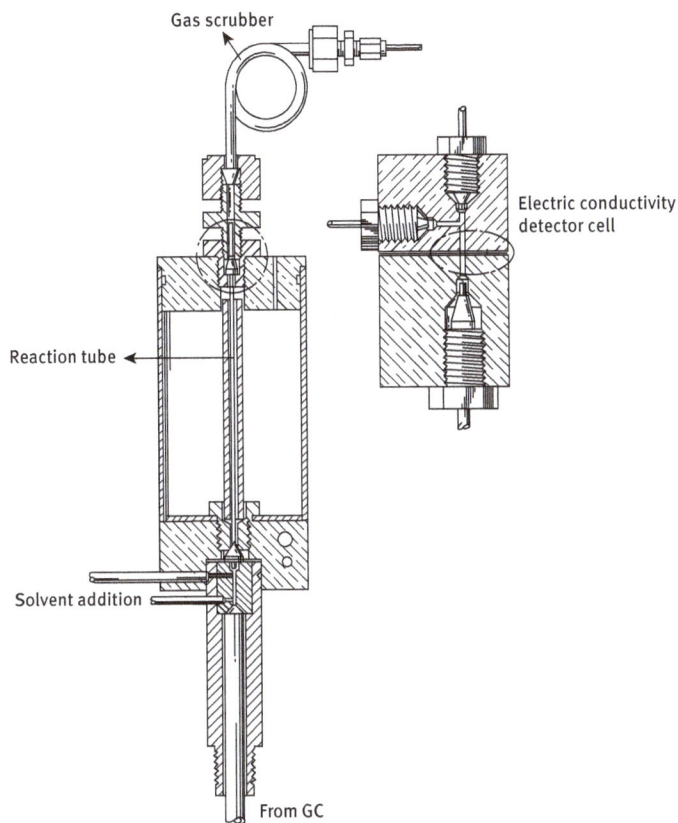

Figure 7.55: Cross section of an electrolytic conductivity detector (ELCD). According to US Patent 4555 383 A, Nov. 26, 1985.

gaseous compounds (See chapter 11.2.3.3). A less expensive version is the combination with *ion-selective mobility* measurement (see chapter 9.4.1).

7.5.9.3 Dedicated GC Detectors
Human Nose as GC Detector   Quantification of aromas, scents or malodors is extremely difficult. But there are many situations where the underlying composition has to be identified and quantified. Usually, an olfactoric panel of several trained persons is used to obtain an impression on odor intensity by comparison of sequential logarithmic dilutions of a gas sample. So, the human nose is used as a detector.

This can be combined with GC separation (see Figures 7.58 and 7.59).

Electroantenna Detector   In the 1950s, entomologists detected that between tip and base of the antenna (olfactory sensilla) of an insect, an electric potential is created when it came into contact with biologically relevant smell (e.g., sex pheromones). The experimental setup is shown in Figure 7.60.

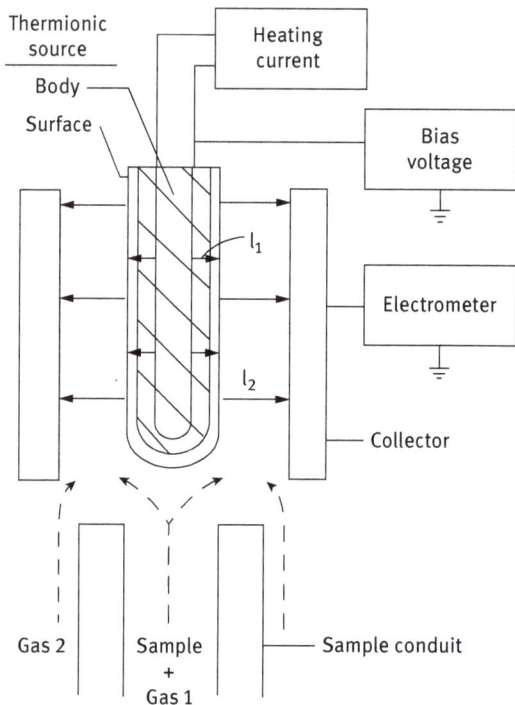

Figure 7.56: Schematic view of a thermionic ionization detector (TID). Carrier gas is air.

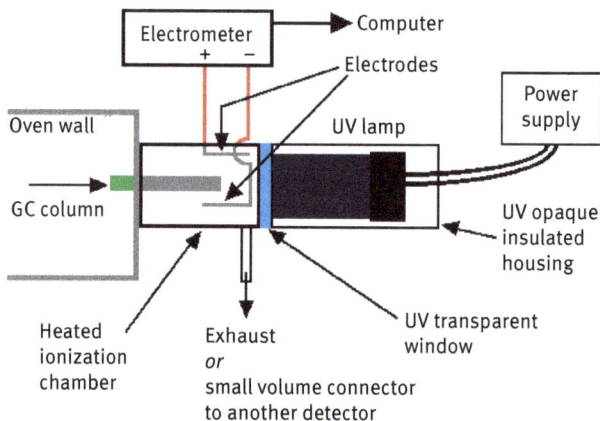

Figure 7.57: Photoionization detector for GC.

Usually, the action potential measured from such excised sensilla is rather noisy and does not show a direct relationship to scent concentration. But recorded in parallel to conventional FID in the same chromatogram, it gives valuable hints on some chemoreceptor functions. An example is given in Figure 7.61. There the smell of a queen bee was analyzed by FID and an electroantenna detector (antennae excised from worker bee !).

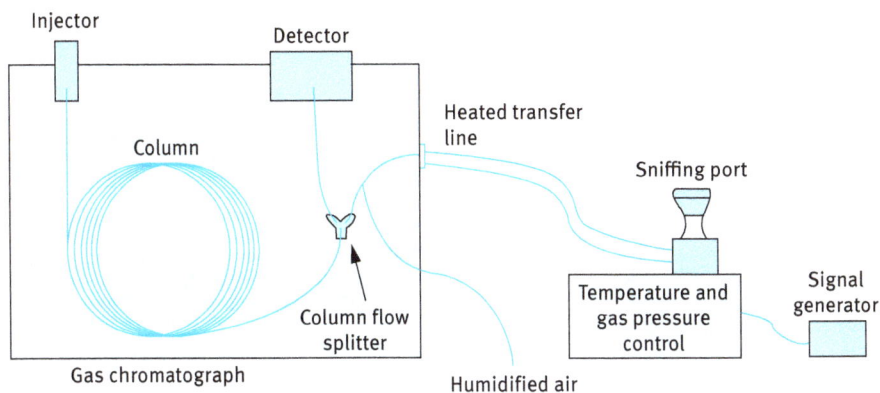

Figure 7.58: Schematic view of a sniffer GC port.

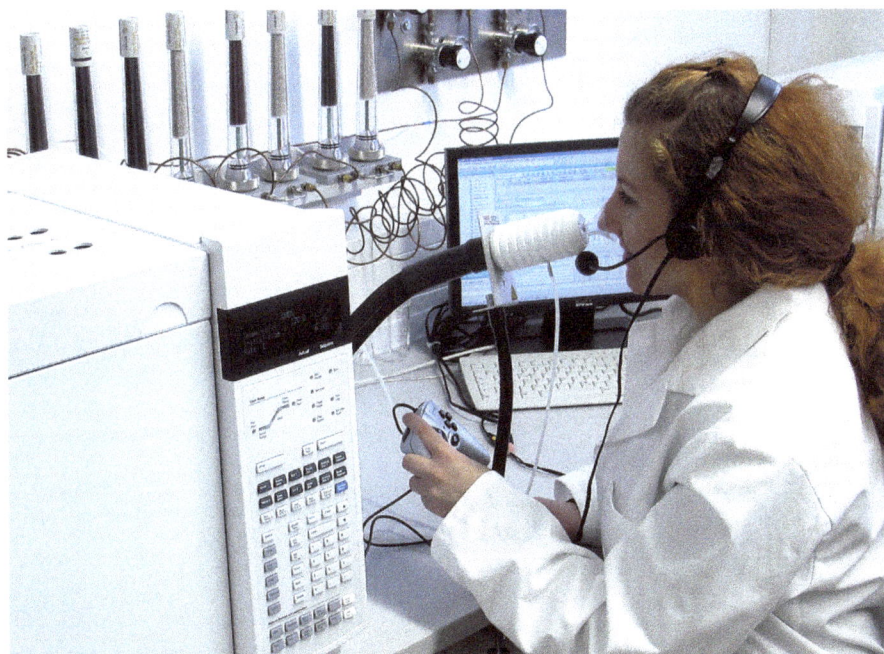

Figure 7.59: Sniffing person acting as detector. Reproduced with permission from Fraunhofer WKI / Manuela Lingnau.

The authors could show that the antennae of the selected worker bee react to this key compound. The compound 3,11-dimethylheptacosane (3,11-diMeC27) is correlated with ovarian activity. Similar numerous studies on the electrophysiological activities of an insect, when confronted with fruit scents, give rise to phantasies for "attractive" perfume development.

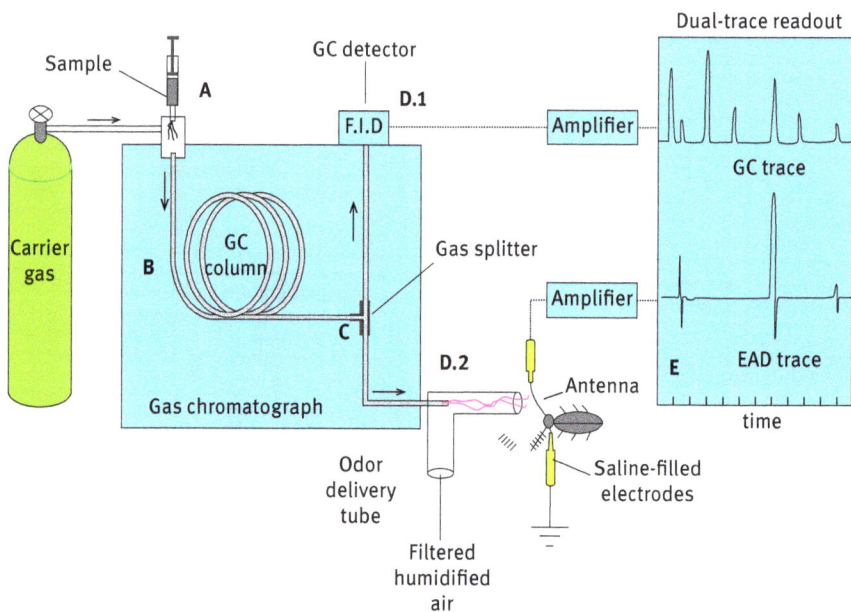

Figure 7.60: Insect's antennae as GC detector.

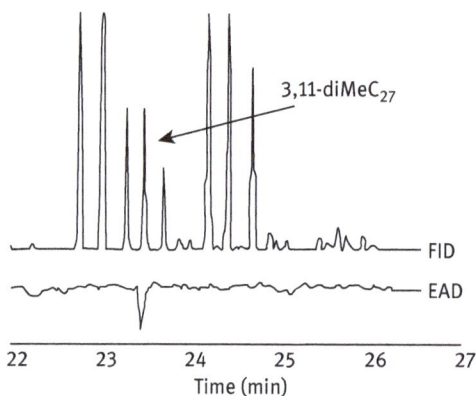

Figure 7.61: Comparison of FID and EAD yields information on electrophysiological activity [21].

7.5.10 Hyphenated Techniques (GC–MS)

7.5.10.1 General Remarks

Hyphenation of GC with other high-performance analytical instrumentation enables the analyst to gain additional knowledge beyond that provided by the commonly used GC detectors. Hirschfeld (1980) was the first who came up with the idea to substitute the "broad-band" sensing detectors by tools delivering a multitude of additional "orthogonal" analytic information. This would help a lot to obtain a much higher reliability for analytically based decisions.

Orthogonality here means the possibility of gaining an additional analytical information value connected to a GC-separated analyte (so far only approved by retention time) through a completely different identification mechanism, for example, by spectroscopy, electrochemical response or MS ion mobility measurement, but not by a subsequent second GC separation. In fact, many disputes about accuracy of analytical results could be resolved if hyphenated techniques instead of low-resolving detectors been applied. This does not exclude the use of multidimensional separations, but one has to be aware of the limited validity in identifying an analyte only by retention time.

The most frequently and highly successfully applied hyphenation is with MS (GC–MS). Upcoming is coupling with ion mobility spectrometry (GC–IMS), since it does not need high vacuum pumping.

The different GC–MS combinations will be presented in Chapter 11, in combination with the different ionization techniques and separation schemes.

In the following, the focus is on the interface between GC exit and MS inlet.

7.5.10.2 GC–MS Interfaces

GC–MS nowadays belongs to routine instrumentation within analytical labs dealing with organic trace analysis. Already in the end of the 1950s, the first coupling of a gas–liquid chromatograph with a mass spectrometer became known. GC–MS in its most common and simplest version is shown in Figure 7.62.

The effluent from the GC is interfaced via a *transfer line* to the high vacuum of a MS system. This causes enormous technical difficulties since the GC is operated with overpressure, whereas the MS needs high vacuum in order to reduce unwanted collisions of analyte ion fragments with residual gas contaminants. By colliding with background molecules, besides the formation of new cluster ions, deflection

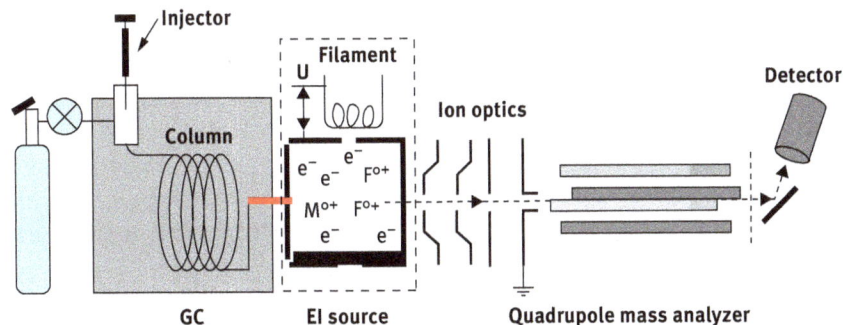

Figure 7.62: GC system and interface to typical mass spectrometric detection.

Table 7.16: Requirements for vacuum inside coupled MS.

Analyzer	Pressure [Torr]
Fourier transform MS	$<10^{-8}$
Magnetic sector field MS	$<10^{-6}$
Time-of-flight MS	$<10^{-6}$
Quadrupole MS	$<10^{-4}$
Ion trap MS	$<10^{-4}$

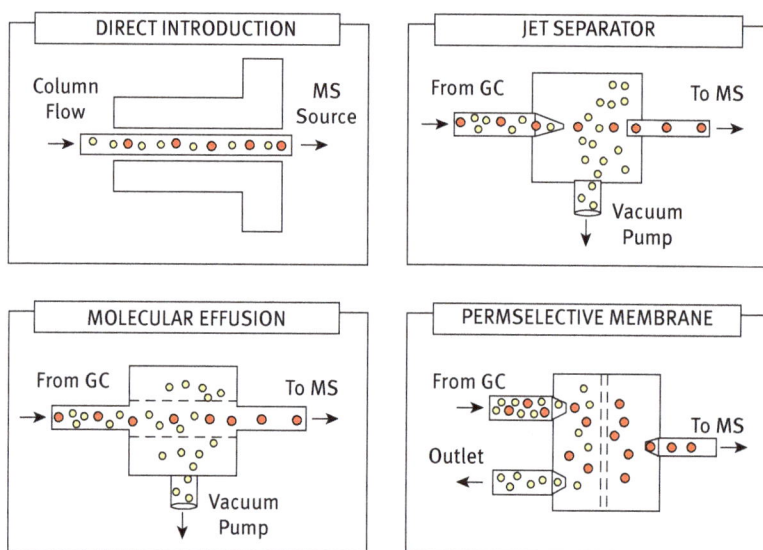

Figure 7.63: Current GC–MS interfaces.

happens or the analyte ions lose their charge, thereby decreasing the sensitivity of the whole determination. In Table 7.16, the required vacuum is compiled for different MS systems.

To meet such requirements, *direct coupling* (see Figure 7.63) would need a very high pumping capacity or the exclusive use of narrow capillary separation columns. Already packed GC columns would introduce an enormous leak to the whole vacuum system with its typical flow rate of 25 mL/min. One way to surpass this at least partially is *open split coupling* of the outlet of a capillary column under helium purging, with the heated transfer line of the MS. By this, even suppressing of unwanted peaks (heart-cut) is possible. Hereby, the pressure is slightly held above atmospheric pressure, so the required pumping capacity is tolerable. Otherwise, this makes the GC–MS system very bulky and heavy. To solve this issue, several other technical solutions are available.

The *jet separator* or *molecular jet interface* (see Figure 7.63) is based on the jet isotope separator of Becker. In this separator, the effluent from the GC column expands through a restrictor capillary into a pre-vacuum space (kept stable by differential pumping), forming a supersonic jet. The smaller molecules of the carrier gas diffuse to the outer part of the jet, whereas the heavier analyte molecules stay in the center of the jet. Therefore, a relative concentration effect occurs. Part of the jet then enters through a skimmer the MS high vacuum stage.

Similarly, transfer is possible by *molecular effusion* (see Figure 7.63). A fritted tube serves as the interface between GC and MS. Around the fritted tube, a slight vacuum serves as a diffusion sink. On time average again due to Brownian motion, the small carrier gas molecules will enter the recipient as sink, thus reducing gas pressure and producing a concentration of heavier molecules within the remaining gas flow entering the MS section. Disadvantageous is a high dead volume which disturbs the chromatographic resolution.

For military purposes (mobile chemical lab in a tank), the *membrane interface* has been developed (Figure 7.63). A permselective membrane (usually prepared of silicone or PTFE) is in contact with the carrier gas of a GC column. Depending on the distribution restriction between gas and polymer, by dissolution of analytes within the polymer and by partitioning between polymer and vacuum, a considerable selection of analytes is achieved. The kinetic limitations make the application of elevated temperatures unavoidable. No vacuum pumping line is needed. A drawback can be memory effects within the membrane.

Table 7.17 gives a comparison of the above mentioned interfaces.

Selection depends clearly on the analytical task. With upcoming miniaturized GC systems and capillary columns in the nm-size, directly coupled MS will be favored in the future.

7.5.11 Multidimensional GC Separations (GC × GC)

7.5.11.1 General Remarks
Currently, the *peak capacity*, which means how many analytes may be resolved within one GC run, is typically within the range of 500. For analytical tasks from the field

Table 7.17: Performance of various GC–MS interfaces.

Interface	Transfer yield [%]	Enrichment [−]	Delay [s]	Carrier gas
Ideal interface	100	∞	0	All
Jet separator	40–70	100	None	All
Membrane	30–95	1,000	Variable	Inorganic
Molecular effusion	20–50	100	1	He, H_2
Direct coupling	100	1	None	All
Open split	<100	1	None	All

of -omics or environment/toxicology/petroleum analysis (metabolites, combustion products, humic substances, waxes, etc.), this displays a serious limitation with the consequence of starting a sequence of different separations. Otherwise, the whole entity of a sample would never be fully characterized. The problem is co-elution of a multitude of isomeric substances or substances different in composition but of the same or similar partition behavior on column. One way is the hyphenation with an orthogonal detection system and high resolving power based on a completely different detection mechanism (as by GC–MS).

7.5.11.2 GC–GC Multidimensional Separation by Heart-Cut

A different option is the serial application of a second separation column but differently sized and operated. One way to achieve a further separation is to cut out (heart-cut) an eluate volume from the first column by multi-port valve switching and subjecting this to a second separation on a differently coated second column. By using an open split interface (see Section 7.5.10.2), part of the split effluent (after the first column) can be recorded by a FID, while most of the effluent becomes now separated under a different temperature and possibly a different flow rate in the second column. This situation is depicted in Figure 7.64.

A certain advantage of this configuration is the possibility to have a full exploitation of the resolving power of the second column.

7.5.11.3 GC × GC Comprehensive Multidimensional Separation by Modulation

The nowadays preferred approach is *modulation* (GC×GC) of the transfer line between two separation columns, for example, by cryogenic treatment, thus performing within ms completely different partitions. The arrangement of the two columns is shown in Figure 7.65.

The peak capacity of the two columns becomes roughly multiplied (total HETP number: N_1 of 1st dimension × N_2 of the 2nd dimension). The first column may consist of a nonpolar one and is operated under best resolving condition like long separation time, whereas the second is a short polar column, where a run is completed within several seconds. The best separations by such arrangement are obtained when a pair

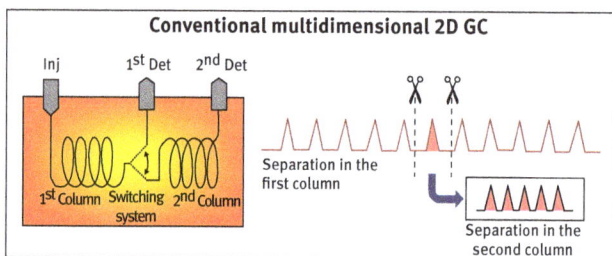

Figure 7.64: Multidimensional GC separation by column switching (GC-GC). Image used courtesy of Thermo Fisher Scientific; copying prohibited.

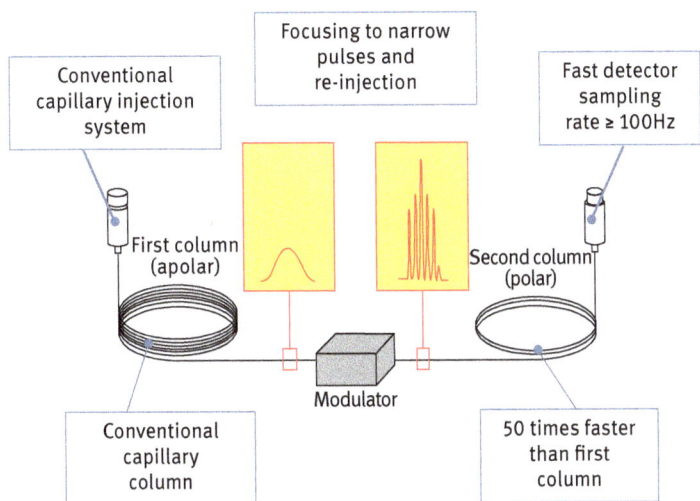

Figure 7.65: Scheme for comprehensive multidimensional GC × GC analysis system.

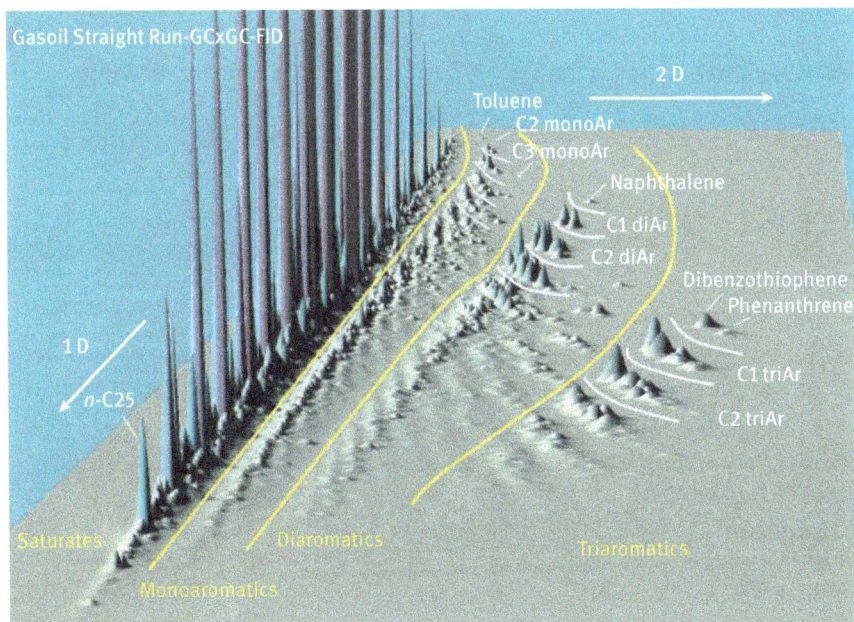

Figure 7.66: GC×GC FID chromatogram of gasoil. Image used courtesy of Thermo Fisher Scientific; copying prohibited.

of analytes A and B is coeluted in the first, but becomes separated by the second column.

The modulator, for example, using a cycled cooling gas nozzle (fed by CO_2 or liquid N_2), is directed against the connecting transfer line and traps (immobilizes) all the components eluting from the first dimension. After a fixed time interval, a hot gas stream pulse is mobilizing (injects) a part of the immobilized compounds again into the second dimension column. The temperatures may cycle between −189 and +475 °C. This also causes a certain refocusing of the separation from the first dimension. Through this, a very high compression of the peaks (peak width <100 ms) is achieved, which in turns asks for very fast detection, usually performed by MS, FID or ECD.

Often claimed is a certain orthogonality of separations in GC × GC, since one separation is dominated by partitioning, in contrast to volatility of the second dimension. But these parameters aren't completely independent of each other. Only detection by MS or other separation-independent observation makes it finally orthogonal.

A typical multidimensional GC × GC FID chromatogram of a gasoil analysis shows the potency of this technique (Figure 7.66).

It seems clear that such combinations can be even more powerful when the detection part is configured also in a multidimensional approach (e.g., by applying wavelength-dependent laser ionization MS; see Chapter 11 for dioxins and PCBs).

Further Reading

Bartle K, Myers P. History of gas chromatography. Trends Anal Chem 2002;21:547–57.

Bhattacharyya L, Rohrer JS. Applications of ion chromatography for pharmaceutical and biological products. Hoboken, NJ: John Wiley & Sons, Inc., 2012.

Blau K, King G. Handbook of derivatives for chromatography. London: Heyden & Sons Ltd., 1979.

Calderon, L. Chromatography-the most versatile method of chemical analysis. Rijeka: InTech, 2012.

Cecchi T. Ion-pair chromatography and related techniques. Boca Raton: CRC Press, 2009.

Corradini D, Phillips TM. Handbook of HPLC. Boca Raton: CRC Press, 2011.

Deng P, Zhan Y, Chen X, Zhong D. Derivatization methods for quantitative bioanalysis by LC-MS/MS. Bioanalysis 20124(1):49–69.

Escrig-Domenech A, Simo-Alfonso EF, Herrero-Martinez JM, Ramis-Ramos G. Derivatization of hydroxyl functional groups for liquid chromatography and capillary electroseparation. J Chromatogr A 2013;1296:140–56.

Fanali S, Haddad PR, Poole C. Liquid chromatography: applications. Waltham: Elsevier Science Publishing Co Inc., 2013.

Fanali S, Haddad PR, Poole CF, Schoenmakers P, Lloyd D. Liquid chromatography: fundamentals and instrumentation. Waltham: Elsevier Inc., 2013.

Gonella NC. LC-NMR: expanding the limits of structure elucidation. Chromatographic science series (Book 105). Boca Raton: CRC Press, 2013.

Grant D. Capillary gas chromatography. New York: Wiley, 1996.

Grob K, Grob G, Grob K. Comprehensive, standardized quality test for glass capillary columns. J Chromatogr 1978;156:1–20.

Grob K. Split and splitless injection in capillary GC. Heidelberg: Hüthig, 2000.

Günther H. NMR spectroscopy: basic principles, concepts, and applications in chemistry. New York: Wiley-VCH, 2013:734.

Guo X, Lankmayr E. Hyphenated techniques in gas chromatography. In Mohd MA, editor. Advanced gas chromatography – progress in agricultural, biomedical and industrial applications. Rijeka: InTech, 2012:3–26.

Hage DS. In Corradini D, Katz E, Eksteen R, Schoenmakers P, Miller N, Affinity chromatography: editors Handbook of HPLC. New York: Marcel Dekker, 1998;361–384.

Handley A, Adlard E. Gas chromatographic techniques and applications. Sheffield: Sheffield Academic Press, 2002.

Hubschmann H-J. Handbook of GC/MS: fundamentals and applications. Weinheim: Wiley-VCH, 2000.

Knapp D. Handbook of analytical derivatization reactions. New York: John Wiley and Sons, 1979.

Kolb B, Ettre L. Static headspace – gas chromatography: theory and practice. New York: Wiley – VCH, 1997.

Komsta L, Waksmundzka-Hajnos M, Sherma J. Thin layer chromatography in drug analysis. Boca Raton: CRC Press, 2014.

Lundanes E, Reubsaet L, Greibrokk T. Chromatography: basic principles, sample preparations and related methods. Hoboken, NJ: Wiley, 2013.

Magdeldin S. Affinity chromatography. Rijeka: InTech, 2012.

McFadden W. Techniques of combined gas chromatography/mass spectrometry: applications in organic analysis. New York: Wiley-Interscience, 1973.

McMaster M, McMaster, C. GC/MS: a practical user's guide. New York: Wiley-VCH, 1998.

McNair H, Miller J. Basic gas chromatography. New York: Wiley, 2009.

Medvedovici A, Farca A, David V. Derivatization reactions in liquid chromatography for drug assaying in biological fluids. In Grushka E, Grinberg N, editors. Advances in chromatography, vol. 47. Boca Raton: CRC Press, 2009.

Meyer V. Practical high-performance liquid chromatography. Chichester: John Wiley & Sons, 2010.

Mondello L, Keith L. Multidimensional chromatography. Chichester: Wiley, 2004.

Poole CF. Gas chromatography. Oxford: Elsevier Ltd., 2012.

Ramos F. Liquid chromatography: principles, technology and applications. Hauppauge, NY: Nova Science Publishers, Inc., 2013.

Rood D. A practical guide to the care, maintenance, and troubleshooting of capillary gas chromatographic systems. Heidelberg: Hüthig, 1991.

Roth M, Uebelhart D. Liquid chromatography with fluorescence detection in the analysis of biological fluids. Anal Lett 2000;33(12):2353–72.

Schomburg G. Gas chromatography. A practical course. Weinheim: VCH, 1991.

Snyder LR, Kirkland JK, Dolan JW. Introduction to modern liquid chromatography. Hoboken, NY: John Wiley & Sons, 2010.

van Deemter JJ, Zuiderweg FJ, Klinkenberg A. Longitudinal diffusion and resistance to mass transfer as causes of non ideality in chromatography. Chem Eng Sci 1956;5:271–89.

Wang PG, He W. Hydrophilic interaction liquid chromatography (HILIC) and advanced applications. (in Chromatogr. Sci. Ser., 103). Boca Raton: CRC Press, 2011.

Xu QA. Ultra-high performance liquid chromatography and its application. Hoboken, NJ: John Wiley & Sons, Inc., 2013.

Zerbe O, Jurt S. Applied NMR spectroscopy for chemists and life scientists. Weinheim: Wiley-VCH Verlag GmbH & Co. KGaA, 2013.

Zuo Y. High-performance liquid chromatography (HPLC): principles, practices and procedures. Hauppauge, NY: Nova Science Publishers, Inc., 2014.

Bibliography

[1] Zheng XW, Li Z, Beeram S, Podariu M, Matsuda R, Pfaunmiller EL, White CI, Carter N, Hage DS. Analysis of biomolecular interactions using affinity microcolumns: a review. J Chromatogr B-Anal Technol Biomed Life Sci 2014;968:49–63.

[2] Zhu Z, Liu GH, Chen YH, Cheng JQ. Assessment of aflatoxins in pigmented rice using a validated immunoaffinity column method with fluorescence HPLC. J Food Compos Anal 2013;31(2):252–8.

[3] Abian J. The coupling of gas and liquid chromatography with mass spectrometry. J Mass Spectrom 1999;34(3):157–68.

[4] Adlard E. Review of detectors for gas chromatography. I. Universal detectors Quick View Other Sources. Crit Rev Anal Chem 1975;5:1–11.

[5] Adlard, E. Review of detectors for gas chromatography. II. Selective detectors. Crit Rev Anal Chem 1975;5:13–36.

[6] Han SY, Liang C, Qiao JQ, Lian HZ, Ge X, Chen HY. A novel evaluation method for extrapolated retention factor in determination of n-octanol/water partition coefficient of halogenated organic pollutants by reversed-phase high performance liquid chromatography. Anal Chim Acta 2012;713:130–5.

[7] Yan MQ, Korshin G, Wang DS, Cai ZX. Characterization of dissolved organic matter using high-performance liquid chromatography (HPLC)-size exclusion chromatography (SEC) with a multiple wavelength absorbance detector. Chemosphere 2012;87(8):879–85.

[8] Warton B, Heitz A, Allpike B, Kagi R. Size-exclusion chromatography with organic carbon detection using a mass spectrometer. J Chromatogr A 2008;1207(1-2):186–9.

[9] Chester TL. Recent developments in high-performance liquid chromatography stationary phases. Anal Chem 2013;85(2):579–89.

[10] Sakuma H, Kamata Y, Sugita-Konishi Y, Kawakami H. Method for determination of aflatoxin M-1 in cheese and butter by HPLC using an immunoaffinity column. Food Hygiene Saf Sci 2011;52(4):220–5.

[11] Wang C, Zhu LZ, Zhang CL. A new speciation scheme of soil polycyclic aromatic hydrocarbons for risk assessment. J Soil Sediment 2015;15(5):1139–49.

[12] Schmid T, Panne U, Niessner R, Haisch C. Optical absorption measurements of opaque liquid samples by pulsed laser photoacoustic spectroscopy. Anal Chem 2009;81(6):2403–9.

[13] Farthing C, Halquist M, Sweet DH. A Simple high-performance liquid chromatographic method for the simultaneous determination of monoamine neurotransmitters and relative metabolites with application in mouse brain tissue. J Liq Chromatogr Rel Technol 2015;38(12):1173–8.

[14] Jones A, Pravadali-Cekic S, Dennis GR, Shalliker RA. Post column derivatisation analyses review. Is post-column derivatisation incompatible with modern HPLC columns? Anal Chim Acta 2015;889:58–70.

[15] Alves EG, Sartori L, Silva LM, Silva BF, Fadini PS, Soong R, Simpson A, Ferreira AG. Non-targeted analyses of organic compounds in urban wastewater. Magn Reson Chem 2015;53(9):704–10.

[16] Godejohann M, Heintz L, Daolio C, Berset JD, Muff D. Comprehensive non-targeted analysis of contaminated groundwater of a former ammunition destruction site using 1H-NMR and HPLC-SPE-NMR/TOF-MS. Environ Sci Technol 2009;43(18):7055–61.

[17] Tranchida PQ, Sciarrone D, Dugo P, Mondello L. Heart-cutting multidimensional gas chromatography: a review of recent evolution, applications, and future prospects. Anal Chim Acta 2012;716:66–75.

[18] Koh IO, Wolfgang RB, Thiemann WH. Analysis of chlorinated paraffins in cutting fluids and sealing materials by carbon skeleton reaction gas chromatography. Chemosphere 2002;47(2):219–27.

[19] Walsh ME. Determination of nitroaromatic, nitramine, and nitrate ester explosives in soil by gas chromatography and an electron capture detector. Talanta 2001;54(3):427–38.

[20] Brody SS, Chaney JE. Flame photometric detector – the application of a specific detector for phosphorus and for sulfur compounds – sensitive to subnanogram quantities. J Gas Chromatogr 1966;4(2):42–46.

[21] D'Ettorre P, Heinze E, Schulz C, Francke W, Ayasse M. Does she smell like a queen? Chemoreception of a cuticular hydrocarbon signal in the ant Pachycondyla inversa. J Exp Biol 2004;207(7):1085–91.

8 Capillary Electrophoresis (CE)

8.1 General Remarks

In capillary electrophoresis (CE), an electric field is used to separate components of a mixture dissolved in an electrolytic buffer solvent which is flowing through a narrow tube, a capillary of sub-mm diameter, typically 25–75 µm. When an electrical field is applied, charged molecules, cations and anions move toward the electrode of opposing charge, that is, cations toward the cathode and anions toward the anode. In CE, even uncharged molecules can be analyzed due to the electroosmotic flow of the buffer system (see below).

Many types of CE have been developed which will briefly be described in the following: capillary zone electrophoresis (CZE), capillary gel electrophoresis (CGE), capillary isoelectric focusing (CIEF), capillary isotachophoresis (CITP) and micellar electrokinetic chromatography.

The advantage of using capillaries in electrophoretic separations is that high electric voltages (10–30 kV) and electrical fields (100–500 V/cm) can be applied with only little heat generation because the high resistance of the capillary limits the current generation; in addition, the large surface-to-volume ratio of the capillary efficiently helps dissipating the generated heat. Such high electrical fields lead to short analysis times. The high resolution in CE separations with typically more than 10^5 theoretical plates is due in part to the electroosmotic plug flow (see below) which also enables the simultaneous analysis of all charged and uncharged solutes. Only small amounts of samples are needed for CE (1–50 nL), but with high enough concentrations depending on the detection method applied which will be presented below. CE can be applied for separations of all kind of analytes, such as amino acids, drugs, vitamins, pesticides, inorganic ions, organic acids, dyes, surfactants, peptides and proteins, carbohydrates and oligonucleotides.

8.2 Working Principles

8.2.1 Interaction of Charged Analyte Species with an External Electric Field

In CE, analytes move through electrolyte solutions under the influence of an electric field. A basic scheme of a CE system is shown in Figure 8.1.

The ends of a narrow-fused silica capillary are placed in buffer reservoirs which contain the electrodes providing the electrical contact between a high-voltage power supply and the capillary. The sample is loaded onto the capillary by replacing one of the reservoirs (usually at the anode) with the sample reservoir and applying either an electric field or an external pressure. After replacing the buffer reservoir, the electric field is applied and separation starts. Optical detection can be made at the

DOI 10.1515/9783110441154-008

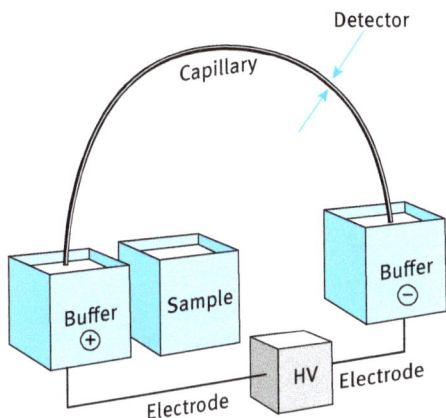

Figure 8.1: Schematic view of a capillary electrophoresis system (HV, high-voltage power supply).

opposite end, directly through the capillary wall; however, other detection modes are available (see below).

The velocity v of an ion in a given electric field E is a function of its electrophoretic mobility μ_e:

$$v = \mu_e \times E \tag{8.1}$$

μ_e is given by

$$\mu_e = \frac{q}{6\pi\eta r} \tag{8.2}$$

with q the charge of the ion, η the viscosity of the electrolyte buffer and r the Stokes radius of the ion. Therefore, small, highly charged species have high mobilities in the electric field, whereas large species of lower charge have lower mobilities.

8.2.2 Electroendosmotic Flow

Electroendosmosis is a fundamental property of CE, resulting in a bulk flow of the electrolyte buffer if an external electric field is applied. This phenomenon is a consequence of the surface charge on the wall of the capillaries used in CE. When using fused silica, the ionizable silanol groups in contact with the buffer in the capillary have an isoelectric point of about 1.5, and the degree of ionization is directly controlled by the pH of the buffer (Figure 8.2).

The negatively charged wall will attract cations that are hydrated from the electrolyte solution, creating an electrical double layer maintaining a charge balance. The formation of a double layer creates a potential difference very close to the wall, known as the zeta potential. In an electrical field, the cations will migrate toward the cathode, dragging water molecules along and creating the EOF. The zeta potential increases with the density of the charge on the surface. For fused silica and many other

Capillary wall

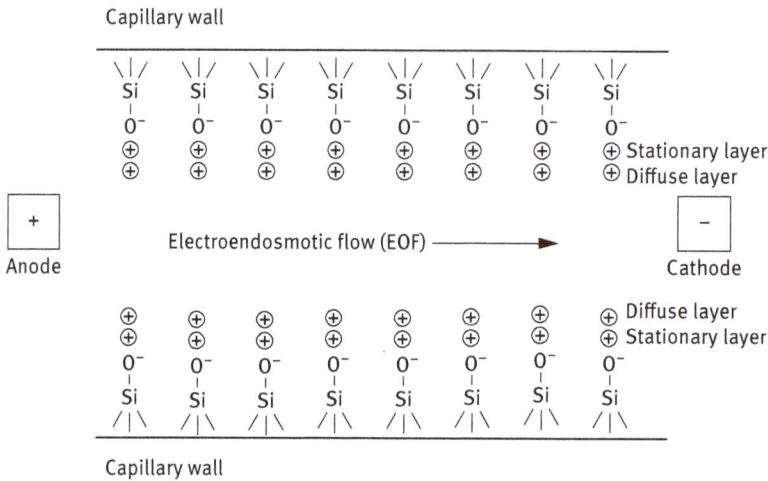

Figure 8.2: Deprotonation of silanyl–OH groups at the surface in fused silica capillaries (pK_a 6.25) and the effect on the electroendosmotic flow (EOF) when using fused silica as capillary material. The charge density varies with pH (with increasing pH more deprotonated Si–OH groups). Other materials are also used for CE capillaries.

materials, charge density will vary with pH. The zeta potential is related to the inverse of the charge per unit surface area and the square root concentration of the electrolyte, that is, increasing the ionic strength of the electrolyte will decrease the EOF (Figure 8.4). The velocity of EOF (v_{EOF}) in a capillary is given by the following equation:

$$v_{EOF} = \frac{\varepsilon \xi}{4\pi \eta} E \tag{8.3}$$

with ε the dielectric constant of the electrolyte, ξ the zeta potential (V), η the viscosity of the electrolyte (P) and E the potential applied (V/cm). The EOF results in a net flow of buffer solution in the direction of the negative electrode with a velocity around 2 mm/s at pH 9 in 20 mM borate buffer. For a 50-mm inner diameter capillary, this equals a volume flow of about 4 nL/s. At pH 3, the EOF is much lower, about 0.5 nL/s.

The mobility μ_{EOF} of the endosmotic flow is independent on the potential E applied:

$$\mu_{EOF} = \frac{\varepsilon}{4\pi \eta} \tag{8.4}$$

and varies depending on the pH of the electrolyte buffer (Figure 8.3).

An advantage of the EOF is that all species – cations, neutrals and anions – are moving in the same direction. If the capillary surface is negatively charged, flow of the species is toward the cathode. Anions will also be dragged toward the negatively charged electrode, the cathode, since the magnitude of EOF is usually greater than any

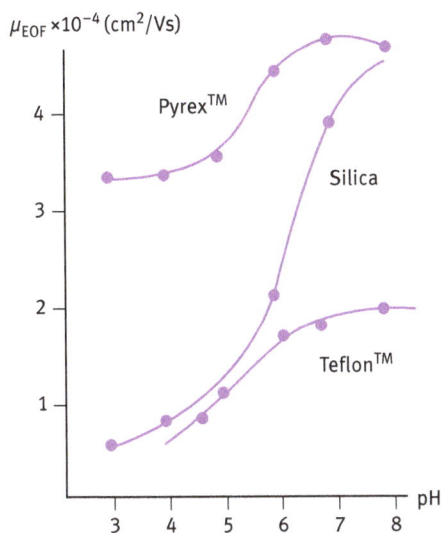

Figure 8.3: Dependence of the electroendosmotic flow (EOF) with pH of the electrolyte for different capillary materials (Agilent).

electrophoretic mobility. Cations migrate fastest since the vectors of electrophoretic attraction toward the cathode and of EOF are aligned. Neutral species are all carried at the velocity of the EOF but are not separated because of the missing interaction with the electric field.

8.2.3 Control of EOF

If EOF is too strong, its dragging influence may overlay the movement of ions due to their charge interactions with the electric field. As a consequence, separation, which is provided only by electrostatic interactions, may be hampered. Therefore, EOF needs to be controlled. Reduction of the electric potential in order to lower the EOF has the disadvantage that analysis time, efficiency and resolution may be negatively affected. At increased potential, Joule heating will be increased. Alternatively, the pH can be changed (Figure 8.3) which however will also affect the solute charge and the mobility: At low pH, the capillary wall will be progressively protonated, that is, not charged, and this may be true also for the analyte, whereas at high pH, both will be deprotonated.

Another parameter controlling the magnitude of EOF is the ionic strength. Increased ionic strength results in double-layer compression, decreased zeta potential and reduced EOF (Figure 8.4). High ionic strength will generate high current but increase Joule heating. In addition, the peak shape of the signal may be distorted if the ionic strength of the electrolyte differs from that of the injected sample.

In order to reduce sorption interactions of the analyte with the capillary wall, it is possible to modify the surface by coatings, either permanently by covalent binding or physically adhered phases or transiently by additives in the electrolyte buffer. Surface modification can reduce or eliminate or even reverse EOF. Covalent modification can

μ_{EOF} ($\times 10^{-4}$ (cm^2/Vs)

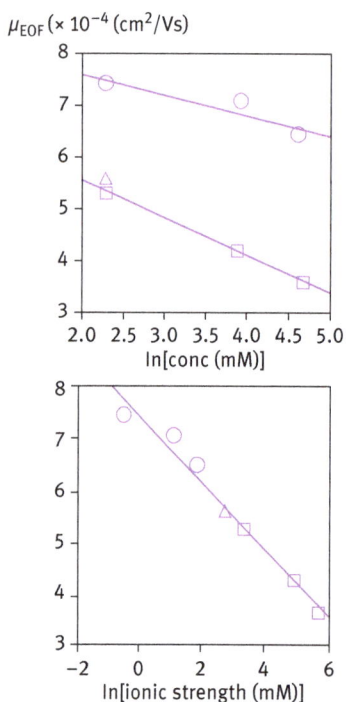

Figure 8.4: Dependence of EOF on the concentration (top) and ionic strength (bottom) of borate buffer (circles), phosphate buffer (squares) and carbonate buffer (triangles) at pH 8 (Agilent).

be obtained by silylation (with various organic silylation agents), polymers like cellulose or polyethyleneimine or the surface modifiers used for the stationary phases in gas chromatography or high-performance liquid chromatography (HPLC). However, many coatings suffer from the limited stability. Buffer additives are for instance hydrophilic polymers such as polyvinyl alcohols or dextrans, surfactants with anionic, cationic, nonionic or zwitterionic properties or quaternary amines acting as ion pairs.

8.2.4 Flow Dynamics, Efficiency and Resolution

The flow profile in CE is different from other separation techniques which are driven by pumping a solvent. In the narrow capillary, the liquid is uniformly moving which results in a "plug flow," unlike the laminary flow profile in pumped systems (Figure 8.5). In CE, the driving force of the flow is uniformly distributed along the capillary, and there is no pressure drop within the capillary. The advantage of plug flow is the minimal broadening of the solute zone leading to extremely sharp signals.

The separation efficiency is, like in other separation techniques, given by the theoretical plate numbers N which in CE equals

$$N = \frac{\mu_e Vl}{2DL} = \frac{\mu_e El}{2D} \tag{8.5}$$

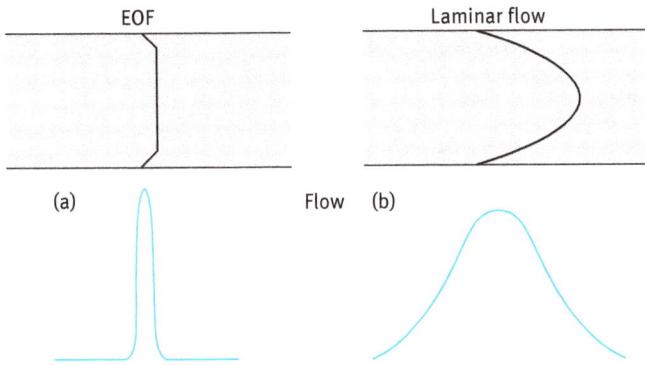

Figure 8.5: Plug flow in CE (a) and laminary flow in pumped systems (b). On the bottom, the solute zones indicate minimal band broadening in CE compared to pumped systems as for instance in HPLC analyses.

with μ_e the electrophoretic mobility of the analyte (see eq. (8.2)), V the voltage, l the effective length of the capillary (cm), D the diffusion coefficient of the analyte (cm^2/s), L the total length of the capillary and E the electric field (V/cm).

Equation (8.5) makes it clear that high voltages (up to about 30 kV) are advantageous since the ionic analyte will move more rapidly, and there is less time to diffuse compared to longer travel times. The problem of Joule heating at high electric field strengths has been discussed. N can also be derived from the migration time t and the half-height width $w_{1/2}$ of an idealized (Gaussian) peak from a CE electropherogram by

$$N = 5.54 \left(\frac{t}{w_{1/2}} \right)$$
(8.6)

Peaks can be distorted if the conductivities (ionic strengths) of the sample zone and the electrolyte do not match. When the sample containing the analyte has a higher mobility, that is, a higher conductivity and lower resistance, than that of the running buffer, the front of the solute experiences a higher voltage drop when entering the buffer zone. This causes the solute, for instance anions if the EOF is toward the cathode, to accelerate away from the sample zone and results in fronting of the solute zone, that is, distortion of the peak. Neutral species are unaffected by such conductivity differences.

The efficiency in separation is affected by many factors, that is, temperature gradients (due to Joule heating), the injection plug length (the length of the injected sample volume in the capillary; it should not be more than about 1–2% of the capillary length) and sorption interactions of the analyte with the capillary wall to name the most important ones.

The resolution R in the separation of two analytes with the migration times t_1 and t_2 can be calculated (eq. (8.7)) by the baseline peak widths w_1 and w_2 or the peak widths at 13.5% of the peak height, that is, covering the data with four times the standard deviation (4s, when assuming a Gaussian peak shape):

$$R = \frac{2(t_2 - t_1)}{W_1 + W_2} = \frac{t_2 - t_1}{4s} \tag{8.7}$$

Another expression of R is

$$R = \frac{1}{4}\left(\frac{\Delta\mu_e \sqrt{N}}{\bar{\mu}}\right) \tag{8.8}$$

with $\Delta\mu_e = \mu_{e2} - \mu_{e1}$, that is, the difference in the electrophoretic mobilities (eq. (8.2)) of two analytes, $\bar{\mu} = 1/2\,(\mu_{e1} + \mu_{e2})$, that is, their average electrophoretic mobility and N is the number of theoretical plates (eq. (8.5)). This can be translated to the final equation:

$$R = \frac{1}{4\sqrt{2}}\Delta\mu\left(\frac{V}{D\,(\bar{\mu} + \mu_{EOF})}\right) \tag{8.9}$$

where V is the voltage, D is the diffusion coefficient and μ_{EOF} is the mobility of the EOF (eq. (8.4)). This equation means that resolution is highest when the average mobilities of the two analytes and that of the EOF are similar but opposite in direction. In this case, however, the analysis time will increase and a compromise has to be found to balance resolution and the time needed for analysis.

8.3 Modes of CE Operation

Several types of capillary electroseparation methods have been developed: We will describe five methods, namely CZE, CGE, CIEF, CITP and micellar electrokinetic capillary chromatography (MECC). They can be classified as shown in Figure 8.6. Methods in continuous systems, in which one electrolyte is in the capillary, comprise CZE, MECC and CGE, if the composition of the electrolyte remains constant, and CIEF, if the composition varies within the capillary. An alternative method, CITP, is based on the migration of analytes in zones that are separated by two different electrolytes. Two-dimensional CE methods combine two of the separation methods, for instance CZE and MECC, but require specific interfacing [1].

CE analysis has been used in many disciplines such as forensic, biological, medical, biotechnological and environmental fields. Nowadays, even portable systems are available which can directly be used in the field [2].

8.3.1 CZE

CZE is the most frequently used method. Its principle is based on the charge and radius of the ionic analyte, the viscosity of the electrolyte and the electric field as described in eqs. (8.1) and (8.2), including the contribution of the EOF (eq. (8.4), Figure 8.2). Separations of both large and small molecules can be accomplished by CZE. The analytes migrate in discrete zones and at different velocities.

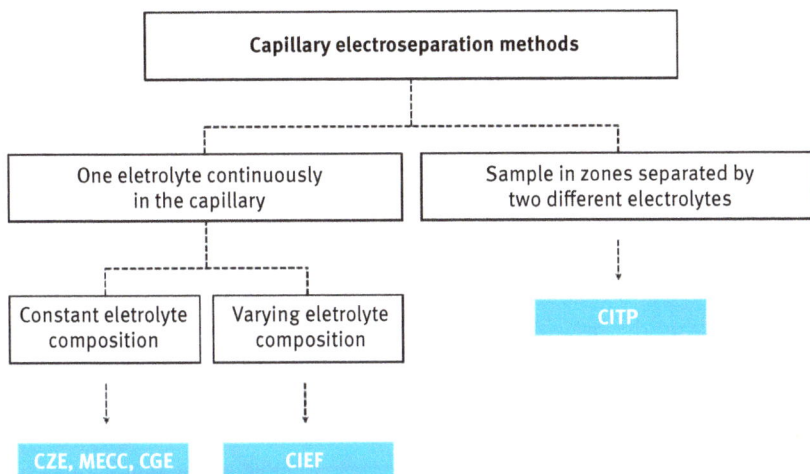

Figure 8.6: Classification of CE methods.

The impact of pH on the analyte and on EOF, as discussed, is critical. As a rule of thumb, a pH at least two units above or below the pK_a of the analyte should be chosen to ensure substantial ionization. At highly alkaline pH, EOF becomes so rapid that the electrophoretic separation efficiency of charged analytes will drop or even cease (if elution occurs before, resolution is achieved), but the principle order of migration in the direction toward the cathode will be: cations with the largest charge-to-mass ratios, then cations with smaller ratios, after that neutral species and later anions with smaller charge-to-mass ratios followed lastly by anions with greater ratios.

Selectivity can be altered through changes in the pH of the electrolyte buffer or by use of buffer additives. Suitable buffers are phosphate (pH ranges 1.1–3.1 and 8.1–10.1), acetate (pH 3.8–5.8), borate (pH 8.1–10.1) and organic zwitterionic substances such as 2-(N-morpholino)ethanesulfonic acid (MES) (pH 5.2–7.2), piperazine-N,N'-bis(2-ethanesulfonic acid) (PIPES) (pH 5.8–7.8), 4-(2-hydroxyethyl)-1-piperazineethanesulfonic acid) (HEPES) (pH 6.6–8.6) or 2-amino-2-(hydroxymethyl)propane-1,3-diol (Tris) (pH 7.3–9.3). The organic buffers can be used at rather high concentrations without generating significant currents and Joule heating; however, their high UV absorptivity has to be taken into account.

8.3.2 Capillary Gel Electrophoresis

CGE separation is based on the different sizes of analytes migrating through a gel as a molecular sieve (Figure 8.7). Besides this function, gels have several advantages: first, minimizing the diffusion of analytes and thus leading to high resolution (sharp peaks); second, they reduce EOF and third, they reduce the adsorption of analytes to the charged capillary walls.

CGE

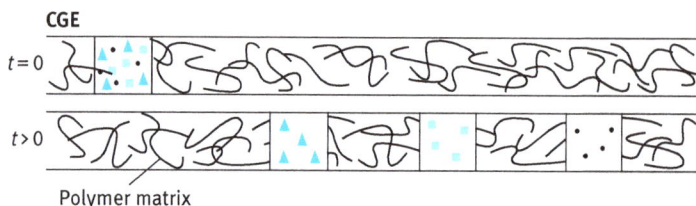

Polymer matrix

Figure 8.7: Principle of size separation in CGE.

Two types of gels are used: physical gels form a porous flexible network by entangling organic polymers; examples are agarose, polyacrylamide and dextran. In chemical gels, covalent binding of polymer strands produces a porous, more rigid structure; an example is bis-polyacrylamide.

CGE is often used for separation of proteins and nucleic acids which can't be separated without a gel since their mass-to-charge ratios do not vary with size. For example, each additional nucleotide added to a DNA chain adds an equivalent unit of mass and charge and does not affect the migration in an electric field free solution.

8.3.3 Micellar Electrokinetic Capillary Chromatography

Like CZE and CGE, MECC is also a "zonal" electrophoresis technique, that is, analytes migrate within discrete zones, differing from CIEF, which is a focusing electrophoresis (see Section 8.3.4), and ITP, a moving boundary technique (see Section 8.3.5). Here, analytes partition between micelles and the electrolyte buffer. Amphiphilic molecules such as surfactants have a hydrophobic molecular part and a hydrophilic head group that may have a charge. Above the critical micellar concentration (cmc) depending on the surfactant type, anionic, cationic, nonionic or zwitterionic aggregates are formed in order to reduce the free (Gibbs) energy of the aqueous system. Both cationic and anionic surfactants can be employed in MECC. Hydrophobic analytes partition to the hydrophobic interior of the aggregates. If the micelles have negative charges at their surface and thus move toward the positively charged electrode, the anode, the EOF moves in the opposite direction, that is, toward the cathode (Figure 8.8). The micelles are also dragged by the EOF (which is strong at neutral and alkaline pH) toward the cathode but move with a lower velocity.

When an analyte has a strong affinity to partitioning into a micelle, the overall migration velocity of the analyte slows down. An uncharged analyte in the aqueous bulk phase (i.e., not inside a micelle) has the same migration velocity as the EOF. Therefore, analytes that have greater affinity for the micelle have slower migration velocities compared to analytes that spend most of their time in the bulk phase. When working with sodium dodecylsulfate (SDS) micelles, the typical order of migration toward the cathode will be anions, neutrals and cations. Anions spend more time in the bulk phase due to electrostatic repulsions from the micelle (and thus move with the EOF). Neutral

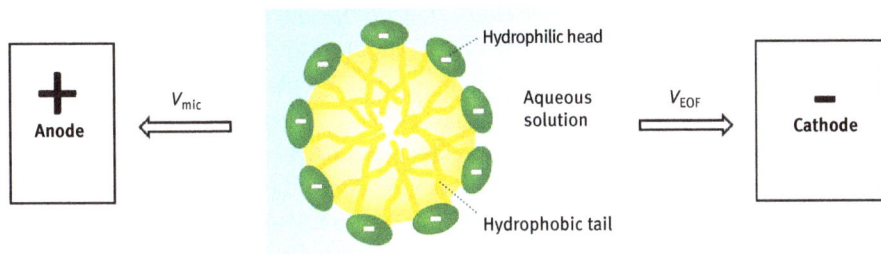

Figure 8.8: A micelle with negative charges at the surface such as sodium dodecyl sulfate (cmc = 8 mM) has an electrophoretic velocity v_{mic} toward the positively charged electrode. If the velocity of the electroendosmotic flow v_{EOF} is larger than v_{mic}, the micelle moves toward the cathode but with a reduced velocity.

molecules are separated based only on hydrophobicity. Cations elute later due to the strong electrostatic attraction to the micelle. However, this general tendency can be reversed: Strong hydrophobic interaction may overcome electrostatic repulsions and attractions and thus the elution order may change.

When using a cationic surfactant such as cetyltrimethylammonium bromide (cmc = 0.9 mM), the micelles migrate in the same direction of the EOF.

Modifiers such as methanol or acetonitrile in concentrations from a few vol % up to 50 vol % (v/v) can lessen hydrophobic interactions between the analytes and the micelles which will influence their elution and separation due to differences in the time that analytes spend in the micellar phase.

MECC has been successfully used to detect polycyclic aromatic hydrocarbons (PAH) in environmental samples [3]. PAH tend to be totally incorporated into the micelles and thus migrate at the same velocity as these. To improve the selectivity in MECC, bile salts are added to the electrolyte buffer, which reduce the hydrophobicity of the micelles, or cyclodextrins or urea. Moreover, organic solvents have to be added to guarantee that the hydrophobic compounds get dissolved.

8.3.4 Capillary Isoelectric Focusing

CIEF can be used for the analysis of zwitterionic analytes, such as peptides, which carry both one or more functional groups with pH-dependent positive charges and negative charges. At low pH, the positive charge will dominate and at alkaline pH, the negative charge, and as long as the analytes are charged, they will move in the electric field. At a certain pH, the positive and negative charges will cancel each other, that is, the molecule carries no net charge: this is called the isoelectric point pI. A pH gradient can be used to separate analytes with different pI from each other. The pH gradient is generated with a series of zwitterionic chemicals, so-called carrier ampholytes. In an electric field, the ampholytes separate: Positively charged ampholytes migrate toward the cathode and that with a negative charge toward the anode. As a consequence, the pH then will decrease at the anodic part of the capillary and increase at the cathodic part. Thus, the buffer medium is discontinuous forming a pH gradient. The ampholyte migration will stop when it has a zero charge and reaches its isoelectric point. An analyte is mixed with suitable ampholytes and loaded onto the capillary (see Section 8.4). Analytes with a net negative charge will migrate toward the anode where the buffer has low pH until it reaches the isoelectric point (Figure 8.9) and vice versa for analytes with positive charges.

The greater the number of ampholytes in the electrolyte, the more even the resulting pH gradient. The pH of the anodic buffer reservoir must be lower than the pI of the most acidic ampholyte to prevent migration into the analyte (the reservoir may be filled, for instance, with phosphoric acid). Likewise, the cathodic buffer reservoir must have a higher pH than the most basic ampholyte (filled, for instance, with sodium hydroxide).

After the analytes stop moving when reaching the pH that corresponds to their pI, they need to be mobilized to pass them through the detector. Mobilization can be accomplished by either application of pressure to the capillary or by addition of salt to one of the reservoirs.

In CIEF, the EOF should be annulled, for example by using coated capillaries, that is, the analytes move in the electric field only due to their charge interactions with the electrodes.

8.3.5 Capillary Isotachophoresis

Like in CIEF, in CITP the EOF has to be annulled, for example by using dynamic or static modifiers. Different to the other CE techniques, two different buffer electrolytes are used here: A so-called leading electrolyte is first introduced to the capillary; it must have a higher mobility than any of the sample components to be determined. For example, if anions are to be analyzed, the leading electrolyte must contain anions with higher mobility than that of the analytes. Then, the sample is injected. The reservoir at

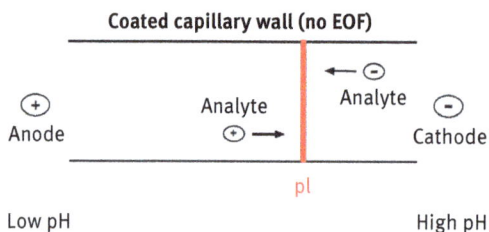

Figure 8.9: The principle of CIEF is based on the electrophoretic movement of charged zwitterionic analytes until they reach their isoelectric point pI within the pH gradient of the electrolyte buffer in the capillary.

Coated capillary wall (no EOF)

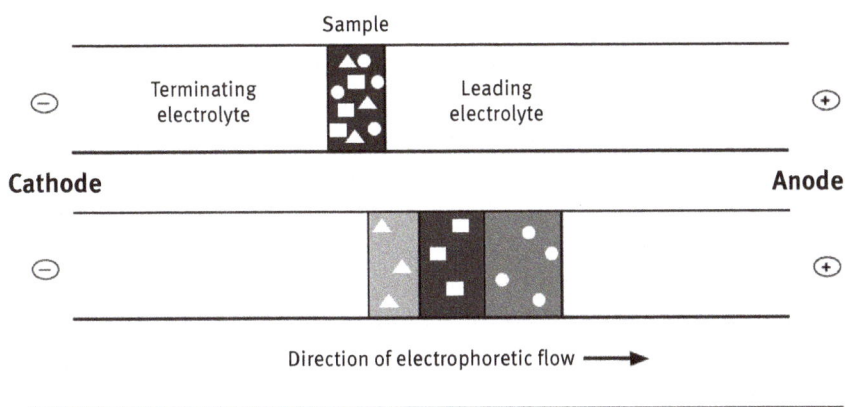

Figure 8.10: Principle of CITP (example for the analysis of anionic analytes). Top: analyte mixture before separation; bottom: separated zones each containing one analyte.

the injection site contains a terminating electrolyte which has an ionic mobility lower than any of the sample components. The analytes are separated in the gap between the leading and terminating electrolytes (Figure 8.10): Stable zone boundaries are forming between the various components of the analyte mixture, all moving at the same speed, resulting in efficient separation. Both anions and cations can be analyzed by CITP, however, only in separate runs.

In CITP, the velocity of the zones remains constant because the electric field varies in each zone: The field is self-adjusting with the lowest field within a zone having the highest mobility according to eq. (8.1). If an ion enters a neighboring zone, its velocity changes and it will return to its original zone. Thus, sharp boundaries between the zones are formed. Moreover, the concentration of analytes in each zone is constant: Since CITP usually runs at a constant current, the ratio between the concentration and the mobility of the analytes remains constant. This means that zones are either sharpened if the analyte concentration is lower than that of the ions in the leading electrolyte or broadened if the analyte concentration in a zone is higher.

8.4 Sample Injection (CE)

In principle, the following steps are performed for introducing a sample to the capillary: (a) insertion of the capillary end from the buffer reservoir to the sample vial, (b) loading the sample by applying either low pressure or voltage across the capillary, (c) reinsertion of the capillary end to the buffer reservoir and (d) applying the voltage for separation of the analytes. The separated sample zones will reach the optical window by the processes described above where they are detected, for example, by a diode array detector (DAD).

Very low volumes of sample are injected into the capillary fitting to the small volumes of the capillaries. If too large sample volumes are injected, the injection plug length may become longer than the diffusion-controlled zones leading to peak broadening. As a rule of thumb, the sample plug length should be not more than about 1–2% of the total length of the capillary which corresponds to an injection plug length of few millimeters (equal to about 1–50 nL), depending on the inner diameter of the capillary. Since the exact injected volume is not precisely determinable, concentrations can only be compared to each other.

The two most commonly applied methods to inject samples are hydrodynamic and electrokinetic loading described in the following.

8.4.1 Hydrodynamic Injection

Samples are loaded into the capillary by (a) applying pressure at the injection end of the capillary, (b) by elevating the injection reservoir relative to the exit reservoir (hydrodynamic pressure) or (c) by setting a vacuum at the elution end of the capillary (Figure 8.11).

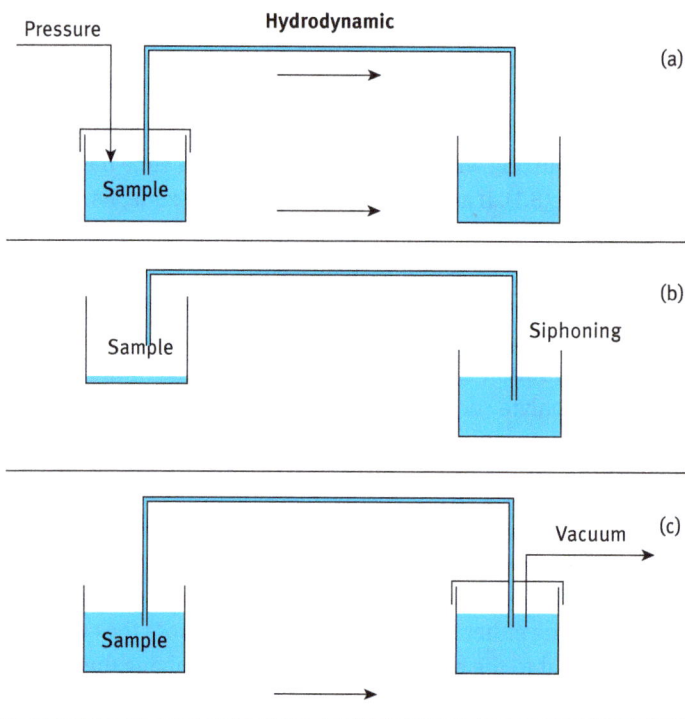

Figure 8.11: Hydrodynamic injection modes (see text).

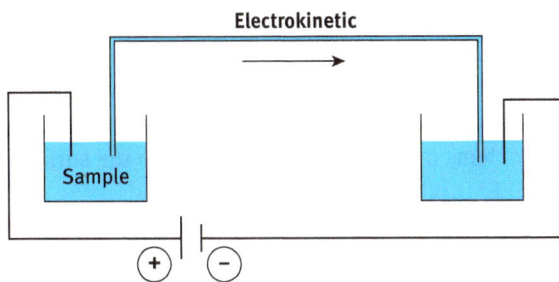

Figure 8.12: Electrokinetic injection.

The volume of sample injected depends on the capillary dimensions, the viscosity of the electrolyte, the applied external or hydrodynamic pressure and the time of loading. Typical injection pressures are below 100 mbar and loading times a few seconds; for hydrodynamical loading, the sample is elevated for up to about half a minute, 5–10 cm higher than the eluent reservoir. It is necessary for reproducibility reasons to keep the temperature of the capillary constant ($\pm 0.1\,^\circ$C) due to its dependence on the viscosity of the buffer and thus on the injection volume.

8.4.2 Electrokinetic Injection

An electric field, usually several times lower than the field that is later used for separation, is applied. In electrokinetic injection, the analytes enter the capillary by migration (if the analytes are charged) and by the EOF acting as a pumping system. Analytes with a higher electrophoretic mobility are loaded to a greater extent than those that are less mobile. Electrokinetic loading is preferred if hydrodynamic loading is ineffective, for instance when working with viscous media or gels in the capillary (Figure 8.12).

8.5 Separation Parameters in CE

Reproducibility in CE analyses depends on the condition of the capillary wall, the composition of the electrolyte, the pH and viscosity of the buffer, the nature of the sample, and the quality of the instrument and the power supply. Several of these experimental parameters have been discussed. Here, we focus on further influencing factors: properties of the capillaries, conditioning of the capillary surface, the temperature and the electric field.

8.5.1 Capillary Properties and Conditioning

Criteria for the choice of capillaries are inertness against chemical erosion and electricity, and flexibility but being robust as well as transparency for the detection method.

Fused silica is often used in CE, which for better stability and handling is coated by an organic polymer such as polyimide. For transparency of the optical window, the organic polymer has to be removed. Another material from which CE capillaries are produced is Teflon®, oder being transparent for UV light, but its disadvantage is that it reduces EOF and may give problems due to sample adsorption to the surface. Moreover, polyether ether ketone (PEEK) capillaries are used.

Internal diameters of fused silica capillaries range from 10 to 200 µm, the effective lengths from 10 cm for gel-filled capillaries up to 100 cm for CZE separations. The total length is generally 5–15 cm longer, depending on the distance from the detector to the exit reservoir.

The condition of the capillary surface is very important because it will directly influence the analytical result. Washing with 1 M NaOH followed by diluted (0.1 M) NaOH and finally the buffer is often used for clearing the surface from material that adsorbed during previous runs. However, for analysis with low pH buffers, hysteresis of the silanol charging may cause irreproducible EOF. Moreover, buffer components and surfactants can sorb to the capillary surface which suggests to use a capillary exclusively for one specific electrolyte system for good reproducibility.

8.5.2 Temperature Effect in CE

Rising temperature will increase the viscosity of the running buffer by about 2–3% with each centigrade Celsius. The Joule heat, generated by the passage of electrical current, depends on the power (=voltage x current) and on the capillary dimensions, the conductivity (ionic strength) of the buffer and the applied voltage. Joule heating is thus a consequence of the resistance of the buffer to the flow of current. Increase in temperature within the capillary will result when the power generation and Joule heating exceed thermal dissipation. While the absolute rise in temperature is not detrimental per se, temperature differences at the capillary wall and its center are problematic. Thermal dissipation of the heat through the capillary walls can result in relatively higher temperatures in the inner part of the capillary which may cause viscosity differences of the electrolyte and peak deformation. Joule heating can be lowered by reducing the electric field (with the disadvantage of reduced resolution power), using narrow capillaries, low ionic strength and a thermostat, the latter also useful for maintaining constant temperature within the capillary.

Effective control of capillary temperature (ideally ±0.1 °C) is therefore most important for reproducible operation. Temperature regulation can be obtained by a strong air stream (ca. 10 cm/s), flushing the capillary from outside or by immersing the capillary in a liquid.

8.5.3 Electric Field in CE

In CE, electric fields of up to about 30 kV and currents of 200–300 mA are typical. It is obvious (eq. (8.2)) that the voltage has to be maintained at a constant level (±0.1%) to maintain a good reproducibility in migration times. Most often, the EOF is in the direction of the cathode and injection is at the anode end of the capillary. However, if EOF is reduced or reversed, for example, by coating the surface, or if gels are used, the polarity of the electrodes should be switched, that is, to have the cathode at the injection end.

Increasing the strength of the electric field during analysis can be useful to decrease the analysis time of complex samples just like the solvent gradients in HPLC. If samples are eluting close to each other reduction of the field prior to, for example, an automatic fraction collection will increase the time window and allow a more precise separation of the fractions.

8.6 Detection of Analytes in CE

Several detection methods have been used in CE; an overview is given in Table 8.1, and some of them will be described in more detail below. Due to the small diameter of the capillary with the embodied optical cell, the concentration of the analytes has be high enough to be detected and often concentration steps are necessary, that is, CE cannot be considered a trace analytical method. An advantage of the small volume in the optical window is that band broadening is negligible unlike in the much larger analytical cells as for instance in HPLC UV–vis cells.

Table 8.1: Detection methods in CE.

Method	LOD (molar)	Comments
UV–vis	10^{-5}–10^{-8}	Universal method; spectral information (diode array)
Fluorescence	10^{-7}–10^{-9}	Sensitive, may require derivatization
Laser-induced fluorescence	10^{-14}–10^{-16}	Very sensitive, may require derivatization
Amperometry	10^{-10}–10^{-11}	Sensitive, only for electroactive analytes
Conductivity	10^{-7}–10^{-8}	Universal
Mass spectrometry	10^{-8}–10^{-9}	Sensitive, structural information
Indirect UV–vis, fluorescence, amperometry	1–10% of the sensitivity of the direct methods	Universal, low sensitivity

8.6.1 UV–Vis Detection in CE

According to Lambert-Beer's law (eq. (7.6)) and due to the on-column measurement in CE, the short pathlength limits the sensitivity in UV–vis detection. In addition, the real pathlength in the curved capillary is less than its inner diameter and, thus, only a fraction of the light passes directly through the center. Since the molar absorptivity of most chemicals increases with decreasing wavelength, an option is to lower the detection wavelength to the range around 200 nm or even below (the UV cutoff of fused silica is around 190 nm). However, it is essential to use low absorbing electrolyte buffers, such as phosphate or borate, whereas organic buffers such as TRIS or HEPES cannot be used. The optical pathlength of the capillary can be effectively multiplied by using mirrors to reflect the incident light inside the capillary prior to detection, for example, by using silver-coated capillaries. The optical pathlength increases while the narrow dimension of the flow cell is maintained [4].

Single wavelength and DADs can be used, the latter valuable for checking the identity of analytes due to the measurement of the complete absorption spectrum, for example, between 200 and 800 nm. DAD optics can yield detection limits, sensitivity and linear detection range that at least equal that of single or multiple wavelength detectors. DAD analysis can rapidly yield information regarding the optimal detector conditions, and in particular the optimal wavelength. Once all peaks have been detected, the diode array can be used to determine the best wavelength for the absorbance maximum for all analytes. With a DAD, the purity of a peak at different times during its elution can be examined if the spectra of the analyte and the contaminant differ. If the spectra match, the peak is most probably pure.

8.6.2 Fluorescence and Chemiluminescence Detection in CE

Fluorescence detectors in CE systems often use lasers producing intense light at a single wavelength for efficient excitation. The monochromatic nature of the laser light allows filtering any stray light which might interfere with the detection of analytes. Analytes will vary in their excitation and fluorescence wavelengths so that the detector will not see all the components in a sample. Analytes may be fluorescent themselves or they can be derivatized with a fluorescent probe and the sensitivity of this type of detector is up to 1,000 times higher than an UV–vis absorbance detector. Nowadays, LED lights are also used for excitation.

Chemiluminescence detection is a sensitive alternative for coupling with CE. Many examples for clinical, environmental and food analysis have been described [5] with several reagents being used such as ruthenium(II) complexes, luminol, peroxyoxalates or acridinium derivatives. On-column, off-column and end-column interfaces have been described for introducing the chemiluminescence reagents to the CE system. Two compounds are necessary: the luminophore, for instance a ruthenium(II) reagent, and the analyte as co-reactant carrying suitable functional groups such as secondary or tertiary amines. CZE is most often used in connection with chemoluminescence detection.

8.6.3 Amperometric and Conductometric Detection in CE

Both techniques offer a high sensitivity and can be applied to a wide variety of analytes, including those without appreciable UV–vis or fluorescence properties. The sensing electrodes are scaled to the dimensions of the electrophoresis capillary. In amperometric detection, an electroactive analyte undergoes an electrochemical reaction (oxidation and reduction) inside a detector cell. In amperometry, the current is measured at a certain voltage when the electroactive analyte is oxidized at the anode or reduced at the cathode. The CE separation is carried out at microampere currents and kilovolt potentials, whereas the detection cell must operate at picoampere currents and millivolt potentials.

Two cylindrical electrodes around the capillary measure the difference in conductivity between the electrolyte buffer and the analytes which should be as high as possible, that is, by using buffers with low conductivity if that of the analyte is high, or vice versa for indirect detection.

8.6.4 Hyphenated Techniques (CE-MS)

CE–mass spectrometry (CE–MS) offers the possibility to extend the information on migration time, quantity and absorbance spectrum (DAD analysis) with structural information.

The predominant MS interface to CE is electrospray MS (for a more detailed presentation, see Chapter 9, Section 9.3.2). The outlet end of the CE capillary is inserted into the electrospray interface. Because the volume of eluent from the capillary is very small, an additional "make-up" liquid is pumped through an axial needle. The liquid is mixed with a flowing gas stream and nebulized into a spray. The spray evaporizes and the ionized analyte particles are carried into the MS detector. Because the peak width in CE is quite small, the MS must rapidly scan the desired m/z range.

Volatile buffers like ammonium formate are preferred to prevent the accumulation of buffer salts inside the MS, even if such buffers may not be ideal for separation of analytes.

8.6.5 Indirect Detection Methods in CE

For compounds that have for instance no UV absorbance at the suitable wavelength range, indirect UV detection can be used, a universal and nonselective detection method. Direct detection methods are based on the presence of an analyte in the detector cell and the proportionality of its concentration to the output signal of the detector within a certain range. In indirect methods, the decrease of the background signal is used for quantification when an analyte passes the cell. As an example for UV detection, the capillary is filled with a buffer that has a significant UV absorbance. When the analyte enters the cell, the chromophore of the buffer is "diluted" giving rise to a negative peak which is proportional to the amount of analyte present. An example is the quantification of sulfate by using a buffer with chromate as the chromophore. The negative signal can be transformed and integrated by a suitable software. The best results and symmetrical peak shapes are obtained if the mobility of the UV probe in the running buffer is the same as that of the analyte.

The principle of indirect detection can also be applied to other detector types, for instance in conductivity measurements.

8.6.6 Increasing the Sensitivity of Detection in CE

Various preconcentration techniques are based on the change in analyte velocity in systems with two or three discontinuous solutions, that is, field amplified stacking, transient isotachophoresis, pH-mediated stacking and sweeping [4]. Such methods have been employed to enrich analytes from environmental samples in CE separations [6]. As an example, in field-amplified stacking, a low-conductivity sample solution is hydrodynamically injected into the capillary filled with an electrolyte buffer with a high conductivity. Because the electric field strength is inversely proportional to the electrical conductivity, the sample zone experiences a higher field relative to the high-conductivity zone. Therefore, the electrophoretic migration of ionic analytes in the sample zone is faster than that in the high-conductivity buffer, which causes the "stacking" of the analytes around the sample-buffer boundary. Significant increases in the detection sensitivity can be obtained by such methods.

Another method to decrease the limit of detection (LOD) in CE analysis is the combination of solid-phase extraction (SPE) [7–10]. SPE can be combined with CE offline, in-line or on-line. In-line SPE–CE systems have either an open-tubular capillary that is coated with the SPE sorbent, or a packed bed material in a capillary, or a thin membrane impregnated with the SPE material positioned between two capillaries inside a sleeve and retained by two membranes. In in-line SPE–CE, the complete desorbed sample from the SPE column is analyzed by CE. In on-line SPE–CE systems, the SPE column is not part of the CE system but is interfaced to it by a flow-switching capability.

As an example for the combination of SPE enrichment and CE analysis, atrazine, terbutylazine and their N-dealkylated chloro- and hydroxyl-metabolites have been analyzed by CZE and MECC with LODs in the low µg/L range [11]. Similarly, 14 different aromatic sulfonates of environmental concern were analyzed by CZE [12] (Figure 8.13).

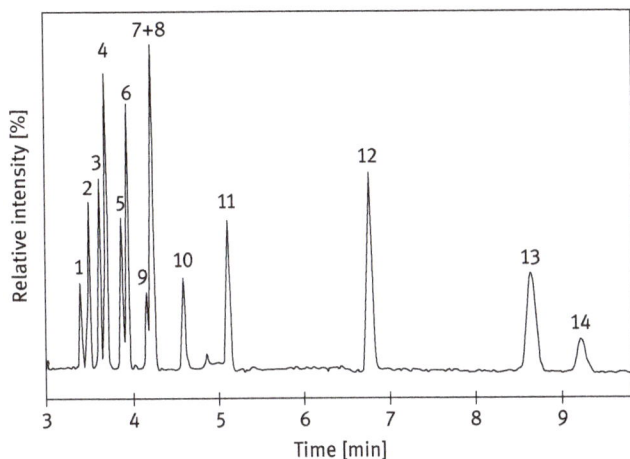

Figure 8.13: Electropherogram (CZE) of a 14-compound aromatic sulfonate mixture (reference substances) containing 5 mg/L of each compound. Conditions: running electrolyte 25 mM sodium borate, pH 9.3, capillary 60 cm, 375 mm ID, voltage 25 kV, temperature 308 °C, pressure injection 50 mbar for 12 s and UV detection at 210 nm [12]. Reproduced with permission from Elsevier.

Further Reading

Weinberger R. Practical capillary electrophoresis. San Diego: Academic Press, 2000.

Schmitt-Kopplin P. Capillary electrophoresis – methods and protocols. Berlin: Springer Verlag, 2016.

Bibliography

[1]   Grochocki W, Markuszewski MJ, Quirino JP. Multidimensional capillary electrophoresis. Electrophoresis 2015;36(1):135–43.
[2]   Lewis AP, Cranny A, Harris NR, Green NG, Wharton JA, Wood RJ, Stokes KR. Review on the development of truly portable and in-situ capillary electrophoresis systems. Meas Sci Technol 2013;24(4):1–20.
[3]   Ferey L, Delaunay N. Capillary and microchip electrophoretic analysis of polycyclic aromatic hydrocarbons. Anal Bioanal Chem 2015;407(10):2727–47.
[4]   Albin M, Grossman PD, Moring SE. Sensitivity enhancement for capillary electrophoresis. Anal Chem 1993;65(10):489A–97A.
[5]   Lara FJ, Airado-Rodriguez D, Moreno-Gonzalez D, Huertas-Perez JF, Garcia-Campana AM. Applications of capillary electrophoresis with chemiluminescence detection in clinical, environmental and food analysis. a review. Anal Chim Acta 2016;913:22–40.
[6]   Kitagawa F, Otsuka K., Recent applications of on-line sample preconcentration techniques in capillary electrophoresis. J Chromatogr A 2014;1335:43–60.
[7]   Ramautar R, de Jong GJ, Somsen GW. Developments in coupled solid-phase extraction-capillary electrophoresis 2009–2011. Electrophoresis 2012;33(1):243–50.
[8]   Ramautar R, Somsen GW, de Jong GJ. Developments in coupled solid-phase extraction-capillary electrophoresis 2013–2015. Electrophoresis 2016;37(1):35–44.

[9]   Ramautar R, Somsen GW, de Jong GJ. Developments in coupled solid-phase extraction-capillary electrophoresis 2011–2013. Electrophoresis 2014;35(1):128–37.

[10] Ramautar R, Somsen GW, de Jong GJ. Recent developments in coupled SPE-CE. Electrophoresis 2010;31(1):44–54.

[11] Loos R, Niessner R. Analysis of atrazine, terbutylazine and their N-dealkylated chloro and hydroxy metabolites by solid-phase extraction and gas chromatography mass spectrometry and capillary electrophoresis-ultraviolet detection. J Chromatogr A 1999;835(1-2):217–29.

[12] Loos R, Niessner R. Analysis of aromatic sulfonates in water by solid-phase extraction and electrophoresis. J Chromatogr A 1998;822(2):291–303.

9 Mass Spectrometry

9.1 Introduction

Mass spectrometry (MS) is certainly the highest performer and best service provider among the detection techniques, combined with chromatography or other separation schemes. Briefly spoken, MS makes use of the determination of electrical mobility of charged species within an electromagnetic field, thus representing an orthogonal separation technique to chromatography (i.e., two essentially different methods used to separate the same analyte). Hyphenation of both generates extremely high resolutions in separating complex samples, required for example in crude oil analysis, protein characterization or nontargeted analysis within the fields of -*omic* techniques [1].

Already 100 years ago, Dempster and Aston (1918/1919) were forerunners of the modern MS technology, after Thompson (1911) had detected the separation of gas ions within an electric field. Furthermore, tremendous technical advances were achieved during World War II, when huge sector-field MS instruments were built and run in parallel for preparative uranium isotope fractionation ("Manhattan project"). With the advent of various energy-selective and intensity-adapted ionization techniques (e.g., by laser, radiation from radioactive decay, inductively coupled plasma, collision-rated ionization or thermal surface interaction, etc.), the applications became numerous [2].

Nowadays, MS is (beyond the still applied isotope fractionation) mainly used for separation and identification of larger organic molecules (up to >100 kDa), derived from environment, industry and life sciences. In contrast to the first very bulky and heavy versions (due to huge magnets and vacuum pumping), nowadays table-top instrumentation with high-resolution capabilities is available, and miniaturized instruments are already a permanent part of space missions for extraterrestrial research [3].

9.2 Basic Equipment

MS is a technique where gaseous ions become accelerated and separated under the influence of an electric and/or magnetic field according to their masses and charges (strictly speaking to their mass-to-charge ratio, m/z). Ions are separated according to their m/z and either collected and quantified on a detector plate or are continuously recorded as an ion current. The latter is mainly used for analytical purposes.

A typical mass spectrometer (see Figure 9.1) consists basically of an ion source where gaseous charged species from the submitted gaseous analyte molecules are produced, followed by an electric and/or magnetic separator stage. Finally, an open photomultiplier tube is acting as ion counter. Mass spectra consist of the relative abundance of detected ions as a function of the m/z. Since not only the ionization process is influenced by competing collisions with other present gas molecules, but

DOI 10.1515/9783110441154-009

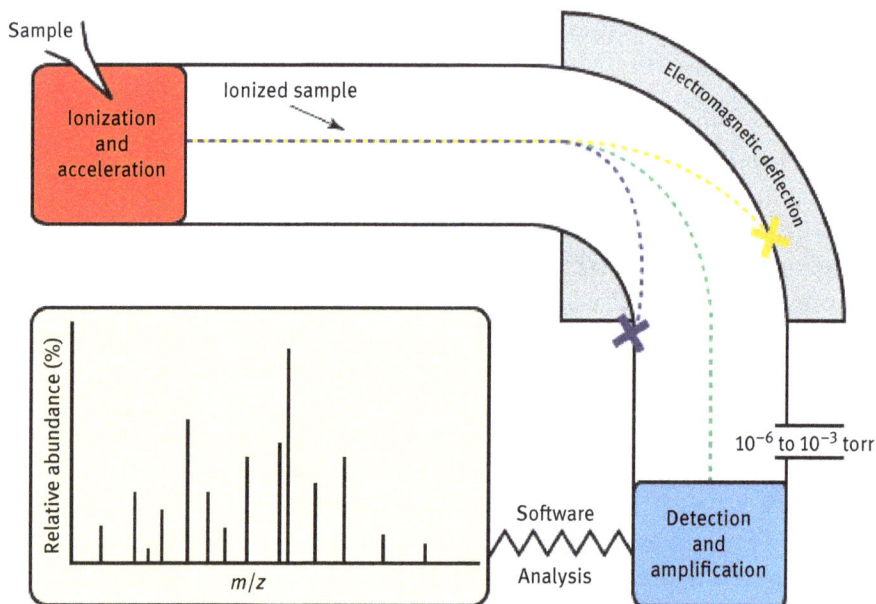

Figure 9.1: Basic setup for mass spectrometry.

also separation of the charged species will be hampered due to mobility losses, the whole process is conducted under vacuum conditions ($<10^{-10}$ torr).

Depending on the molecular structure of an analyte, the accompanying matrix and the applied ionization conditions, one or a series of characteristic charged fragments or adducts is the detected result of such an interaction. Based on this, MS became one of the most powerful tools for structure determination in chemistry. For qualitative and quantitative purposes, fragmentation patterns of numerous substances are compiled in accessible mass spectral libraries (e.g., from Wiley, NIST or AIST), supporting identification of unknown substances in a sample [4].

It is beyond the scope of this textbook to introduce structural elucidation by MS, and we would like to refer to the respective literature, for example, Ref. [5].

Rather, we have to learn the different ionization possibilities with their strengths and weaknesses for organic trace analysis. Furthermore, the typical MS separation configuration with its intrinsic resolution power needs to be communicated and discussed, in order to have an idea on the appropriate selection of instrumentation (chromatographic separation and MS embodiment) for a given analytical problem.

9.3 Ionization in MS

In analytical MS, the sample is subjected to ionization. To do so, the sample is usually in the gaseous state, as it is already the case when leaving a GC unit. Liquid samples as obtained from LC or CE separation need to become transformed into the gaseous state by dispersion, evaporation or ablation. There are several ways of achieving ionization. By principle, they can be divided into hard or soft ionization techniques, which describes the amount of energy transferred to the analyte molecule.

9.3.1 Hard Ionization by Electron Impact

Hard ionization means bombardment with highly energetic particles like accelerated thermal electrons (*electron impact ionization*, EI) [6]. This is done by crossing an electron beam under vacuum conditions (see Figure 9.2).

Depending on pressure (i.e. the presence of other carrier gas molecules) and the kinetic energy of the colliding electrons, analyte molecules not only become charged but also fragmented. Typical kinetic energies in the range of 50–100 eV are applied. The simplified process is

$$M + e^- \rightarrow M^+ + 2e^-$$

where M is the analyte molecule, e^- is the colliding electron and M^+ is the resulting molecular ion. The high energy transferred by the electrons leads to ionization and after bond dissociation to fragmentation.

After charging and rearrangement of the fragments, the stable ion clusters reach the separator stage and are recorded by the subsequent ion detector. Typically, 70 eV is applied for EI fragmentation in MS. EI often creates a variety of typical fragments, which can be used for structural elucidation of unknown molecules. A complex sample ionized by EI would provide a tremendous mixture of fragments; therefore, a pre-separation is needed. This is the reason why EI is often combined with GC. An EI mass spectrum of 2,3,7,8-tetrachloro-p-dibenzodioxine is exemplarily shown in Figure 9.3.

The cluster of ions presented at $m/z = 322$ is at the molecular weight of TCDD. The pattern is unique for a substance with four chlorines. The mass spectrum has lower mass ions due to the decomposition of the molecular ion in the ion source. This demonstrates that TCDD within a congener mixture can only be determined in

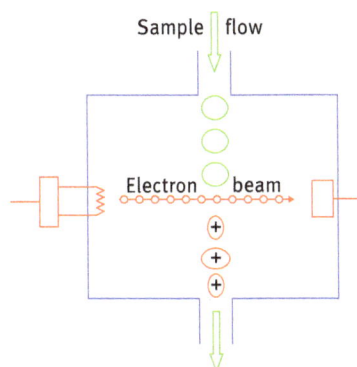

Figure 9.2: Electron impact ionization source.

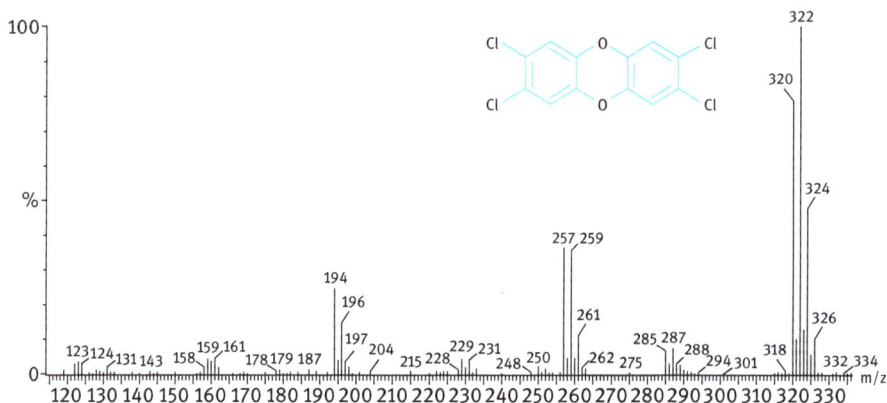

Figure 9.3: EI mass spectrum of 2,3,7,8-tetrachloro-p-dibenzodioxin (TCDD). Ionization energy: 70 eV.

combination with an effective high-resolution separation, for example, by capillary GC. Often stable-isotope labeling makes the assignment of fragments to the searched compound possible.

9.3.2 Soft Ionization

Soft ionization means application of far less energy in order to achieve only a few and possibly selective bond breakings. Figure 9.4 shows the variation of fragmentation as a function of EI energy for diphenylmethanone.

In this case, optimal sensitivity was achieved at 70 eV, but here, use of a lower ionization energy reduces fragmentation and increases the intensity of the molecular ion. So, for analytical purposes, soft ionization with 14 or 16 eV is favorable. As shown, reduced fragmentation leads to an increase of the molecular ion peak M^+.

Chemical ionization (CI) happens when reactant gases, like methane, water, higher alkanes or ammonia, become admixed to the gaseous analyte molecules within an electron beam (150 eV), causing plasma conditions with ionic adduct formation [7]. This leads to transiently instable protonated species and adducts/radicals, for example, for methane as reactant gas

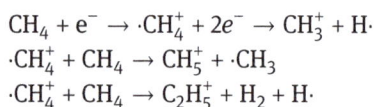

$$CH_4 + e^- \rightarrow \cdot CH_4^+ + 2e^- \rightarrow CH_3^+ + H\cdot$$
$$\cdot CH_4^+ + CH_4 \rightarrow CH_5^+ + \cdot CH_3$$
$$\cdot CH_4^+ + CH_4 \rightarrow C_2H_5^+ + H_2 + H\cdot$$

The formed instable carbonium ions $\cdot CH_4^+$ or CH_5^+ are then ionizing in a next step the analyte molecule M by charge exchange

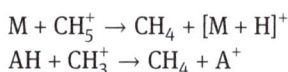

$$M + CH_5^+ \rightarrow CH_4 + [M + H]^+$$
$$AH + CH_3^+ \rightarrow CH_4 + A^+$$

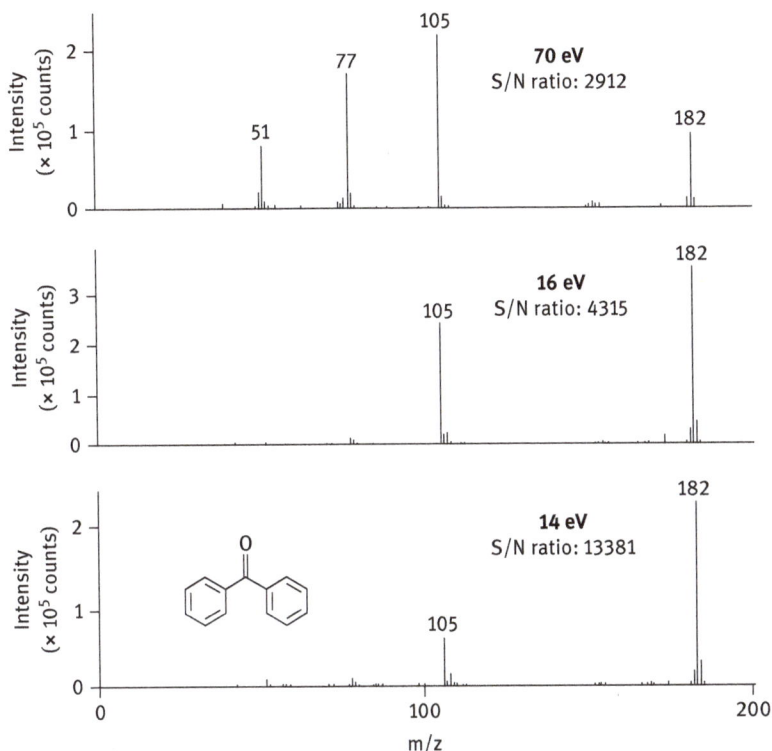

Figure 9.4: Comparison of mass spectra for diphenylmethanone at 70 eV, 16 eV and 14 eV.

Since the proton affinity decreases in the order $CH_5^+ > C_2H_5^+ > H_3O^+ > C_4H_9^+ > NH_4^+$, only energy to a lesser extent is liberated in course of the proton transfer, and hence, molecule fragmentation is negligible. Therefore, such spectra are dominated by the molecular ion, which in turn can be used for molecular mass determination. The extent of collision-induced ionization can be directed by appropriate selection of admixed gas and pressure conditions. Under an appropriately configured potential, also negative ionization can be favored (*negative chemical ionization*, NCI). A paramount example is the determination of endocrine disruptors in environmental matrices by GC–NCI–MS [8].

Often, CI is performed under atmospheric pressure (APCI) conditions. Many applications are reported, especially after LC separations of analytes which need a gentle fragmentation, for example, lipids or polar PAH metabolites [9].

A special case is the direct ionization within a liquid, for example, from LC or CE separation. Figure 9.5 visualizes the working principle of the *atmospheric pressure chemical ionization* (APCI) process [10].

In APCI, a liquid is pumped through a capillary and nebulized at the tip converting the eluent stream into an ultrafine aerosol flow. Since the leaving aerosol

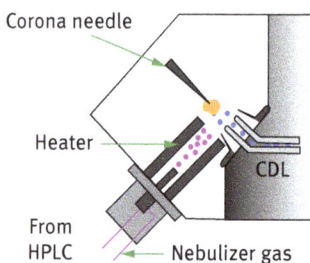

Corona needle

Heater

CDL

From HPLC

Nebulizer gas

Figure 9.5: Schematic view of atmospheric pressure chemical ionization (APCI).

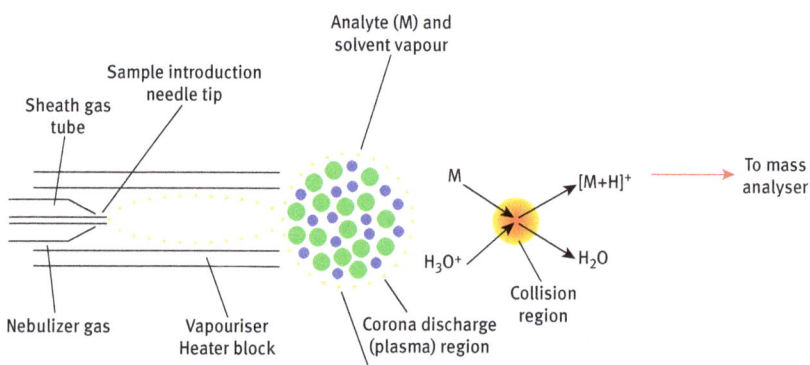

Analyte (M) and solvent vapour

Sample introduction needle tip

Sheath gas tube

M

[M+H]+

To mass analyser

H_3O^+

H_2O

Collision region

Nebulizer gas

Vapouriser Heater block

Corona discharge (plasma) region

Figure 9.6: Scheme of an atmospheric pressure ionization HPLC–MS interface. Copyright permission granted by Dr. Gates, University of Bristol.

crosses a heated desolvation chamber surrounding a corona charger (3–6 kV), the dissolved analyte molecules become intensely mixed with electrons produced from the corona charger. This results in negatively or positively charged adducts or radical ions. These ions then ionize the analyte via charge transfer (Figure 9.6). The difference in electrospray ionization (ESI, see later) and APCI is that in the latter the analyte solution is nebulized and desolvated before interacting with the corona discharge creating ions, whereas in ESI, ionization occurs by the potential difference between the spray needle and the counter electrode and simultaneous desolvation. The corona discharge produces primary reactive ions like N_2^+ and N_4^+ which produce secondary ions by reaction with the vaporized solvent molecules and finally the ionized analyte. This aspect is similar to the old CI technique in MS, although in APCI, the efficiency of ionization is higher because it occurs under atmospheric pressure (more collisions occur) compared to the vacuum used for CI. Fragmentation of the analyte is of little relevance.

The technique is useful for small, thermally stable, semipolar molecules that are not well ionized by ESI. Unlike ESI, ions do not carry multiple charges.

APCI is usually run at flow rates of a few hundreds µL/min up to mL/min.

Even larger analyte molecules can be easily ionized in a soft way by ESI [11]. ESI is suitable for the analysis of moderately polar molecules such as many xenobiotics, metabolites and peptides. The liquid HPLC eluent is directed through a metal capillary surrounded with a nitrogen flow (the atmospheric pressure aperture plate of the mass spectrometer is the counter electrode maintained at 3–5 kV); eluent flow rates are low, typically in the μL/min range. Typical solvents for ESI contain volatile organic compounds, for example, methanol or acetonitrile mixed with some water. Compounds increasing the conductivity, for example acetic acid or ammonium salts, are often added to the solution. The eluent is nebulized at the end of the capillary forming a fine spray of charged droplets in the nm range. Since the charge of the droplets has the same polarity as the needle, they are repelled. The vapor pressure of the solvent increases exponentially in the small droplets, with the result of fast shrinking sizes. Once the droplet size becomes so small that the intrinsic charges exceed the so-called Rayleigh limit, a Coulombic explosion happens producing many ultrafine charged particles or molecules. Under the influence of vacuum conditions, finally individual charged analyte molecules enter the separation stage of the MS (Figure 9.7). Usually, no fragmentation happens during this process. The ion source and subsequent ion optics can be operated to detect either positive or negative ions.

ESI is a rather soft ionization method, that is, little fragmentation of the analyte ion will occur, unlike other MS ionization methods such as EI used in GC–MS. By increasing voltages within the source, fragmentation in ESI can be increased. Coupling ESI with tandem MS may also overcome the problem of rather little structural information due to limited fragmentation. Other fragmentation methods, such

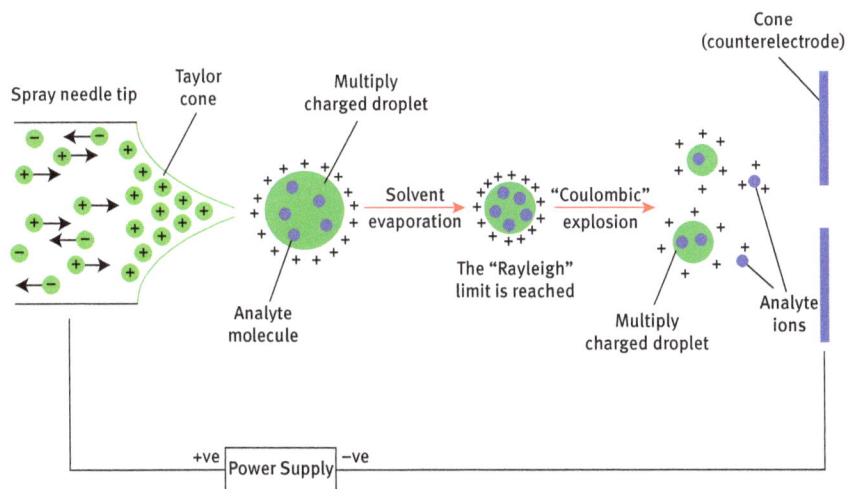

Figure 9.7: Electrospray ionization of dissolved analyte molecules under a strong electric field when leaving a capillary. Copyright permission granted by Dr. Gates, University of Bristol.

as collision-induced dissociation, are used to produce stronger fragmentation of the analyte for structural studies.

Small molecules with a single functional group capable of carrying an electrical charge give predominantly singly charged ions by addition of a proton $[M + H]^+$ when the ion source is operated in positive ion mode or by the loss of a proton when in negative ion mode $[M - H]^-$. Larger molecules with several charge-carrying functional groups, such as proteins and peptides, can exhibit multiple charging resulting in masses $[M + 2H]^{2+}$, $[M + 3H]^{3+}$ and $[M + nH]^{n+}$.

Desorption electrospray ionization (DESI) is an ambient ionization technique in which a solvent electrospray is directed at a sample to which a voltage is applied. The analyte is extracted into the solvent which is nebulized to charged droplets that evaporate to form highly charged ions that are directed to the atmospheric pressure interface of the mass spectrometer.

Currently of high interest is the ionization under ambient conditions in the DESI mode (see Figure 9.8) [12] or, when using He^+ gas molecules as surface colliding and ionizing sputterer, named DART (*direct analysis in real time*).

For DESI, clean solvent nanodroplets leave the capillary and strike the analytes deposited at a surface. The charged solvent molecules partly sputter the adsorbed analyte molecules and then enter the MS entrance. Fast monitoring of organic contaminants (e.g., drugs, explosives and biopolymers) by this technique has been demonstrated. The ionization and collection efficiency is depending on the geometric arrangement of the sprayer nozzle and the MS inlet, surface properties (contact potential) and chemical composition of the dispersed solvent, as well.

Proton-transfer reaction-MS makes use of adduct formation of free hydronium (H_3O^+) ions with gaseous organic compounds with higher proton affinity than free water molecules. Volatile organic compounds fulfill this requirement. The hydronium ions are produced within a hollow cathode discharge from water vapor and become directly mixed with the sample gas. Ambient air or exhaled breath gas can be quantitatively analyzed for VOCs without pre-separation at pptv level [13].

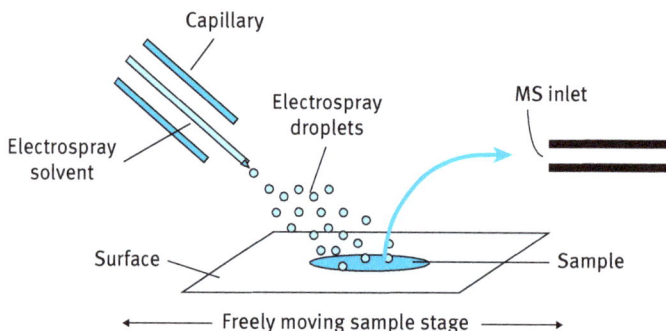

Figure 9.8: Desorption electrospray ionization (DESI) process.

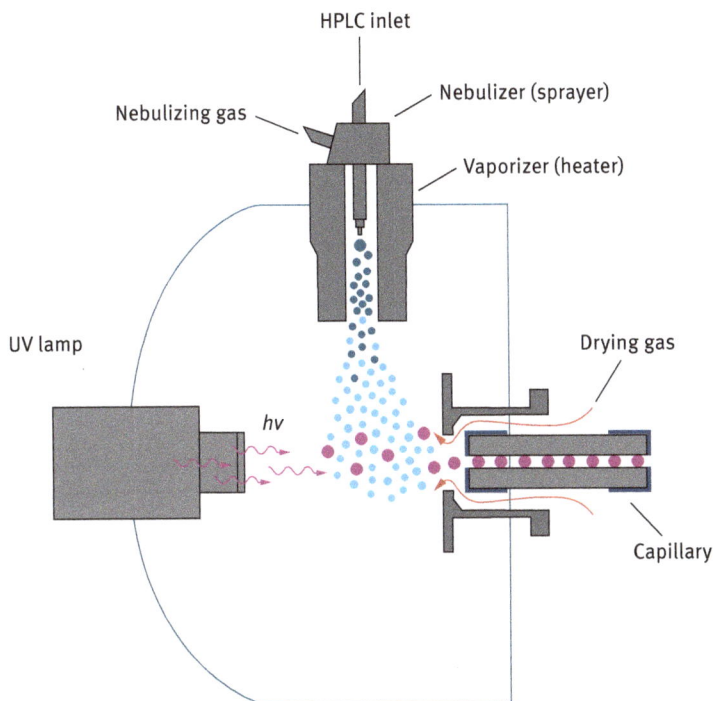

Figure 9.9: Principle of atmospheric pressure photoionization.

Atmospheric pressure photoionization uses photons to excite and ionize molecules after nebulization (Figure 9.9). The energy of the photons is chosen to minimize concurrent ionization of solvents and ion source gases, for instance, by use of a krypton lamp emitting photons around 10 eV. By adding a so-called dopant, the percentage of ionized analytes can be increased. Dopants are molecules with ionization energies below the photon energy of the lamp used. For krypton lamps, for example, toluene, benzene or acetone may be used as dopant.

The technique also gives predominantly singly charged ions and can be applied for compounds that ionize poorly by ESI and APCI.

Photoionization (PI) offers a wide range of adjustable energies for selective ionization. The basic mechanism of PI is $M + hv \rightarrow M^+ + e^-$. However, the predominant ion observed is typically $[M + H]^+$. Especially, *resonance-enhanced multiphoton ionization* (REMPI) with tunable laser sources has found applications for complex mixture analysis. Ionization is caused by absorption of photons (see Figure 9.10). Applications are quasi-online analyses of dioxin precursors or PAHs in waste incineration flue gas [14].

Since the ionization energy is in a similar range (5–12 eV), only small excess energy is transferred to the analyte molecule, hence avoiding much fragmentation. Using

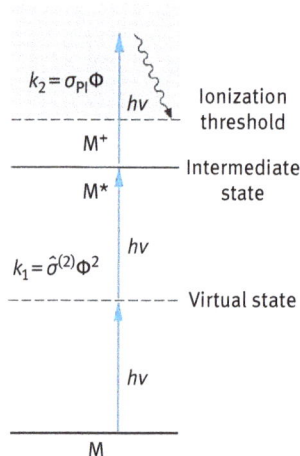

Figure 9.10: Resonance multiphoton ionization process ($n = 2$).

high photon densities by short-pulsed laser irradiation, photon energies become additive within the absorption process. Therefore, ionization is caused at wavelengths usually not appropriate for direct ionization by single photons. Applications toward direct process analysis of very complex mixtures (e.g., combustion exhaust) or even optically selective ionization (separation of optical isomers) have been reported [15].

A special variant represents the matrix-assisted laser desorption ionization (MALDI) [16]. Here, the analytes of interest are either co-crystallized or surface-absorbed to a photon-absorbing "matrix" compound (e.g., sinapinic or cinnamic acid or dihydroxybenzoic acid). This matrix becomes ablated by a pulsed UV laser beam. Within the small plasma formed, ionization happens, causing a very soft ionization without much fragmentation. Beneficial is the predominance of the molecular ion peak M^+ in the produced spectrum. Therefore, MALDI–MS (often as time-of-flight [TOF] configuration) is perfect for qualitative molecular mass characterization of large biomolecules (proteins, peptides etc.), but since the plasma conditions depend on the photon/solid-phase interaction, and the per se unknown charging equilibrium within the laser plasma, a quantitative determination so far is not possible.

Increasing distribution is seen with *dielectric-barrier-discharge ionization* (DBDI). DBDI is the formation of a low-temperature plasma between two insulated electrodes [17]. Figure 9.11 depicts the simple geometry of a capillary-based ionizer.

DBDI is applied in GC–MS- or LC–MS-coupled systems. Optimization of the ionization parameters for the detection of analytes was carried out by varying the discharge power, the kind of gas, the gas flow rate and the dielectric material between the electrodes.

A summary of the three most often used HPLC–MS interfaces with respect to the types of analytes for which the techniques can be used is given in Figure 9.12.

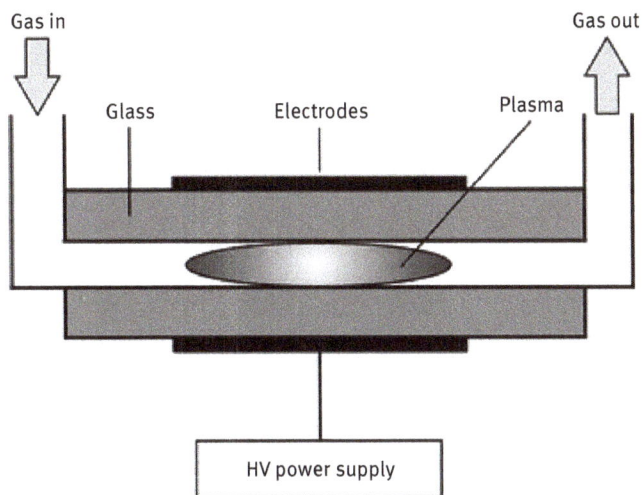

Figure 9.11: Dielectric-barrier-discharge ionization of a gas flow within a glass capillary.

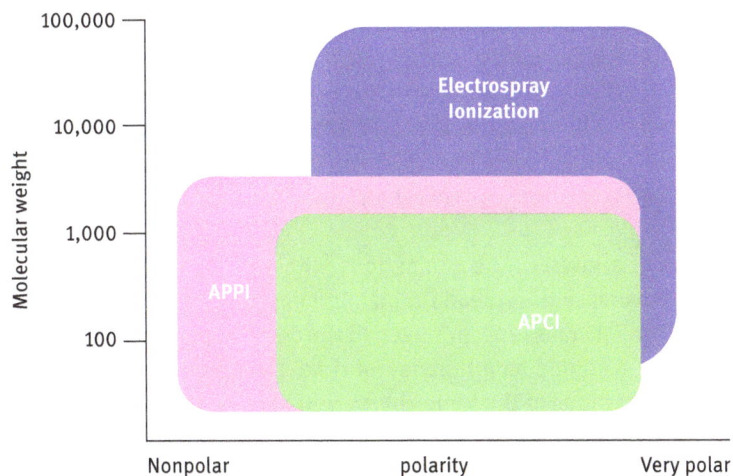

Figure 9.12: Three HPLC–MS interface techniques in view of analyte properties.

9.4 Separation of m/z Species

9.4.1 One-Dimensional Mass Separation

Once a molecule has been ionized, there are in principle two ways to characterize the m/z ratio: (a) trajectory analysis of the charged fragment within an electric and/or magnetic field or (b) after the charging process measurement of the achieved linear

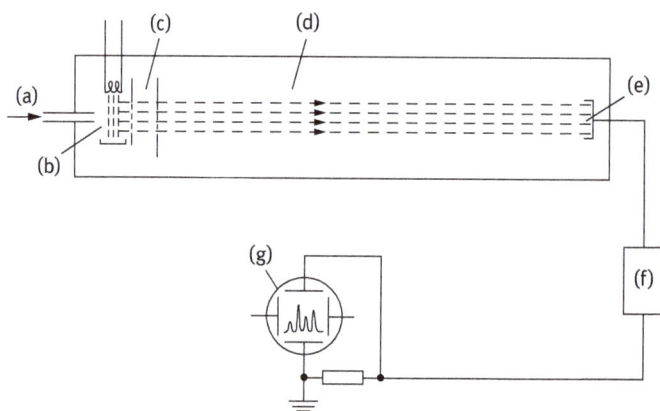

Figure 9.13: Scheme of TOF-MS. (a) Gas inlet, (b) ion source, (c) acceleration, (d) drift chamber, (e) ion detector, (f) amplifier and (g) visual display unit.

velocity of the charged fragment ion under influence of an electric field is performed. The latter is the widely used TOF-MS configuration (see Figure 9.13) [18].

An ion package becomes accelerated first by a voltage pulse and then travels ("drifts") through a field-free evacuated space toward the collection electrode. The ions leaving the ion source of a TOF mass spectrometer do not exhibit monoenergetic behavior; so, various TOF mass spectrometer designs have been developed to compensate for these differences. Often applied is an ion optic device in which ions in a TOF-MS pass through a "mirror" or "*reflectron*" and their flight is reversed (see Figure 9.14) [19].

If a packet of ions of a given m/z contains ions with varying kinetic energies, then the reflectron will decrease the spread in the ion flight times, and therefore improve the resolution of the TOF mass spectrometer. At the detector plate, the incoming charges are continuously recorded as a function of time. Due to their different m/z ratios, differently sized mass ions of the same charge unit are recorded after different traveling times. Nowadays, these TOF-MS units can be built in a very compact manner. The resolution can be still quite high (>100,000). TOF-MS is very fast and therefore best fitting for hybridization with GC. Disadvantageous is the need of pulsed ionization (e.g., by a pulsed laser in REMPI–TOF-MS).

Ion mobility spectrometry (IMS) represents a similar technique [20]. The difference to TOF-MS is the ambient pressure within the drift chamber and the ion source.

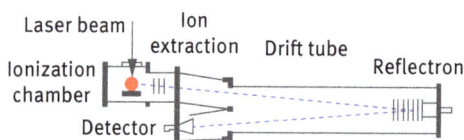

Figure 9.14: Scheme of ion repelling by a reflectron in a TOF-MS.

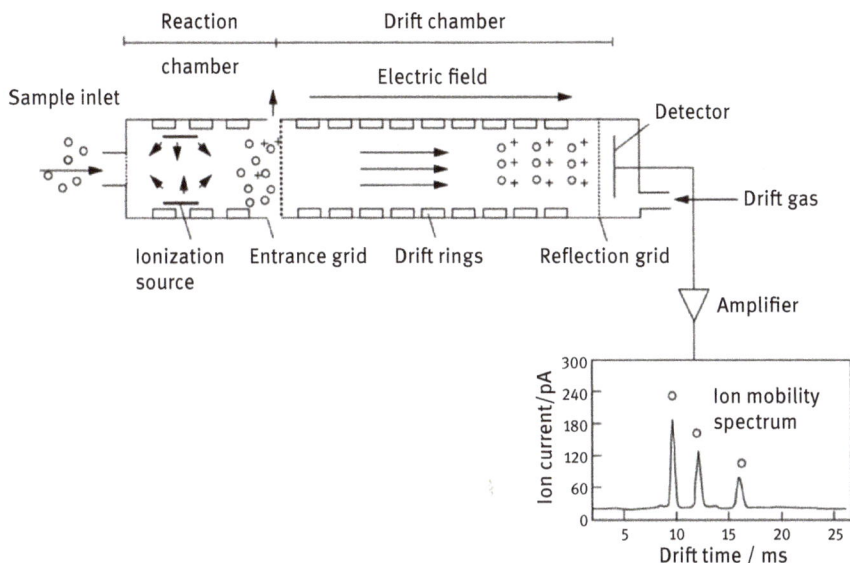

Figure 9.15: Principle of an IMS arrangement.

This principle has been in use for gas sensing for about 50 years (e.g., for chemical weapon monitoring). The principle is shown in Figure 9.15.

The analyte molecules become unipolarly ionized by ion cluster formation at ambient pressure and sometimes with the help of an admixed reactant gas within the ion source by different means (e.g., β rays from radioactive decay, PI or dielectric barrier discharge). At one moment, the gate to the drift tube becomes electronically opened and the differently mobile ion clusters start at the same time to drift against a drift gas stream toward the collector electrode, accelerated by a staggered potential gradient along the drift tube. At the collection electrode, the incoming charges form the ion current, which is amplified and used for mobility characterization. IMS is currently also used for pre-separation to subsequent high-resolution MS (IMS–MS tandem configuration). Also miniaturized versions (<5 cm length) have been published [21].

9.4.2 Two-Dimensional Mass Separation

When a monoenergetic ion beam traverses a magnetic sector field, ions of different *m/z* ratio experience different deviations. At the same time, a focusing happens similarly to optical photon beams. The magnetic sector field acts like a wavelength-selecting prism (see Figure 9.16), sorting and focusing the ion sorts at different locations.

The sector angle can be selected differently. In Figure 9.17, an often used 180° version is shown. By variation of the magnetic field strength, a mass spectrum is

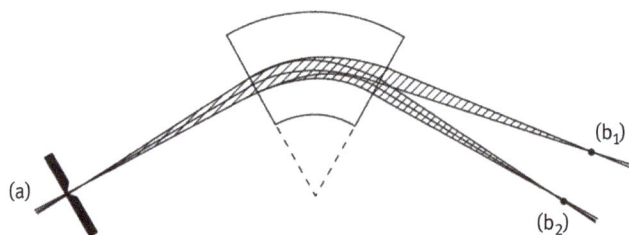

Figure 9.16: Separation of an ion beam consisting of two different m/z crossing a magnetic sector field. (a) injection of analyte mixture; (b_1 and b_2) focal points for separated charged analyte beams.

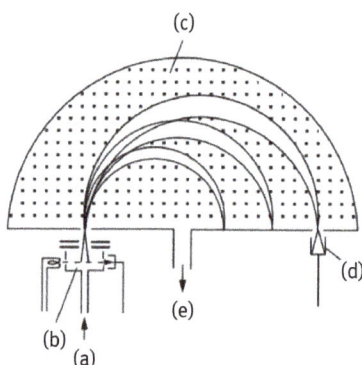

Figure 9.17: Configuration of a 180° magnetic sector field MS. (a) Gas inlet, (b) ion source, (c) magnetic field perpendicular to picture plane, (d) ion detector and (e) to vacuum pumps.

obtained at the detector location. The resolution for such a configuration is not high (<2,000). Magnetic sector field mass spectrometers can be changed from one target mass to another instead of scanning over a given mass range. This *selected ion monitoring* (SIM) method is used to improve sensitivity compared to total ion monitoring for quantitative analysis by focusing only on the masses of interest for specific compounds.

In case the ions leaving the ion source are not monoenergetic, separation of different ion mobilities at one detector location becomes difficult, since focusing becomes dependent on energy and m/z, both. Sorting first by energy (by an electric field) and in a second step by m/z (by a magnetic field) leads to tremendous ion losses. The classical *Mattauch and Herzog MS–MS* configuration represents up to now the best solution for this problem (see Figure 9.18). Focal-plane (array) detectors can detect a range of masses simultaneously.

Using a magnetic sector field angle of 90° and one of 31.49° for the electric sector field, the slit width in between both sectors can be made wide, thus avoiding serious ion losses. Such MS–MS configuration offers excellent resolution (>100,000) but needs bulky and very heavy instrumentation (magnets and vacuum line pumps). The sensitivity depends on the transfer losses at point (d). This configuration shows excellent

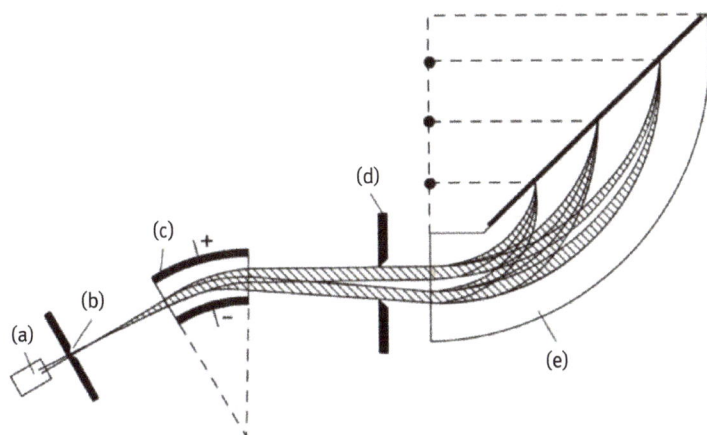

Figure 9.18: Double-focusing Mattauch–Herzog MS–MS configuration. (a) Ion source, (b) entrance slit, (c) electric sector field, (d) slit and (e) magnetic sector field.

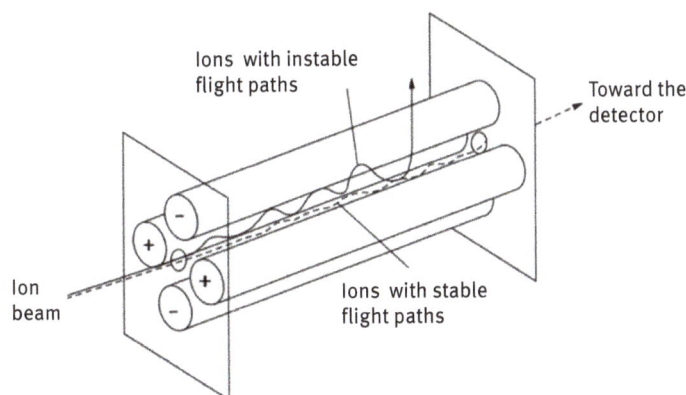

Figure 9.19: Quadrupole mass spectrometer (schematic view).

reproducibility and is best performing for all quantitative tasks, inclusive isotope ratio measurements [22].

Quadrupole MS (QMS) is frequently used in small and rugged MS versions, where only moderate resolution is needed. Ion separation is achieved here by transversal oscillation perpendicular to the direction of movement of ions when traveling along the field of four oppositely charged electrodes (see Figure 9.19).

The oscillations are induced by an appropriate superposition of radiofrequency (RF)-modulated voltage between the electrode rods, which are already held on a direct current voltage. Depending on amplitude, frequency, flow rate and electrical mobility of the charged fragments, only certain ions of a distinct *m*/*z* ratio can penetrate the configuration, hence acting as a mass filter. All other ions collide with the rods and are neutralized, that is, lost for analysis. To acquire a mass spectrum, the potentials must be scanned in constant ratio across a range. Therefore, acquisition of a broader mass range requires a longer scan time.

Advantageous is the low cost of fabrication and the gained sensitivity in the single ion monitoring mode. In contrast to this is the moderate range of observable masses up to 3,500 Da and a lower acquisition rate capability. The acquisition rate determines the possibility to deconvolute neighbored peaks, for example, delivered from an attached GC.

For confirmatory purposes, the SIM mode has disadvantages, since several identification points are needed, which is not achievable in one run. Here, a so-called *triple quadrupole MS/MS* configuration helps (see Figure 9.20).

A triple quadrupole system consists of three quadrupoles. Q1 and Q3 are working as mass filters, while Q2 is acting as collision cell. The precursor ion, selected in SIM mode, becomes reacting in Q2 with collision gas (e.g., N_2), forming a series of new ions which are then separated in Q3. Only one ion of a specific m/z is allowed to pass. All other product ions are filtered out in Q3. An advantage of this *multi-reaction mode* is that it works like a double mass filter which drastically reduces noise and increases selectivity. An introduction to the combination of quadrupole time spectrometry with a flight mass detector was given by Chernushevich and Loboda (2001) [23].

9.4.3 Three-Dimensional Mass Separation by Ion Trap MS

A logic extension of the QMS is the 3D quadrupole ion trap MS [24], where the ions are introduced into a ring-shaped potential cave (see Figure 9.21). Electric fields are used for ion suspension.

Within this trap, the RF field forms a 3D potential well. Since the trap is flooded with helium, incoming ions loose some energy due to collisions with He molecules and become accumulated within the trap. By changing the enveloping RF field strength and symmetry for a short time, part of the accumulated ions can leak the trap and can be measured by a secondary electron multiplier.

By introducing other reactant gases and changing the RF field conditions, fragmentation can be initiated within such trap. By preselection, well-defined

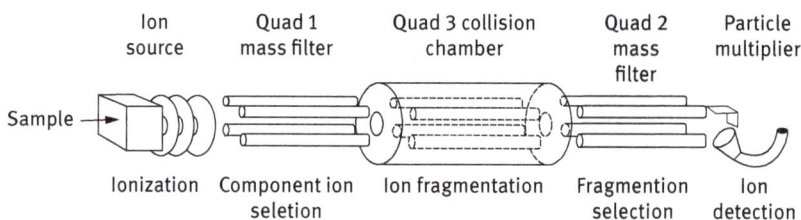

Figure 9.20: Schematic view of a triple quadrupole mass spectrometer.

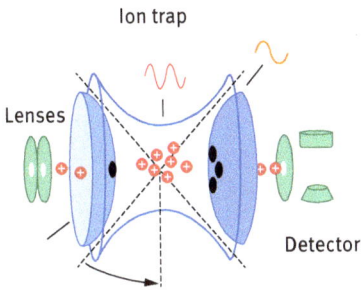

Ion trap

Lenses

Detector   Figure 9.21: Quadrupole ion trap accumulating and ejecting ions.

fragmentation experiments within such trap can be produced, and if wanted, repeated ($n < 7$). After reaching optimal fragmentation, the fragments are ejected from the trap and counted by the secondary electron multiplier. These MS^n experiments are advantageous for protein characterization or peptide sequencing [25].

An improved resolution is obtained with a similar configuration, but using a homogeneous magnetic field. The circulating ions generate electromagnetic RF radiation whose image current is recorded by detector plates in the time domain as Fourier-transformed signal, which then is converted into a mass spectrum (see Figure 9.22). This arrangement is called *Fourier-transform ion cyclotron* MS (FT-ICR-MS) [26].

The trapped ions can remain for days within the trap and then be released to the detector. FT-ICR-MS produces highest mass spectral resolution (>1,000,000) and is also applicable to fragmentation studies. The instrumentation is extremely bulky and expensive.

The *Orbitrap* MS has been developed by Makarov (1999), and its principle is shown in Figure 9.23 [27].

Once injected into the rotational symmetric Orbitrap cell, the ions move back and forth in helical orbits around the spindle-shaped electrode under the influence

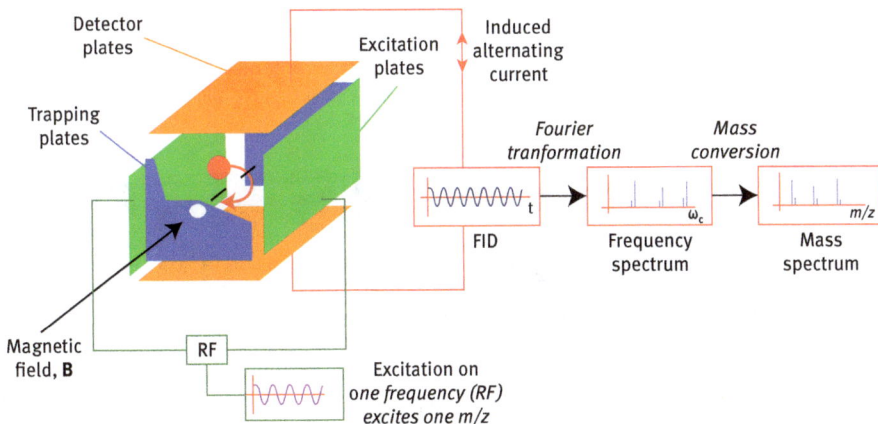

Figure 9.22: Setup of Fourier-transformation ion cyclotron MS.

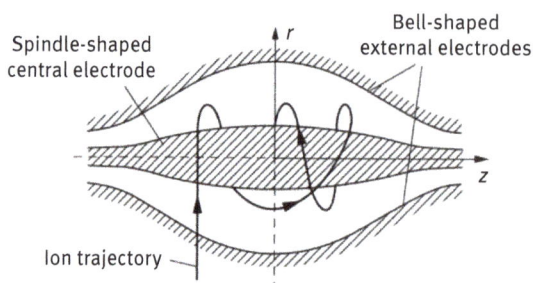

Figure 9.23: Trajectory of an ion within the Orbitrap separation cell.

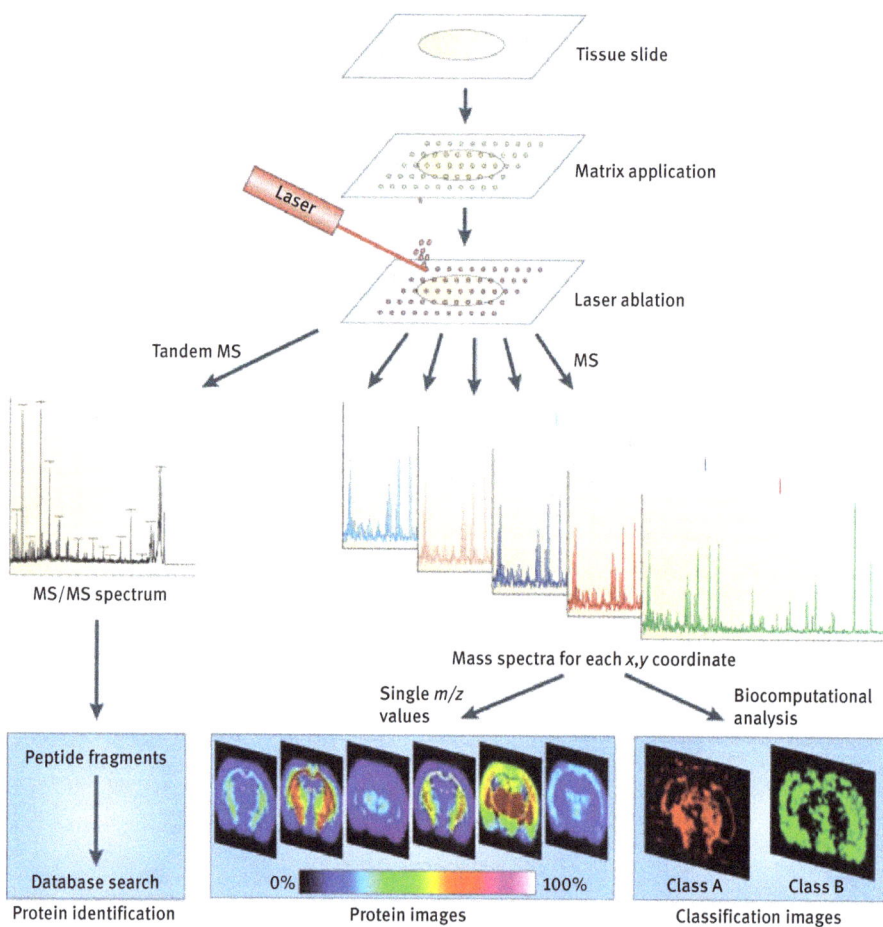

Figure 9.24: The way to chemical imaging by rastering a surface in a MALDI–MS process. Reproduced with permission from Nature Publishing Group.

of an electrostatic field. Ions of different m/z ratios move with different rotational frequencies but the same axial symmetry around the electrode. Axial oscillations of ion rings are detected by their image current induced on the outer electrode. The readout procedure therefore is similar to FT-ICR-MS. Resolving powers can be as high as 1,000,000. Advantageous is the very compact buildup. Orbitrap-based mass spectrometers are used in proteomics, metabolomics, environmental, food and safety analysis [28].

9.5 Molecular Imaging by MS

Chemical (molecular) imaging is one of the most pressing tasks nowadays. Especially, quickly accessible information on the distribution of organic molecules at a surface is often needed. This can be contamination deposited at any surface (e.g., explosives on the skin of fingers or a bacterial infection of tissue), but in medicine the recognition and characterization of the extent of a growing tumor is of utmost importance.

Currently, MALDI–MS technique is applied for this [29]. In Figure 9.24, the steps to gain such 2D distribution are depicted.

The tissue sample is subjected first to a deposition of a photon-absorbing matrix layer. Next, by rastering the surface shot-by-shot with an UV laser pulse, consecutively, the surface-enriched material becomes ablated, ionized and characterized by an MS/MS configuration. Finally, the obtained MS spectra are used to produce 2D information, that is, a 2D picture, about the protein distribution.

Further Reading

Cole, R. Electrospray ionization mass spectrometry: fundamentals, instrumentation, and applications. New York: Wiley, 1997.

Domin, M, Cody R. Ambient ionization mass spectrometry. Cambridge: RSC, 2015.

Ellis A, Mayhew C. Proton transfer reaction mass spectrometry: principles and applications. Hoboken: Wiley, 2014.

Gross J. Mass spectrometry: a textbook. Heidelberg: Springer, 2004.

Gross M, Caprioli R. Encyclopedia of mass spectrometry. Cambridge: Elsevier, 2013.

Harrison A. Chemical ionization mass spectrometry. Boca Raton: CRC Press, 1983.

March R, Todd, J. Practical aspects of ion trap mass spectrometry; volume III: chemical, environmental and biomedical applications. Boca Raton: CRC Press, 1995.

Marshal A. Fourier transform ion cyclotron resonance mass spectrometry. Amsterdam: Elsevier, 1996.

McLuckey S. Quadrupole ion trap mass spectrometry, in Honor of John FJ Todd and Raymond E. Amsterdam: Elsevier, 1999.

Pramanik B, Ganguly A, Gross M. Applied electrospray mass spectrometry. New York: Marcel Dekker, 2002.

Schlag E. Time-of-flight mass spectrometry and its applications. Amsterdam: Elsevier, 1994.

Wilkins C, Trimpin S. Ion mobility spectrometry – mass spectrometry: theory and applications. Boca Raton: CRC Press, 2011.

Bibliography

[1]   Chen CH. Review of a current role of mass spectrometry for proteome research. Anal Chim Acta 2008;624(1):16–36.

[2]   Griffiths J. A brief history of mass spectrometry. Anal Chem 2008;80(15):5678–83.

[3]   Pietrogrande MC. Enantioselective separation of amino acids as biomarkers indicating life in extraterrestrial environments. Anal Bioanal Chem 2013;405(25):7931–40.

[4]   Milman B, Zhurkovitch I. Mass spectral libraries: a statistical review of the visible use. Trends Anal Chem 2016;80:636–40.

[5]   Eberlin MN. Structurally diagnostic ion/molecule reactions: class and functional-group identification by mass spectrometry. J Mass Spectrom 2006;41(2):141–56.

[6]   Portoles T, Pitarch E, Lopez FJ, Hernandez F, Niessen WM. Use of soft and hard ionization techniques for elucidation of unknown compounds by gas chromatography/time-of-flight mass spectrometry. Rapid Commun Mass Spectrom 2011;25(11):1589–99.

[7]   Munson MS, Field FH. Chemical ionization mass spectrometry. I. General introduction. J Am Chem Soc 1966;88(12):2621–30.

[8]   Kuch HM, Ballschmiter K. Determination of endocrine-disrupting phenolic compounds and estrogens in surface and drinking water by HRGC-(NCI)-MS in the picogram per liter range. Environ Sci Technol 2001;35(15):3201–6.

[9]   Letzel T, Poschl U, Rosenberg E, Grasserbauer M, Niessner R. In-source fragmentation of partially oxidized mono- and polycyclic aromatic hydrocarbons in atmospheric pressure chemical ionization mass spectrometry coupled to liquid chromatography. Rapid Commun Mass Spectrom 1999;13(24):2456–68.

[10] Hopmans EC, Schouten S, Pancost RD, van der Meer MT, Damste JS. Analysis of intact tetraether lipids in archaeal cell material and sediments by high performance liquid chromatography/atmospheric pressure chemical ionization mass spectrometry. Rapid Commun Mass Spectrom 2000;14(7):585–9.

[11] Fenn JB, Mann M, Meng CK, Wong SF, Whitehouse CM. Electrospray ionization for mass spectrometry of large biomolecules. Science 1989;246(4926):64–71.

[12] Manikandan M, Kazibwe Z, Hasan N, Deenadayalan A, Gopal J, Pradeep T, Chun S. Biological desorption electrospray ionization mass spectrometry (DESI MS) – unequivocal role of crucial ionization factors, solvent system and substrates. TRAC Trends Anal Chem 2016;78:109–19.

[13] Lindinger W, Hansel A, Jordan A. On-line monitoring of volatile organic compounds at pptv levels by means of proton-transfer-reaction mass spectrometry (PTR-MS) – Medical applications, food control and environmental research. Int J Mass Spectrom 1998;173(3):191–241.

[14] Heger HJ, Zimmermann R, Dorfner R, Beckmann M, Griebel H, Kettrup A, Boesl U. On-line emission analysis of polycyclic aromatic hydrocarbons down to pptv concentration levels in the flue gas of an incineration pilot plant with a mobile resonance enhanced multiphoton ionization time-of-flight mass spectrometer. Anal Chem 1999;71(1):46–57.

[15] Boesl U, Kartouzian A. Mass selective chiral analysis. Ann Rev Anal Chem 2016;9:343–64.

[16] Caprioli RM, Farmer TB, Gile J. Molecular imaging of biological samples: localization of peptides and proteins using MALDI-TOF MS. Anal Chem 1997;69(23):4751–60.

[17] Hayen H, Michels A, Franzke J. Dielectric barrier discharge ionization for liquid chromatography/mass spectrometry. Anal Chem 2009;81(24):10239–45.

[18] Cotter R. Time-of-flight mass spectrometry: instrumentation and applications in biological research. Washington: American Chemical Society, 1997.

[19] Cotter RJ, Griffith W, Jelinek C. Tandem time-of-flight (TOF/TOF) mass spectrometry and the curved-field reflectron. J Chromatogr B-Anal Technol Biomed Life Sci 2007;855(1):2–13.

[20] Kanu AB, Dwivedi P, Tam M, Matz L, Hill HH. Ion mobility-mass spectrometry. J Mass Spectrom 2008;43(1):1–22.
[21] Wilks A, Hart M, Koehl A, Sommerville J, Boyle B, Ruiz-Alonso D. Characterization of a miniature, ultra-high-field, ion mobility spectrometer. Int J Ion Mobility Spectrom 2012;15:199–222.
[22] Feldmann I, Jakubowski N, Thomas C, Stuewer D. Application of a hexapole collision and reaction cell in ICP-MS Part II: Analytical figures of merit and first applications. Fresenius J Anal Chem 1999;365(5):422–8.
[23] Chernushevich IV, Loboda AV, Thomson BA. An introduction to quadrupole-time-of-flight mass spectrometry. J Mass Spectrom 2001;36(8):849–65.
[24] March RE. An introduction to quadrupole ion trap mass spectrometry. Journal of Mass Spectrometry 1997;32(4):351–69.
[25] Schroeder MJ, Shabanowitz J, Schwartz JC, Hunt DF, Coon JJ. A neutral loss activation method for improved phosphopeptide sequence analysis by quadrupole ion trap mass spectrometry. Anal Chem 2004;76(13):3590–8.
[26] Marshall AG, Hendrickson CL, Jackson GS. Fourier transform ion cyclotron resonance mass spectrometry: a primer. Mass Spectrom Rev 1998;17(1):1–35.
[27] Hu QZ, Noll RJ, Li HY, Makarov A, Hardman M, Cooks RG. The Orbitrap: a new mass spectrometer. J Mass Spectrom 2005;40(4)430–43.
[28] Perry RH, Cooks RG, Noll RJ. Orbitrap mass spectrometry: instrumentation, ion motion and applications. Mass Spectrom Rev 2008;27(6):661–99.
[29] Schwamborn K, Caprioli RM. Innovation molecular imaging by mass spectrometry – looking beyond classical histology. Nat Rev Cancer 2010;10(9):639–46.

10 Receptor-based Bioanalysis for Mass Screening

10.1 General Remarks

High-throughput and multiplexed analyte screening is often required, whether it is in clinical chemistry, where hundreds of patient serum samples have to be screened for biomarker presence within few minutes to hours [1], strategies for high-throughput drug screening [2], or for analysis of a limited number of specific compounds (e.g., toxins, pesticides) in food or water samples [3, 4].

A good example for the latter is the mobility determination of pesticides in water from hydrogeological dispersion or breakthrough experiments [5]. After having injected a pesticide into a water well (or on top of a lysimeter), water samples have to be collected at several observation wells around the respective aquifer (or at the exit of the lysimeter). This may yield up to hundreds of water samples within a predetermined observation period. From the results of the pesticide analyses, so-called breakthrough curves for this specific hydrogeological situation are gained and will be used for characterization of the retention function of the soil. This helps to establish the necessary protection areas for the water supplier of a community. Usually, chromatography with mass spectrometry would be the first choice to meet such task. But the unavoidable sample pretreatment and enrichment of analytes present in the ppt range would cause enormous efforts in terms of manpower and costs.

The only way out is high-throughput screening (HTS) by highly parallel analysis and usage of technologies which are applicable to the real sample without much pretreatment [6]. A look to clinical chemistry, where a huge sample frequency is a part of the daily workload, gives an answer, that is, the combination of automatization and receptor-based analysis. Principally, flow injection technologies would be applicable, if the sensitivity and selectivity of electrochemical or optical detection was in the ppt range. So far, for example for selective pesticide analysis, such a scenario is not known. Therefore, combination of a highly affine recognition principle with a highly sensitive detection principle would enable the requested analysis. If such an analysis can be performed in parallel, a high multitude of samples' HTS is possible.

Receptor-based analysis means mainly bioanalysis. In the course of evolution, nature has tailored wonderful tools for this purpose. *Enzymes* (E) and *antibodies* (Ab) are natural, highly selective receptors, nowadays in use for automated high-throughput analysis [7]. Whether they are able to replace the classical analytical instrumentation, like chromatography and/or spectroscopy or spectrometry, is a question of demand. Of course, one has to live with a compromise. Hence, screening analyses serve for a first overview on the presence of harmful substances but need to be validated by classical techniques. But sorting out those samples which are uncontaminated or above a certain level exonerates tremendously the available analytical capacity. In case of presence of only a few analytes, such techniques are equivalent to the classical techniques (e.g., LC, GC, MS).

DOI 10.1515/9783110441154-010

Antibodies and enzymes have been used during the last decades as blueprints for developing stable synthetical analogs, like *aptamers* [8] or *molecularly imprinted polymers* [9]. Currently, only a few examples have been successfully applied toward real-world samples.

In the following, the basics for Ab and E and some applications of them will be discussed.

10.2 Natural Receptors (Enzymes and Antibodies)

Enzymes and antibodies are proteins with the ability to bind appropriate ligands very strongly and with high selectivity. The catalytic activity of enzymes provides an enormous range of analytical techniques. The ability of a single enzyme molecule to catalyze the reaction of numerous substrate molecules also generates an amplification effect which enhances the sensitivity of the analyses. A further advantage is that most enzyme-catalyzed reactions can be followed by simple, widely available spectroscopic or electrochemical methods. Enzymes are normally active only under mild conditions (aqueous solutions at moderate temperatures and controlled pH) and this restricts the circumstances in which they can be used. By immobilization of enzymes to suitable solid matrices [10, 11] or by mutagenesis [12, 13], the enzymatic activity can be preserved even under unfavorable conditions, such as the presence of organic co-solvents or nonphysiological pH. Enzymes are generally classified on the basis of the type of reactions that they catalyze. Six groups of enzymes can be recognized on this basis. Table 10.1 lists the six groups of enzymes along with examples.

Antibodies have been used in analysis for over 70 years and offer a wide range of techniques and applications. In some cases, the specific combination of an antibody

Table 10.1: Classification of enzymes.

Group of Enzyme	Reaction catalyzed	Examples
Oxidoreductases	Transfer of hydrogen and oxygen atoms or electrons from one substrate to another.	Dehydrogenases, Oxidases
Transferases	Transfer of a specific group (a phosphate or methyl etc.) from one substrate to another.	Transaminase, Kinases
Hydrolases	Hydrolysis of a substrate.	Esterases, Digestive enzymes
Isomerases	Change of the molecular form of the substrate.	Phosphohexose isomerase, Fumarase
Lyases	Nonhydrolytic removal of a group or addition of a group to a substrate.	Decarboxylases, Aldolases
Ligases (Synthetases)	Joining of two molecules by the formation of new bonds.	Citrate synthetase

with the corresponding antigen or hapten can be detected directly (e.g., by nephelometry), but more often such reactions are monitored by a characteristic label such as an enzyme, fluorophore and so on. Since antibody reactions do not have a built-in amplification effect, these labels are frequently necessary to provide sufficient analytical sensitivity. Enzyme immunoassays (IAs), in which enzymes act as labels, are very well established. Antibodies of different classes vary greatly in stability, but some are relatively robust proteins, and this contributes significantly to the range of methods available.

10.3 Working Principle of Enzymes (Enzymatic Catalysis, Enzymatic Inhibition)

The basic mechanism for *enzymatic catalysis*, described by Michaelis and Menten, proposes that the enzyme E reacts reversibly with the substrate S to form a complex ES, which subsequently decomposes to release the free enzyme E and the product P:

$$E + S \rightarrow ES \rightarrow E + P$$

In this model, the combination reaction of E and S, with a rate constant k_1, occurs at the catalytically active site of the enzyme (see Figure 10.1).

The reverse process, dissociation of the **ES** complex, has a rate constant k_2, and the decomposition of ES to give E and P has a rate constant k_3. These three rate constants can be used to calculate the Michaelis constant K_M:

$$K_M = \frac{(k_2 + k_3)}{k_1}$$

The Michaelis constant has units in mol/L and in practice values between 10^{-1} and 10^{-7} mol/L.

An important feature of enzymes is that their catalytic (=active) sites can often be occupied by, or react with, molecules other than the substrate, leading to *inhibition*

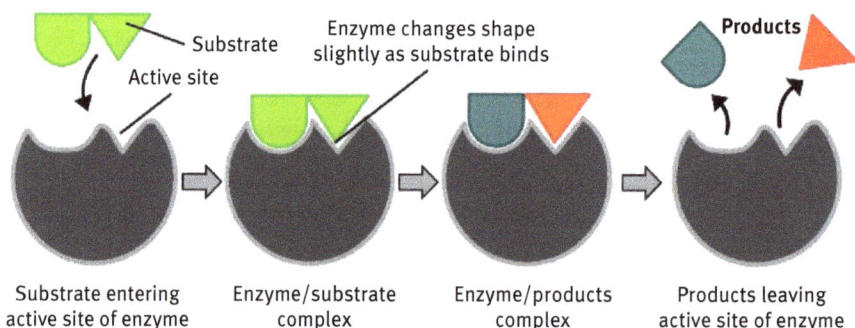

Figure 10.1: Schematic view of enzymatic catalysis process.

of enzyme activity. Several inhibition mechanisms are known, but it is necessary only to distinguish between irreversible and reversible inhibition. Irreversible inhibition arises when the inhibitor molecule *I* dissociates very slowly or not at all from the active site of the enzyme. The best known examples occur when the inhibitor binds covalently to a critical residue in the active site. Inhibition of cholinesterase enzymes by the reaction of organophosphorus compounds with a serine residue is such an example (see Figure 10.2). The chemical warfare compound Soman (*o*-pinacolyl methylphosphonofluoridate) is extremely toxic for mammals due to irreversible blocking of the active site of acetylcholinesterase (AChE) (see Chapter 11, Section 11.3).

This type of inhibition is said to be noncompetitive, as the enzyme activity cannot be restored by addition of excess substrate. Inhibition assays can be rather sensitive down to the ppt level.

Enzyme assays based on inhibition effects are not as commonly employed as substrate determinations, but some are very important. Prominent is the determination of organophosphorus compounds by using their inhibitory effect on cholinesterase enzymes (E.C. 3.1.1.8: the first digit signifies a hydrolase enzyme, the second means that the compounds hydrolyzed are esters and the third that they are phosphoric monoesters). Cholinesterases catalyze the conversion of acetylcholines to choline and the corresponding acid:

$$\text{Acetylcholine} + H_2O \rightarrow \text{choline} + \text{acid}$$

This reaction is crucial to many living systems, so the organophosphorus compounds (pesticides, chemical warfare agents) that inhibit it by irreversible binding to the active site are often highly toxic. AChE assays provide a convenient method for the detection of AChE activity [14]. It uses the substrate 5,5′-dithiobis-2-nitrobenzoic acid (DTNB) to quantify the thiocholine produced from the hydrolysis of acetylthiocholine by AChE in blood, in cell extracts and in other solutions. The optical absorption intensity of the DTNB adduct is used to measure the amount of thiocholine formed, which

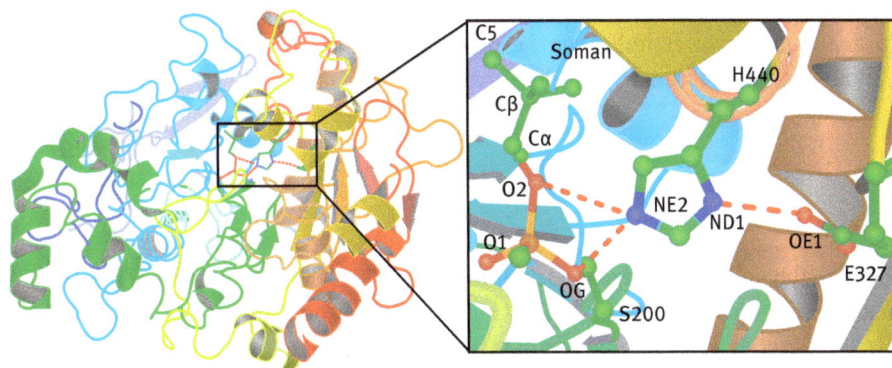

Figure 10.2: Active site of AChE blocked by the organofluorophosphate molecule Soman.

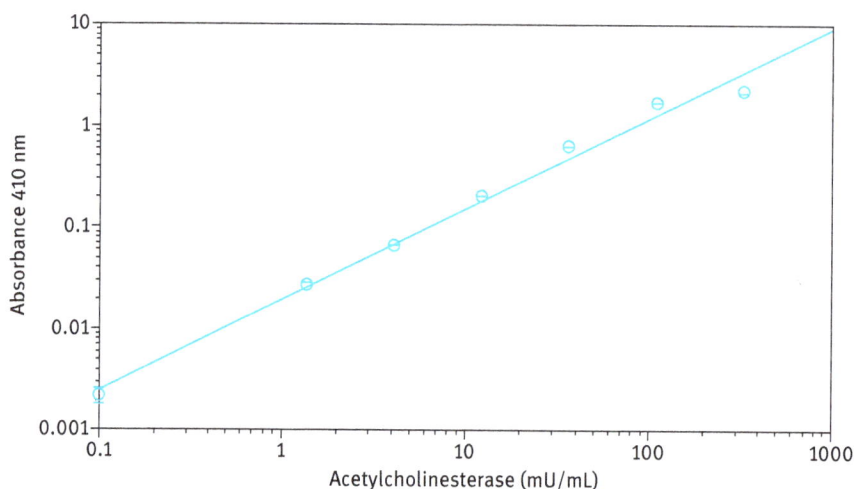

Figure 10.3: Acetylcholinesterase dose response.

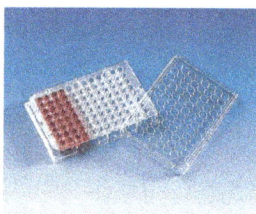

Figure 10.4: Polystyrene 96 microtiter plate (MTP).

is proportional to the AChE activity (see Figure 10.3). The optical measurement is performed within the microplate wells by a microplate reader (see Figure 10.4) within some minutes automatically. This analytical task needs 30 min incubation time and can process 96 samples in parallel. Nowadays, already 1536 well plates can be handled on a routine base [15].

A number of companies have developed robots to specifically handle microplates. Up to thousands of samples can be processed within hours, inclusive the optical readout (e.g., fluorescence or absorbance).

10.4 Working Principle of Antibodies (IA, Test Format, Microarray, Dip Stick)

The probably most applied mass-screening principle is the immunoassay (IA). IAs are based on the formation of an antibody–antigen immunocomplex. IAs are especially interesting when high sample frequencies are expected, extreme sensitivity *without*

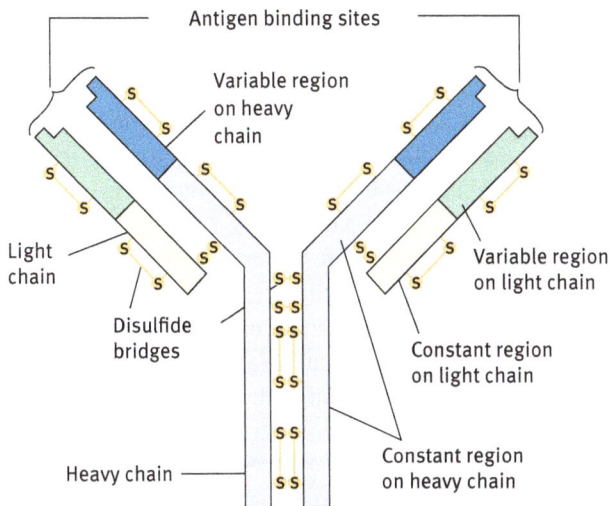

Figure 10.5: Schematic view of an antibody structure.

enrichment is needed or expensive equipment like LC–MS is not available. Today, IAs are widely used in clinical chemistry.

Prerequisite for a successful IA is the availability of highly affine and selective antibodies, which are raised in vertebrates. Abs are Y-shaped glycoproteins, also known as immunoglobulins (Ig) (see Figure 10.5). The short tips of the variable regions (paratopes) serve as recognition sites for its counterpart at the antigen, the epitope.

In organic trace analysis, mainly polar substances of a molecular weight >200 Da might be analyzable by an IA technique. Below 1,000 Da, these analyte molecules do not act as antigens. Therefore, the production of Ab against so-called haptens needs special attention. Usually, large proteins like keyhole limpet hemocyanin, human or bovine serum albumin (BSA) or thyroglobulin are used as carrier proteins. The small analyte is covalently bound to the carrier.

Commonly in use for analytical purposes are *polyclonal* or *monoclonal* antibodies (pAb or mAb). For the production of pAb, the immunogen (hapten linked to a carrier protein) is applied to vertebrates, often with the help by adjuvants. Polyclonal immune sera from an immunization of this type consist of a spectrum of antibodies, which are active against the diverse epitopes of the same antigen and, therefore, possess various affinity constants. The affinity can be described by the dissociation constant K_d of the immunocomplex, that is, how tightly an antigen is bound to the Ab, and may range up to 10^{-12} M. The higher the affinity or the lower K_d, the more sensitive an IA can be configurated. Further, the *cross-reactivity* of the antibodies is of utmost importance. Cross-reactivity refers to the reaction of an antibody with molecules which are structurally very similar to the desired analyte (hapten); it determines the sensitivity to interfering substances in the final test.

The acceptance of IAs depends on the availability of antibodies. Since the amount of polyclonal antibodies is always limited, and even repeated immunizations of the same experimental animal species result in nonreproducible characteristics (specificity, affinity), the production of suitable mAb (they are product of a single cell alone) is most important. The method was developed by Köhler and Milstein [16] and encompasses the immortalization of a single-antibody producing cell (B-lymphocyte) by fusion to a myeloma cell. The resulting hybridoma produces just one species of antibody which can be grown indefinitely and, therefore, offers a means to advise a consistent product in unlimited quantities. For this reason, tests based on mAb principally can become accepted as standardized methods. In recent years, the production of genetically engineered recombinant antibodies has been reported [17].

IAs are nowadays performed in many configurations. One can principally distinguish homogeneous and heterogeneous IAs. The latter uses either surface-bound Ab or antigens on microtiter plates (MTPs), or the reaction takes place only within the liquid phase. Homogeneous IAs can be automatized more easily, in contrast to the heterogeneous format, where different separation/washing steps are needed. This is accomplished with the help of an automatized MTP washing station and a MTP reader.

Furthermore, IAs can be subdivided into competitive or noncompetitive assays. The most applied IA is the so-called enzyme-linked immunosorbent assay (ELISA). Its principle is shown in Figure 10.6.

This assay is the most important in food and environmental analysis. First, antibodies are immobilized on a solid carrier. γ-Irradiation is applied, to make the carrier sorption-active. Standard 96 well MTPs, polystyrene beads or test tubes can be used. In a typical ELISA test on a MTP (Figure 10.4), the carrier is first coated with analyte-binding antibody. In case of the indirect-competitive assay, the antigen (the analyte of interest) becomes attached.

The sample or standard is then added after washing off the excess antibody. The volume added is typically 100–200 μL. After preincubation, a constant quantity of tracer, for example, an enzyme-labeled hapten, is added to the sample or standard. This initiates a competitive reaction, because only a limited number of antibodies are available for binding. After the tracer incubation period, the sample (or standard) and excess reagents are washed away. The bound tracer concentration is inversely proportional to that of the analyte. The amount of bound tracer can be determined via a common enzyme-substrate reaction with a chromogenic substrate. After a reaction time long enough to produce sufficient dye, the enzyme reaction is stopped by addition of acid. Subsequently, the depth of color formed in the individual wells of the MTP is automatically determined with a MTP reader (photometer).

The application of a competitive heterogeneous IA is very common. Some important IAs for trace amounts of environmentally relevant compounds are presented in

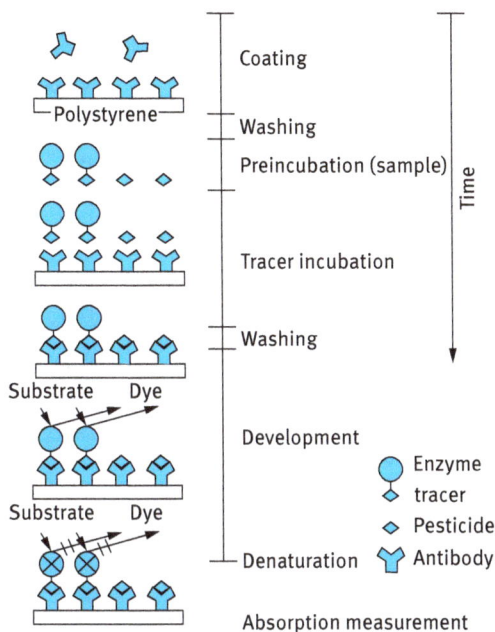

Figure 10.6: Principle of a direct-competitive ELISA with photometric readout.

Table 10.2. Especially for agrochemicals, a large number of determination procedures have been established, because of the legislation on drinking water with a limiting value of 100 ng/L for a single pesticide. Aside these applications, especially for clinical purposes, a manifold of tests is practiced.

IAs are of special value, if a predetermined analytical task has to be fulfilled, and the identity of analytes is known already. For such targeted analysis, not only the use of prefabricated microtitration plates is possible, rather for semiquantitative purposes, often a dipstick format for single-use and single-analyte determination is available.

Recently, the parallel readout principle of a MTP format has been merged with the advantage of high-throughput flow-injection analysis. An example is the microarray platform MCR3 [31] (see Figure 10.7a). Its working principle is the automatized ELISA of liquid samples with trace contaminants within an optically transparent flow-through microarray cell in opposite a CCD camera [32] (see Figure 10.7b). Detection is achieved by a chemiluminescence reaction observed within the flow cell.

The whole analytical protocol (sample introduction, mixing with antibodies, incubation, washing steps, substrate addition, optical readout of produced chemical luminescence, chip regeneration, data handling) happens computer assisted. The heart is the microarray flow cell (ca. 100 µL volume), consisting of a glass plate base, where with the help of a spotting machine, the different antibodies or antigens have been placed in a known order. Usually, a chemical architecture assures a covalent

Table 10.2: Selected ELISA tests using monoclonal antibodies for environmental analysis.

Substance	Detection limit [ng/L]	Reference
Pesticides		
Atrazine	4	[18]
Terbuthylazine	30	[19]
2,4-Dichlorophenoxyacetic acid	30	[20]
Aromatic compounds		
Benzo[a]pyrene	20	[21]
Trinitrotoluene	60	[22]
Halogenated aromatic compounds		
PCB80	150	[23]
2,3,7,8-Tetrachlorodibenzodioxin	10	[24]
Surfactants		
Alkyl ethoxylate ($C_{12}EO_7$)	5,000	[25]
Linear alkylbenzene sulfonates (C_9–C_{13})	20,000	[26]
Toxins		
Microcystin LR	6	[27]
Microcystins (ADDA)	70	[28]
(Pharmaceutical) drugs		
Diclofenac	8	[29]
Caffeine	1	[30]

linkage to the transparent glass surface. One spot has a diameter of ca. 450 μm. Once the sample has experienced the competitive ELISA, the optical CCD readout observes location of a spot and its luminescence intensity (see Figure 10.8).

Since a series of redundant parallel spots for the same analyte is in one row, outliers are recognized and can be easily removed. The number of analytes to be accomplished by one chip depends on the availability of suitable antibodies. A typical example is the analysis of antibiotics in fresh milk. Here, up to 13 different antibiotics can be analyzed directly within the fresh milk [33]. Analysis time is 6 min, incl. regeneration of the microarray by chaotropic reagent addition (a chaotropic agent in aqueous solution disrupts the hydrogen-bonding network between water molecules and weakens hydrophobic effects; examples are ethanol, guanidinium chloride or thiourea). More than 50 regenerations are feasible with 1 chip. This reduces analysis costs considerably and gives the opportunity for intermittent blank and standard measurements. Such microarrays may be produced by mass fabrication; hence, certification is basically an option.

Different tests based on the same microarray format and chemiluminescence readout have been reported for a variety of analytes [antibiotics, polycyclic aromatic hydrocarbons (PAHs), toxins, pharmaceuticals, pesticides, microorganisms, etc.]. It has to be mentioned that beyond a luminescence readout, also electrochemical or photon emission detection after laser stimulation (fluorescence, Raman emission) is known.

(a)

(b)

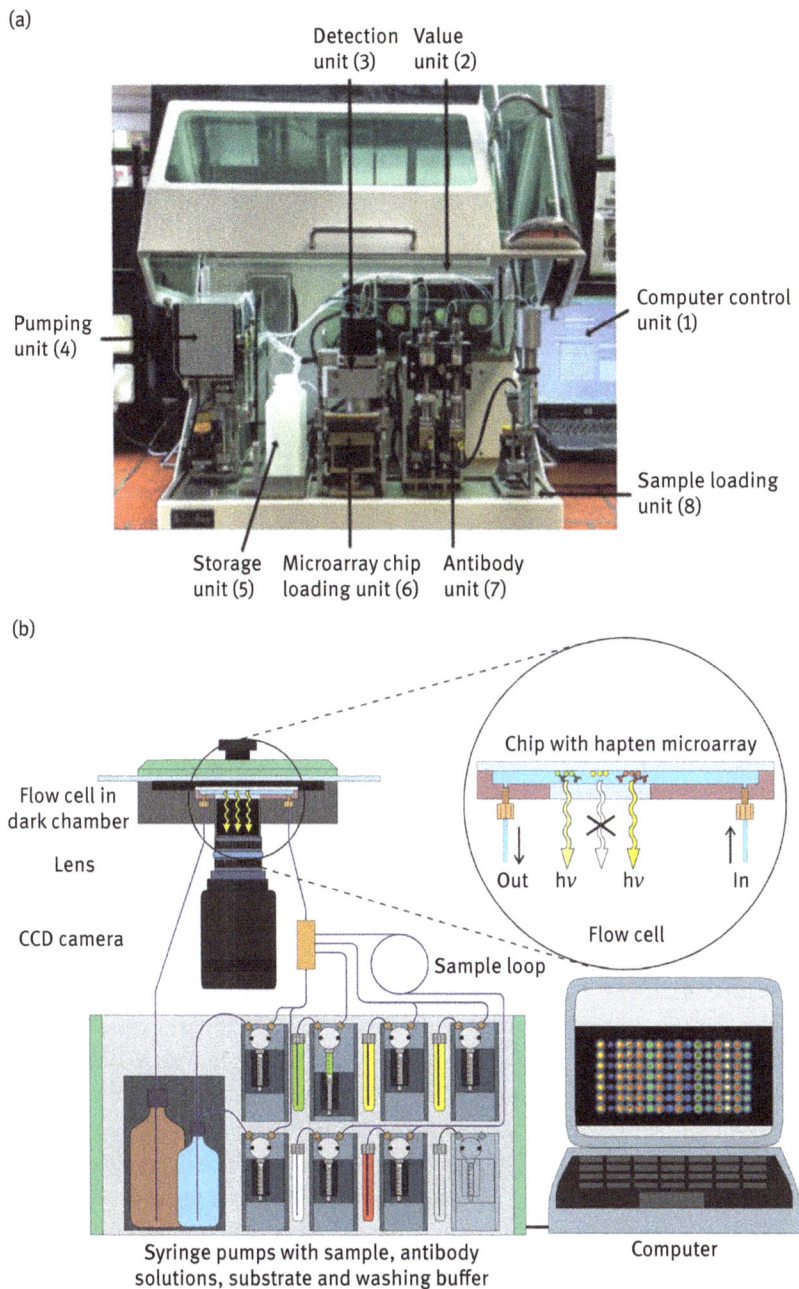

Figure 10.7: (a) Microarray platform MCR3. Reproduced with permission from Elsevier. (b) Functional parts of a microarray platform and chemiluminescence readout.

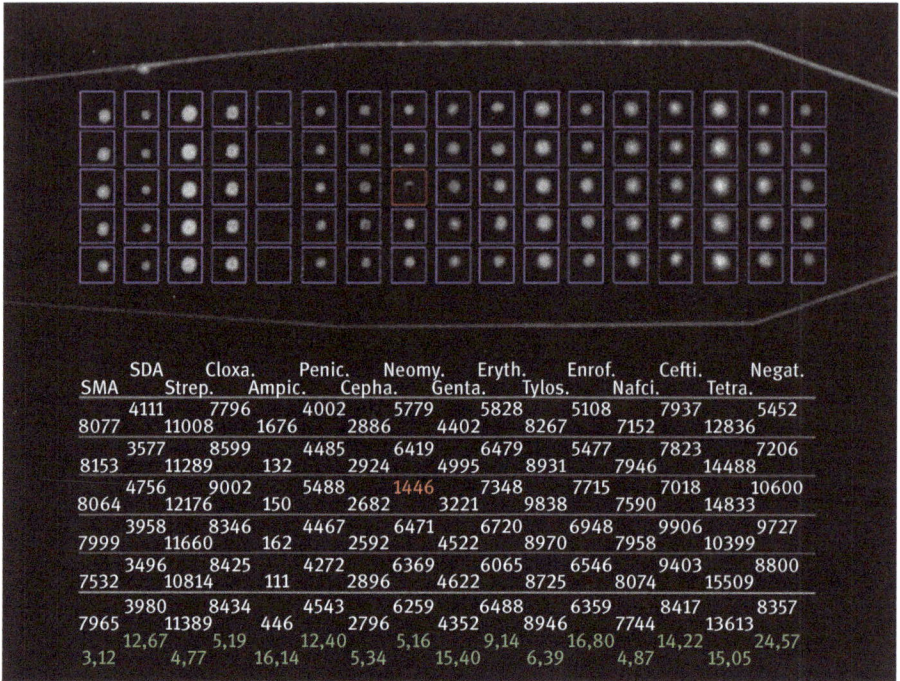

| SDA | Cloxa. | Penic. | Neomy. | Eryth. | Enrof. | Cefti. | Negat. |
SMA	Strep.	Ampic.	Cepha.	Genta.	Tylos.	Nafci.	Tetra.
4111 7796	4002	5779	5828	5108	7937	5452	
8077 11008	1676	2886	4402	8267	7152	12836	
3577 8599	4485	6419	6479	5477	7823	7206	
8153 11289	132	2924	4995	8931	7946	14488	
4756 9002	5488	1446	7348	7715	7018	10600	
8064 12176	150	2682	3221	9838	7590	14833	
3958 8346	4467	6471	6720	6948	9906	9727	
7999 11660	162	2592	4522	8970	7958	10399	
3496 8425	4272	6369	6065	6546	9403	8800	
7532 10814	111	2896	4622	8725	8074	15509	
3980 8434	4543	6259	6488	6359	8417	8357	
7965 11389	446	2796	4352	8946	7744	13613	
12,67 5,19	12,40	5,16	9,14	16,80	14,22	24,57	
3,12 4,77	16,14	5,34	15,40	6,39	4,87	15,05	

Figure 10.8: Automatic data processing of a microarray view.

In contrast to microarray formats, which need to be embedded in a complex microfluidic handling of reagents, incubation followed by washing steps and a detection reaction, test strips or dip sticks allow a quick semiquantitative and fairly cheap information on the presence of an antigen or antibody in an aqueous environment. This format has especial advantages when only one analyte has to be screened. Nowadays, many variants are commercialized. They are mainly used for clinical or self-diagnostics, for example, pregnancy testing for human chorionic gonadotropin in the mM range. The general principle of such dipsticks is shown in Figure 10.9. As the dipstick is dipped in the biological fluid (urine, blood) to be tested, antigen–antibody complexes are formed with the first antibody, which is conjugated to colloidal gold nanoparticles. The complexes then migrate by capillarity until they reach the line of immobilized capture antibody, which becomes materialized as a red line. A second line of immobilized antibodies catches the gold-conjugated Ab that did not bind to the first line and serves as a control for dipstick functioning. Evaluation can be done visually or by reflectometry.

For applications in the trace-concentration range, this principle has to be combined with extremely affine antibodies and further separation means for preconcentration. This is shown in Figure 10.10 for a membrane-based lateral-flow immunodipstick assay developed for the fast screening of aflatoxin B_2 (AFB_2) in food samples

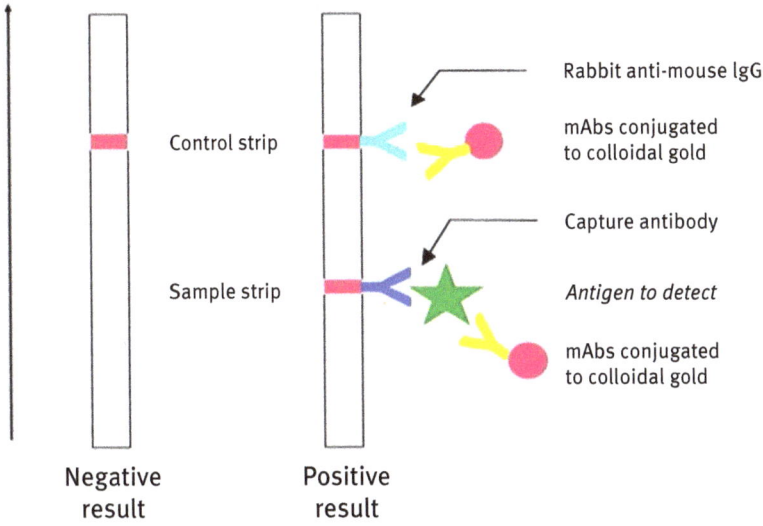

Figure 10.9: Working principle of a dipstick.

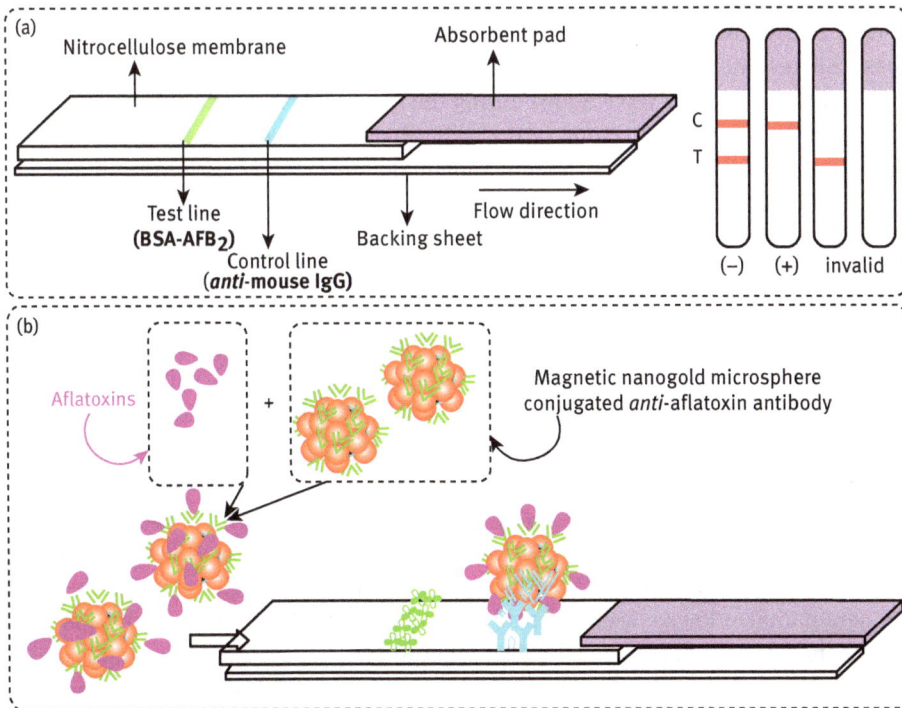

Figure 10.10: (a) The schematic illustration of the immunodipstick (left) and readout (right) and (b) principle of the detection method. Reproduced with permission from Elsevier.

[34]. The detector reagent consisted of magnetic nanogold microspheres (MnGMs) with nano-Fe_2O_3 particles as core and gold nanoparticles as shell and biofunction-alized with monoclonal anti-AFB_2 antibodies. Spotted AFB_2–BSA conjugates and goat antimouse IgG on nitrocellulose membrane were used as test and control lines, respectively. For usage, the dipstick was put into the well, and the liquid moved up to the membrane with the aid of absorbent pad. When reaching the test line, anti-AFB_2–MnGMs react with the AFB_2–BSA conjugate on the membrane and form a pink line. When the AFB_2 concentration in the sample solution is above a certain value, all of the anti-AFB_2 antibodies on the surface of the nanocomposites are occupied and therefore, MnGMs pass through the test line. After reaching the control line, the immobilized antimouse IgG reacts with the anti-AFB_2–MnGMs to form a second pink line. The color is recorded ~10 min after the dipstick was inserted into the well and yields a yes/no answer. The MnGM-based dipstick IA shows a limit of detection of 0.9 ng/mL AFB_2, which is about threefold lower compared to a conventional immunodipstick test using gold nanoparticles as detection reagent.

10.5 Principles of Effect-Directed Analysis

Often, environmental pollutants are present as complex mixtures [35], for instance pesticide residues in agricultural soils due to tank-mix applications with several ingredients, subsequent applications during the growing season or due to applications from last years and stabilization of the residues after sorption to soil. Moreover, sediments accumulate more and more contaminants over the years and residues may be preserved for long times. Sometimes, even thousands of compounds are detectable [36]. Such mixtures cannot be addressed with target analyses alone. Instead, bioassays may be used for detecting the combined effects of complex mixtures which can be fractionated by chromatographic separations. Biotesting means the analysis of the biological activity of a sample using responses of molecular and cellular systems or whole organisms. The fractions can then be reassayed by suitable biological assays to detect active components which subsequently can be identified, for example, by high sensitivity mass-spectrometric analysis (Figure 10.11). The combination of fractionation, biotesting and chemical analysis is called effect-directed analysis (EDA) [37].

In the future, chemical monitoring programs may be more and more complemented by bioassays as this combination can give rapid indications of chemical contamination followed by chemical identification of predominant toxicants [38–40]. For chemicals with known effect concentrations (EC_x), that is, concentrations above which a defined effect is observed for a given percentage (x) of exposed organisms, toxic units (TU) can be calculated as the ratio of detected concentrations and EC [41]. Such TU can be summed up in a mixture toxicity approach (concentration addition) [42] or hazard quotients (HQ) can be obtained by using predicted no effect concentrations (PNECs) instead of EC values [43]. In environmental risk assessment, the PNEC

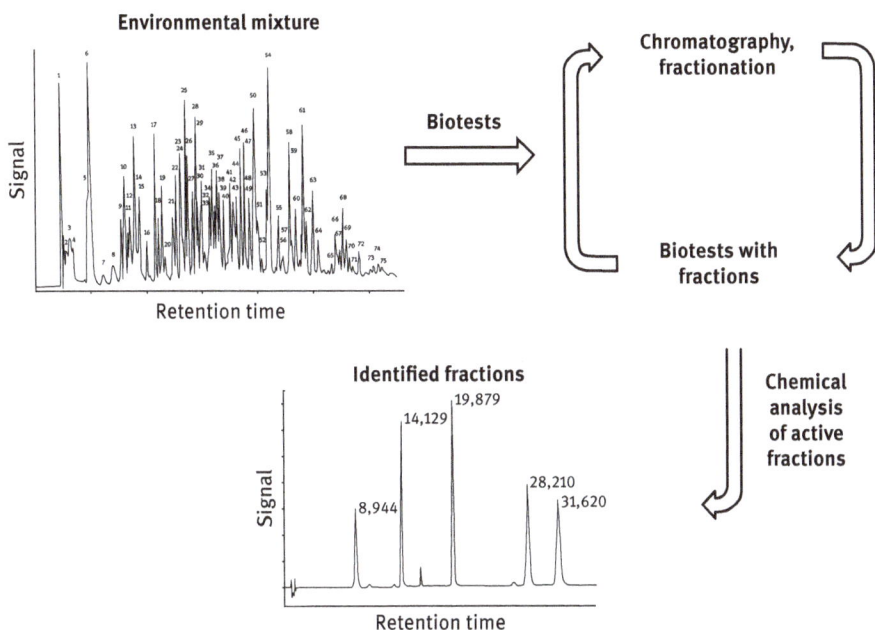

Figure 10.11: Principle of effect-directed analysis of environmental pollutants in complex mixtures.

is compared to the predicted environmental concentration (PEC), that is, by the ratio of PEC/PNEC. If the ratio is below 1, no risk is assumed.

The choice of the bioassays as toxicity indicators is an important step in EDA since ideally it should be integrated in a high-throughput approach with low-volume requirements (in the higher µL up to the lower mL range) [36]. Many different bioassays are at hand ranging from in vitro and reporter gene tests to complex in vivo responses (Table 10.3).

One of the most often used reactive toxicological endpoints in EDA is genotoxicity (DNA damage or DNA repair activity) or mutagenicity. This approach has been successful for identifying mutagens in sediments [44] and atmospheric particulate matter fractions [45, 46]. Specific mode-of-action-based assays are most suitable for EDA, that is, binding assays to specific receptors such as the estrogen or androgen receptor, photosystem II assays in phototrophic systems or specific enzyme assays [47].

Conventionally, an extract or a fraction in water often supported by low amounts (up to about 1 vol%) of an organic solvent for solubilization like DMSO or methanol is used for EDA analysis. Alternatively, passive dosing can be applied for lipophilic compounds in which the cells or organisms are exposed to a solid, inert polymer (e.g., polydimethylsiloxane) loaded with the environmental mixture of lipophilic compounds from which an equilibrium concentration in the water phase can be established.

Sampling, clean-up and enrichment procedures prior to chemical analysis have been discussed in Chapters 4 and 6, respectively. A time-integrated approach for

Table 10.3: Choice of bioassays for effect directed analysis (modified according to Brack et al. [36]).

	Assays on the cellular level				Assays with organisms	Assays with populations
	Metabolic assays	Target interactions	Cellular assays	Cytotoxicity assays		
In vitro assays	Activation of aryl-hydrocarbon receptor	Reporter gene assays, enzymatic assays	Activation of DNA repair, reporter gene assays for adaptive stress response, mutagenicity assays	Cell viability assays, mutagenicity assays		
In vivo assays	EROD assay*	Inhibition of algal photosynthesis			Fish embryo test, population growth of algae, daphnia immobilization	

*EROD assay: 7-ethoxy-resorufin-O-deethylase, a cytochrome P450 monooxygenase 1A (CYP1A)-dependent enzyme, via fluorescence measurement.

large volumes of water sampling (50–1,000 L) by in situ enrichment and solid-phase extraction was recently described [48].

After the first bioassays with the complex mixture, fractionation is typically based on preparative liquid chromatography with all types of interaction mechanisms such as normal-phase adsorption [49], reversed phase [50], ion exchange [51], size-exclusion chromatography [52] and affinity chromatography [53, 54]. Criteria for efficient fractionation procedures in EDA have been published [36]. Separation can be achieved by combining different techniques in an orthogonal approach, for instance by combination of normal phase and reversed phase separation, or gas chromatography and liquid chromatography involving mobile and stationary phases with different properties.

Identification of the fractionated toxicants is performed by GC–MS and LC–MS analyses either by target or suspect screening or nontarget screening approaches by use of spectral reference databases and in-silico fragmentation approaches by ranking a set of candidate structures based on the fragmentation information in the mass-spectrometric data.

Further Reading

Bender M, Bergeron R, Komiyama, M. The bioorganic chemistry of enzymatic catalysis. New York: John Wiley and Sons, 1984.

Diamandis E, Christopoulos T. Immunoassays. San Diego: Academic Press, 1996.

Law B. Immunoassay – a practical guide. London: Taylor and Francis, 2002.

Mayforth R. Designing antibodies. San Diego: Academic Press, 2012.

Miller J, Niessner R, Knopp D. Enzyme and immunoassays. Ullmann Biotechnol Biochem Eng 2007;2:585–612.

Sharma R. Enzyme inhibition and bioapplications. Rijeka: InTech, 2012.

Suelter C, Kricka L. Bioanalytical applications of enzymes. New York: Wiley, 1992.

Bibliography

[1]   Kingsmore SF. Multiplexed protein measurement: technologies and applications of protein and antibody arrays. Nat Rev Drug Discovery 2006;5(4):310–20.

[2]   Dittrich PS, Manz A. Lab-on-a-chip: microfluidics in drug discovery. Nat Rev Drug Discovery 2006;5(3):210–218.

[3]   Zhang K, Wong JW, Wang PG. A perspective on high-throughput analysis of pesticide residues in foods. Se Pu 2011;29(7):587–93.

[4]   Araoz R, Vilarino N, Botana LM, Molgo J. Ligand-binding assays for cyanobacterial neurotoxins targeting cholinergic receptors. Anal Bioanal Chem 2010;397(5):1695–704.

[5]   Burauel P, Führ F. Formation and long-term fate of non-extractable residues in outdoor lysimeter studies. Environ Poll 2000;108(1):45–52.

[6]   Hertzberg RP, Pope, AJ. High-throughput screening: new technology for the 21st century. Curr Opin Chem Biol 2000;4(4):445–51.

[7]   Inglese J, Johnson RL, Simeonov A, Xia MH, Zheng W, Austin CP, Auld, DS. High-throughput screening assays for the identification of chemical probes. Nat Chem Biol 2007;3(8):466–79.

[8]   Mascini M. Aptamers in Bioanalysis. Hoboken: Wiley-Blackwell, 2009:314.

[9]   Haupt K, Linares AV, Bompart M, Bernadette, TSB. Molecularly Imprinted Polymers. In Haupt K, editor. Molecular imprinting. Heidelberg: Springer, vol. 325, 2012:1–28.

[10] Dong J, Kun Z, Tang TT, Ai SY. Enzyme-catalyzed removal of bisphenol A by using horseradish, peroxidase immobilized on magnetic silk fibroin microspheres. Res J Chem Environ 2011;15(3):13–18.

[11] Yang XN, Huang XB, Hang RQ, Zhang XY, Qin L, Tang B. Improved catalytic performance of porcine pancreas lipase immobilized onto nanoporous gold via covalent coupling. J Mat Sci 2016;51(13):6428–35.

[12] Zhang LJ, Tang XM, Cui DB, Yao ZQ, Gao B, Jiang SQ, Yin B, Yuan YA, Wei, DZ. A method to rationally increase protein stability based on the charge-charge interaction, with application to lipase LipK107. Protein Sci 2014;23(1):110–16.

[13] Peters RJ, Shiau AK, Sohl JL, Anderson DE, Tang G, Silen JL, Agard, DA. Pro region C-terminus: Protease active site interactions are critical in catalyzing the folding of alpha-lytic protease. Biochemistry 1998;37(35):12058–67.

[14] Miao YQ, He NY, Zhu, JJ. History and new developments of assays for cholinesterase activity and inhibition. Chem Rev 2010;110(9):5216–34.

[15] Giera M, Heus F, Janssen L, Kool J, Lingeman H, Irth H. Microfractionation revisited: a 1536 well high resolution screening assay. Anal Chem 2009;81(13):5460–6.

[16] Kohler G, Milstein C. Continuous cultures of fused cells secreting antibody of predefined specificity. Nature 1975;256(5517):495–7.

[17] de Marco A. Recombinant antibody production evolves into multiple options aimed at yielding reagents suitable for application-specific needs. Microb Cell Fact 2015;14:1–17.

[18] Winklmair M, Weller MG, Mangler J, Schlosshauer B, Niessner R. Development of a highly sensitive enzyme-immunoassay for the determination of triazine herbicides. Fresenius J Anal Chem 1997;358(5):614–22.

[19] Schneider P, Hammock, BD. Influence of the ELISA format and the hapten-enzyme conjugate on the sensitivity of an immunoassay for S-triazine herbicides using monoclonal antibodies. J Agric Food Chem 1992;40(3):525–30.

[20] Drgoev AB, Gazaryan IG, Lagrimini LM, Ramanathan K, Danielsson B. High-sensitivity assay for pesticide using a peroxidase as chemiluminescent label. Anal Chem 1999;71(22):5258–61.

[21] Matschulat D, Deng AP, Niessner R, Knopp D. Development of a highly sensitive monoclonal antibody based ELISA for detection of benzo[a]pyrene in potable water. Analyst 2005;130(7):1078–86.

[22] Zeck A, Weller MG, Niessner R. Characterization of a monoclonal TNT-antibody by measurement of the cross-reactivities of nitroaromatic compounds. Fresenius J Anal Chem 1999;364(1-2):113–20.

[23] Inui H, Takeuchi T, Uesugi A, Doi F, Takai M, Nishi K, Miyake S, Ohkawa H. Enzyme-linked immunosorbent assay with monoclonal and single-chain variable fragment antibodies selective to coplanar polychlorinated biphenyls. J Agric Food Chem 2012;60(7):1605–12.

[24] Okuyama M, Kobayashi N, Takeda W, Anjo T, Matsuki Y, Goto J, Kambegawa A, Hod S. Enzyme-linked immunosorbent assay for monitoring toxic dioxin congeners in milk based on a newly generated monoclonal anti-dioxin antibody. Anal Chem 2004;76(7):1948–56.

[25] Goda Y, Hirobe M, Kobayashi A, Fujimoto S, Ike M, Fujita M. Production of a monoclonal antibody and development of enzyme-linked immunosorbent assay for alkyl ethoxylates. Anal Chim Acta 2005;528(1):47–54.

[26] Fujita M, Ike M, Goda Y, Fujimoto S, Toyoda Y, Miyagawa, KI. An enzyme-linked immunosorbent assay for detection of linear alkylbenzene sulfonate: development and field studies. Environ Sci Technol 1998;32(8):1143–6.

[27] Zeck A, Eikenberg A, Weller MG, Niessner R. Highly sensitive immunoassay based on a monoclonal antibody specific for 4-arginine microcystins. Anal Chim Acta 2001;441(1):1–13.

[28] Zeck A, Weller MG, Bursill D, Niessner R. Generic microcystin immunoassay based on monoclonal antibodies against Adda. Analyst 2001;126(11):2002–7.

[29] Huebner M, Weber E, Niessner R, Boujday S, Knopp D. Rapid analysis of diclofenac in freshwater and wastewater by a monoclonal antibody-based highly sensitive ELISA. Anal Bioanal Chem 2015;407(29):8873–82.

[30] Carvalho JJ, Weller MG, Panne U, Schneider, RJ. A highly sensitive caffeine immunoassay based on a monoclonal antibody. Anal Bioanal Chem 2010;396(7):2617–28.

[31] Kloth K, Niessner R, Seidel M. Development of an open stand-alone platform for regenerable automated microarrays. Biosens Bioelectron 2009;24(7):2106–12.

[32] Weller MG, Schuetz AJ, Winklmair M, Niessner R. Highly parallel affinity sensor for the detection of environmental contaminants in water. Anal Chim Acta 1999;393(1-3):29–41.

[33] Kloth K, Rye-Johnsen M, Didier A, Dietrich R, Martlbauer E, Niessner R, Seidel M. A regenerable immunochip for the rapid determination of 13 different antibiotics in raw milk. Analyst 2009;134(7):1433–9.

[34] Tang D, Sauceda JC, Lin Z, Ott S, Basova E, Goryacheva I, Biselli S, Lin J, Niessner R, Knopp D. Magnetic nanogold microspheres-based lateral-flow immunodipstick for rapid detection of aflatoxin B-2 in food. Biosens Bioelectron 2009;25(2):514–18.

[35] Malaj E, von der Ohe PC, Grote M, Kuehne R, Mondy CP, Usseglio-Polatera P, Brack W, Schaefer, RB. Organic chemicals jeopardize the health of freshwater ecosystems on the continental scale. Proc Natl Acad Sci USA 2014;111(26):9549–54.

[36] Brack W, Ait-Aissa S, Burgess RM, Busch W, Creusot N, Di Paolo C, Escher BI, Hewitt LM, Hilscherova K, Hollender J, Hollert H, Jonker W, Kool J, Lamoree M, Muschket M, Neumann S,

Rostkowski P, Ruttkies C, Schollee J, Schymanski EL, Schulze T, Seiler T.-B, Tindall AJ, Umbuzeiro GD, Vrana B, Krauss M. Effect-directed analysis supporting monitoring of aquatic environments – An in-depth overview. Sci Total Environ 2016;544:1073–118.

[37] Brack W, Schirmer K, Kind T, Schrader S, Schuurmann G. Effect-directed fractionation and identification of cytochrome P4501A-inducing halogenated aromatic hydrocarbons in a contaminated sediment. Environ Toxicol Chem 2002;21(12):2654–62.

[38] Wernersson A.-S, Carere M, Maggi C, Tusil P, Soldan P, James A, Sanchez W, Dulio V, Broeg K, Reifferscheid G, Buchinger S, Maas H, Van Der Grinten E, O'Toole S, Ausili A, Manfra L, Marziali L, Polesello S, Lacchetti I, Mancini L, Lilja K, Linderoth M, Lundeberg T, Fjallborg B, Porsbring T, Larsson DG, Bengtsson-Palme J, Forlin L, Kienle C, Kunz P, Vermeirssen E, Werner I, Robinson CD, Lyons B, Katsiadaki I, Whalley C, den Haan K, Messiaen M, Clayton H, Lettieri T, Carvalho RN, Gawlik BM, Hollert H, Di Paolo C, Brack W, Kammann U, Kase R. The European technical report on aquatic effect-based monitoring tools under the water framework directive. Environ Sci Eur 2015;27:1–11.

[39] Altenburger R, Ait-Aissa S, Antczak P, Backhaus T, Barcelo D, Seiler T.-B, Brion F, Busch W, Chipman K, Lopez de Alda M, de Aragao Umbuzeiro G, Escher BI, Falciani F, Faust M, Focks A, Hilscherova K, Hollender J, Hollert H, Jaeger F, Jahnke A, Kortenkamp A, Krauss M, Lemkine GF, Munthe J, Neumann S, Schymanski EL, Scrimshaw M, Segner H, Slobodnik J, Smedes F, Kughathas S, Teodorovic I, Tindall AJ, Tollefsen KE, Walz K.-H, Williams TD, Van den Brink PJ, van Gils J, Vrana B, Zhang X, Brack W. Future water quality monitoring – Adapting tools to deal with mixtures of pollutants in water resource management. Sci Total Environ 2015;512:540–51.

[40] Kortenkamp A, Scholze M, Ermler S. Mind the gap: can we explain declining male reproductive health with known antiandrogens? Reproduction 2014;147(4):515–27.

[41] von der Ohe PC, Dulio V, Slobodnik J, De Deckere E, Kuhne R, Ebert RU, Ginebreda A, De Cooman W, Schuurmann G, Brack W. A new risk assessment approach for the prioritization of 500 classical and emerging organic microcontaminants as potential river basin specific pollutants under the European Water Framework Directive. Sci Total Environ 2011;409(11):2064–77.

[42] Swartz RC, Schults DW, Ozretich RJ, Lamberson JO, Cole FA, Dewitt TH, Redmond MS, Ferraro, SP. Sigma-PAH – a model to predict the toxicity of polynuclear aromatic hydrocarbon mixtures in field-collected sediments. Environ Toxicol Chem 1995;14(11):1977–87.

[43] Kuzmanovic M, Ginebreda A, Petrovic M, Barcelo D. Risk assessment based prioritization of 200 organic micropollutants in 4 Iberian rivers. Sci Total Environ 2015;503:289–99.

[44] Lubcke-von Varel U, Machala M, Ciganek M, Neca J, Pencikova K, Palkova L, Vondracek J, Loffler I, Streck G, Reifferscheid G, Fluckiger-Isler S, Weiss JM, Lamoree M, Brack W. Polar compounds dominate in vitro effects of sediment extracts. Environ Sci Technol 2011;45(6):2384–90.

[45] Pedersen DU, Durant JL, Penman BW, Crespi CL, Hemond HF, Lafleur AL, Cass, GR. Human-cell mutagens in respirable airborne particles in the Northeastern United States. 1. Mutagenicity of fractionated samples. Environ Sci Technol 2004;38(3):682–9.

[46] Pedersen DU, Durant JL, Taghizadeh K, Hemond HF, Lafleur AL, Cass, GR. Human cell mutagens in respirable airborne particles from the Northeastern United States. 2. Quantification of mutagens and other organic compounds. Environ Sci Technol 2005;39(24):9547–60.

[47] Escher BI, Allinson M, Altenburger R, Bain PA, Balaguer P, Busch W, Crago J, Denslow ND, Dopp E, Hilscherova K, Humpage AR, Kumar A, Grimaldi M, Jayasinghe BS, Jarosova B, Jia A, Makarov S, Maruya KA, Medvedev A, Mehinto AC, Mendez JE, Poulsen A, Prochazka E, Richard J, Schifferli A, Schlenk D, Scholz S, Shiraish F, Snyder S, Su GY, Tang JY, van der Burg B, van der Linden SC, Werner I, Westerheide SD, Wong CK, Yang M, Yeung BH, Zhang XW, Leusch, FD. Benchmarking organic micropollutants in wastewater, recycled water and drinking water with in vitro bioassays. Environ Sci Technol 2014;48(3):1940–56.

[48] Schulze T, Krauss M, Bahlmann A, Hug C, Walz KH, Brack W. Onsite large volume solid phase extraction — how to get 1,000 litres of water into the laboratory? Poster, Society of environmental chemistray and toxicology (SETAC) Europe, 2014.

[49] Lubcke-von Varel U, Streck G, Brack W. Automated fractionation procedure for polycyclic aromatic compounds in sediment extracts on three coupled normal-phase high-performance liquid chromatography columns. J Chromatogr A 2008;1185(1):31–42.

[50] Thomas KV, Balaam J, Hurst MR, Thain, JE. Identification of in vitro estrogen and androgen receptor agonists in North Sea offshore produced water discharges. Environ Toxicol Chem 2004;23(5):1156–63.

[51] Gallampois CM, Schymanski EL, Bataineh M, Buchinger S, Krauss M, Reifferscheid G, Brack W. Integrated biological-chemical approach for the isolation and selection of polyaromatic mutagens in surface waters. Anal Bioanal Chem 2013;405(28):9101–12.

[52] Brack W, Kind T, Hollert H, Schrader S, Moder M. Sequential fractionation procedure for the identification of potentially cytochrome P4501A-inducing compounds. J Chromatogr A 2003;986(1):55–66.

[53] Jonker N, Kretschmer A, Kool J, Fernandez A, Kloos D, Krabbe JG, Lingeman H, Irth H. Online magnetic bead dynamic protein-affinity selection coupled to LC-MS for the screening of pharmacologically active compounds. Anal Chem 2009;81(11):4263–70.

[54] Pochet L, Heus F, Jonker N, Lingeman H, Smit AB, Niessen WM, Kool J. Online magnetic bead based dynamic protein affinity selection coupled to LC-MS for the screening of acetylcholine binding protein ligands. J Chromatogr B-Anal Technol Biomed Life Sci 2011;879(20):1781–8.

11 Selected Applications

11.1 Trace Analysis of Polycyclic Aromatic Hydrocarbons (PAHs)

11.1.1 General Remarks (Occurrence, Physical and Chemical Properties)

Polycyclic Aromatic Hydrocarbons (PAHs) represent one of the most analyzed substance classes. This is due to the proven mutagenicity and/or carcinogenicity of some prominent representatives of this class [1]. Since they are found everywhere (even in interstellar space, as identified by IR emission spectra), the concomitant matrices (exhaust gas, water, soil, food, ambient air, etc.) often pose unexpected problems to the analyst.

Chemically, PAHs are hydrocarbons without functional groups, characterized by multiple condensed aromatic rings, where π-electrons are delocalized. This sometimes leads to a remarkable thermodynamic stability and low vapor pressure, especially when the molecule has a flat nature and a high symmetry. Even the presence of genuine PAHs together with minerals has been known for more than 150 years: The mineral *Pendletonite* has been identified as nearly pure coronene, a seven-ring PAH with a melting point above 430 °C [2].

According to US-EPA the following 16 members are of special concern (see Figure 11.1) and belong to priority pollutants [3]. This has been adopted by many regulating agencies worldwide, and therefore most analytical procedures need to quantify them all. But there is an ongoing discussion to extend these priority substances by others. Three groups of polycyclic aromatic compounds are missing: larger and highly relevant PAHs, alkylated PAHs and compounds containing heteroatoms.

As hydrocarbons they show less tendency to become solubilized in water. Furthermore, the high π-electron density gives rise to the formation of stable stacked π–π sandwich complexes (donor–acceptor complexes). This is of importance when humic substances from soil, or graphene systems in soot particles, are present. Due to this, separation through extraction or vaporization isn't always easy. One should also recognize that there is a range of derivatives possible, for example, alkylated PAHs, or C atoms of the aromatic ring(s) are partially substituted by S-, N- or O-atoms. Substitution or radical reactions create numerous derivatives or metabolites. This has enormous consequences, as for the partition behavior between water and nonaqueous solvents (K_{ow} value), or the extreme wide range of vapor pressure (see Figure 11.2) [4].

Incomplete combustion is often reported as the main direct anthropogenic source for PAH pollution (house heating or energy production by wood, coal or oil; coke production; exhaust from internal combustion engines). Soot formation is tightly linked to the presence of PAH structures; therefore, PAHs are always found surface bound to soot particles. Smouldering fires (like cigarette smoking) are responsible for PAH formation by dehydration, cyclization, aromatization and ring expansion of leaf phytosterols. Hence, cigarette smoke contains PAHs, but not adsorbed to a soot

DOI 10.1515/9783110441154-011

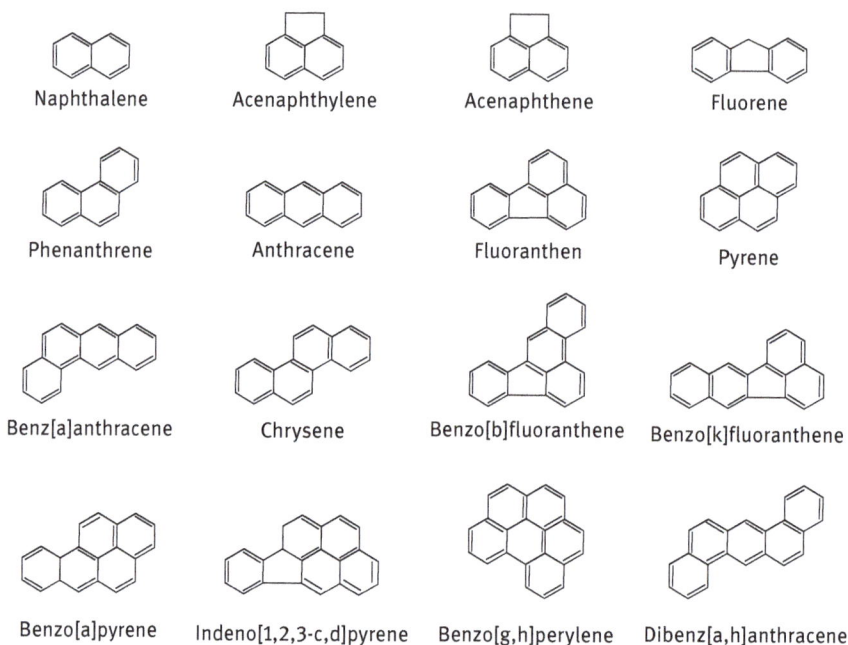

Naphthalene Acenaphthylene Acenaphthene Fluorene

Phenanthrene Anthracene Fluoranthen Pyrene

Benz[a]anthracene Chrysene Benzo[b]fluoranthene Benzo[k]fluoranthene

Benzo[a]pyrene Indeno[1,2,3-c,d]pyrene Benzo[g,h]perylene Dibenz[a,h]anthracene

Figure 11.1: Chemical structures of the 16 US-EPA PAHs.

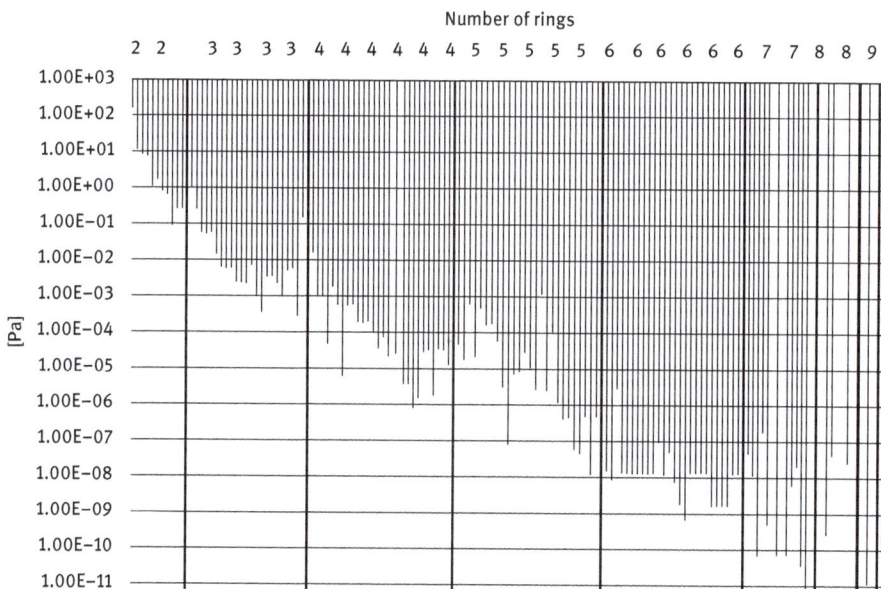

Figure 11.2: Vapor pressures of PAHs (including alkylated PAHs) versus number of rings.

nucleus. Beyond this, there are several other sources known. Considerable amounts are of natural origin (oil spills, forest fires), even termite mounds are hot spots for a naphthalene contamination, produced by associated microorganisms. Indirect pollution sources exist a lot. Sometimes, even the drinking water distribution lines themselves serve as a PAH source. Last century, most iron pipes used tar coatings as liner to prevent corrosion. When drinking water is chlorinated, the protecting biofilm is destroyed and PAHs are released from the tar liner into the drinking water. Street or roof run-off water is frequently polluted by PAHs. Curing meat, roasting, grilling and cooking also serve as food pollution source.

Toxicologically, it is generally accepted that especially benzo[*a*]pyrene (BaP) acts as a strong cancerogen. Other PAHs are mutagenic and/or teratogenic. This is also the reason for the low threshold limit set by several regulations, for example, 10 ng BaP/L in drinking water or 1 ng/m^3 in ambient air.

On the basis of this, consequences for analytical procedures are:

– enrichment is needed in most cases in order to reach a low ppt range;
– in aqueous phases, PAHs are surface-adsorbed to colloidal matter (humic substances or other interfaces) and not dissolved, except oilfield water with a lot of two- or three-ring PAHs dissolved;
– airborne particles (soot) are PAH – submonolayer covered; low-molecular-weight compounds form the most outer layer, in contrast to the carbonaceous nucleus;
– if not particle bound, low-molecular-weight PAHs (two–four-ring PAHs) and especially uneven PAH molecules (e.g., fluorene, fluoranthene, indeno[1,2,3-c,d]pyrene, etc.) are partitioned to a considerable amount within the gas phase. This equilibrium is strongly dependent on temperature. They represent a semi-volatile fraction.

These peculiarities require special techniques for sampling, sample pretreatment and analysis. Ideal would be an online and *in situ* technique.

11.1.2 Sampling Strategies for PAHs (Water, Air, Soil, Body Fluids)

11.1.2.1 Water
PAH sampling from the aqueous matrix has many peculiarities. Most water-related guidelines ask for a pre-filtration of water sampled (e.g., by a 0.45-μm pore-sized membrane filter), in order to remove suspended silt particles or biological debris. But most ultrafine hydrocolloids with surface-enriched PAHs will pass this step and will remain in the liquid phase.

Already, the sample container material poses several risks. Light-protected quartz glass or electropolished stainless steel containers show the lowest adsorption losses. The highest losses occur with plastics like PTFE [5]. Nowadays *solid-phase microextraction* (*SPME*) is preferred for sampling, but one has to be aware of competing adsorption of PAHs to other solid surfaces. *SPME* (e.g., polydimethylsiloxane [PDMS] fibers) is applicable if no strong interaction of PAHs with always present humic substances or other colloidal matter is competing [6]. *Liquid–liquid extraction* (LLE) with acetonitrile or hexane therefore is still the best technique, with the disadvantage of handling larger solvent volumes. When LLE is used, volume reduction by evaporation under slight vacuum, enhanced temperature or under nitrogen gas flow has to be applied. Losses of the light-weight PAHs (especially fluoranthene) are the consequence. Sometimes, an alternative is the use of very small amounts of organic solvent, for example performance of *single-drop microextraction* or *dispersive liquid–liquid micro-extraction (DLLME)*. The latter is based on an appropriate mixture of extraction solvent (e.g., C_2Cl_4) and disperser solvent (e.g., acetone) injected into the aqueous sample by a syringe. Thereupon, a turbid solution is formed consisting of fine droplets of extraction solvent which is dispersed entirely into the aqueous phase. After centrifugation, the fine droplets of extraction solvent are sedimented to the bottom of the conical test tube, and this separated solvent phase gets ready for transfer to chromatography.

Micellar extraction (also named *cloud-point extraction*) by addition of tensides and formation of a micellar phase helps to reduce solvent consumption but is only feasible when tensides do not interfere with the subsequent determination step. Details on such enrichment techniques are presented in Chapter 6.

In any case, standard addition of deuterated PAHs is strongly recommended for mass spectrometric (MS) analysis. For HPLC determination, the spiking of rare PAHs is used as internal standard and is added to the sample, before any procedure is started.

Further complications occur with chlorinated or ozonized water causing oxidation losses during storage [7]. Sometimes, sulfite addition is recommended to avoid these artifacts.

Generally, parallel analysis of a PAH containing reference material for validation of the selected technique is greatly advisable. At the present time this is not possible for the matrix water. Several attempts have been made years ago, for example, by controlled release of surface-adsorbed PAHs at constant temperature under a fixed water flow rate. The no longer available standard of the National Bureau of Standards (US) SRM 1644 consisted of separate generator columns for anthracene, benz[a]anthracene and BaP [8]. These columns were 50 cm × 0.6 cm stainless steel tubes packed with 80–100 mesh sea sand and coated with 0.5% (w/w) of the PAH compound of interest. Saturated solutions were generated by pumping HPLC grade water through the generator column at flow rates between 0.1 and 5 mL/min. Since the aqueous solubility is a well-defined thermodynamic quantity, saturated solutions are standard solutions.

11.1.2.2 Air (Ambient Air; Combustion Aerosol)

PAHs in air are found in very different concentration ranges: Combustion exhaust may contain PAHs in the $\mu g/m^3$ range, whereas clean air masses from remote areas are contaminated only in the pg/m^3 range. This leads to different sampling strategies.

PAH monitoring in ambient, moderately contaminated air needs preconcentration. As PAHs are distributed among the gas–particle phase, sampling has to

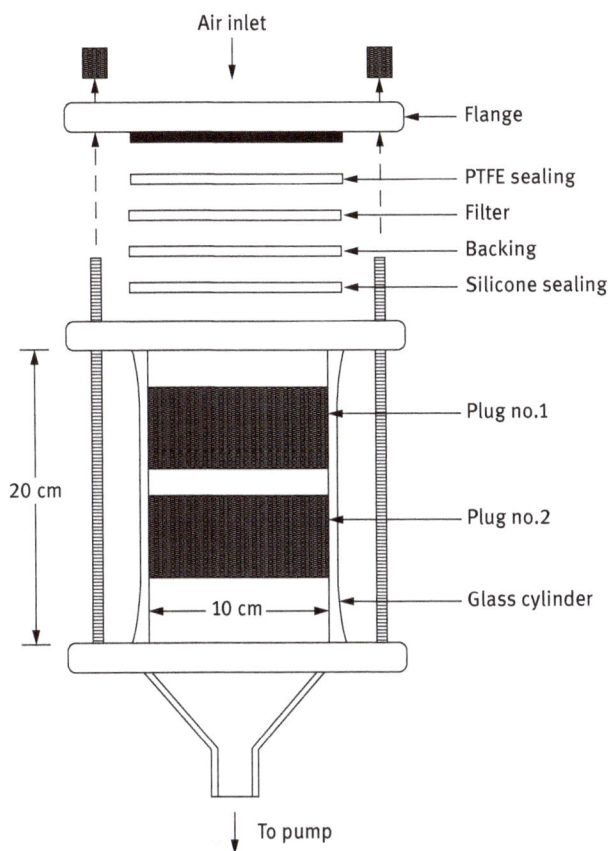

Schematic diagram of a PAH aerosol/gas sampler [9].

be adapted to this. The most frequent separation tool is particle separation by filtration during high-volume sampling. Due to the partitioning of the various PAHs between gas and particle phase, a sequential sampling has been introduced already in 1981. A sampling system for the collection of PAH in air is shown schematically in Figure 11.3 [9].

The particles are collected on a glass fiber filter. The volatile compounds are trapped by two cylindrical plugs of polyurethane foam located behind the filter in a glass cylinder. The glass fiber membranes have to be cleaned by preheating for several hours at 400 °C. The polyurethane foam plugs (used as hydrophobic adsorbent for gaseous PAHs) are thoroughly cleaned by 24-h Soxhlet extraction first under acetone, followed by the same procedure under cyclohexane. Sampling time is 24 h, at a flow rate of 25 m^3/h. Before extraction of the charged filter, two internal standards are added to the samples. Then, Soxhlet extraction by cyclohexane is applied. Concentration by evaporation down to a final volume of 100 µL is the next step, followed by

Table 11.1: Partitioning of gaseous and particulate PAHs in ambient aerosol at different locations [9].

Data on the reproducibility of the sampling system. Unit: ng/m³. Part: particulate phase. Gas: gaseous phase.

Set no.:	1						2					
Sampler:	a			b			a			b		
	Part	Gas	Total	Part	Gas	Total	Part	Gas	Total	Part	Gas	Total
Naphthalene		10.1	10.1		6.6	6.6		6.1	6.1		4.4	4.4
2-Methylnaphthalene		7.3	7.3		4.1	4.7		4.4	4.4		2.8	2.8
1-Methylnaphthalene		4.6	4.6		3.1	3.1		2.5	2.5		1.7	1.7
Biphenyl		4.2	4.2		2.8	2.8		1.9	1.9		1.4	1.4
1,8-Dimethyl naphthalene		3.6	3.6		4.2	4.2		0.8	0.8		0.8	0.8
Acenaphthene		29.5	29.5	0.2	19.6	19.8		3.1	3.1		2.1	2.1
Fluorene		80.0	80.0	0.3	68.6	68.9		15.5	15.5		13.8	13.8
Dibenzothiophene		30.7	30.7	0.2	32.6	32.8		4.3	4.3		3.4	3.4
Phenanthrene	1.8	244.0	246.0	2.5	238.0	241.0	< 0.1	34.3	34.3	< 0.1	39.0	39.0
Anthracene	0.2	20.5	20.7	0.4	21.7	22.1		3.0	3.0		3.7	3.7
1-Methylphenanthrene	0.1	20.2	20.3	0.4	32.8	33.2		3.0	3.0		3.7	3.7
Fluoranthene	4.5	112.0	117.0	4.8	106.0	111.0	0.2	14.1	14.3	0.2	15.3	15.5
Pyrene	4.0	72.9	76.9	4.2	70.1	74.3	0.3	8.9	9.2	0.3	9.4	9.7
Benzo(a)fluorene	2.8	16.5	19.3	2.8	17.5	20.3	0.1	0.8	0.9	0.1	0.9	1.0
Benzo(b)fluorene	2.2	10.7	12.9	2.2	10.5	12.7	< 0.1	0.8	0.8	< 0.1	0.9	0.9
Benzo(a)anthracene	13.0	19.7	32.7	12.8	22.4	35.2	0.1	0.4	0.5	0.2	0.4	0.6
Chrysene/triphenylene	26.1	26.3	50.4	25.6	26.6	52.5	0.5	1.0	1.5	0.6	1.0	1.6
Benzo(b/j/k)fluoranthene	60.9	4.7	65.6	58.1	6.5	64.6	1.2		1.2	1.5		1.5
Benzo(e)pyrene	26.6		26.6	25.5	0.6	25.5	0.6		0.6	0.7		0.7
Benzo(a)pyrene	13.0		13.0	12.9		12.9	0.3		0.3	0.4		0.4
Perylene	2.5		2.5	2.6		2.6	0.3		0.3	0.2		0.2
o-Phenylenepyrene	13.2		13.2	13.1		13.1	0.8		0.8	1.0		1.0
Dibenzo(ac/ah)anthracene	4.7		4.7	4.9		4.9	0.5		0.5	0.3		0.3
Benzo(ghi)perylene	15.7		15.7	15.8		15.8	1.5		1.5	1.4		1.4
Anthanthrene	0.5		0.5	1.6		1.6						
Coronene	4.2		4.2	4.3		4.3	1.8		1.8	1.4		1.4

gas chromatography GC/FID analysis (FID = Flame Ionization Detector). The analyzed PAHs typically ranged from naphthalene (two-ring system) up to coronene (seven-ring system). As can be seen from Table 11.1, PAHs are mainly found in the gas phase (according to their vapor pressures up to four-ring-PAHs), whereas larger PAHs, especially BaP as key PAH, exist in the particulate matter.

It should be kept in mind that during sampling, there are tremendous losses by sublimation of the already deposited PAHs. The longer the sampling periods, the more such desorption losses will happen. This was studied by Miguel et al. [10] by exposing preloaded HighVol filters at ambient temperatures from a heavy traffic situation over a sampling period of up to 9 h. Under prevailing atmospheric conditions, the analyzed PAHs experienced losses up to 100% which (for most of them) followed first-order kinetics.

Chemical transformation by ozone and nitrogen dioxide aside thermal instability has been identified for being responsible for further losses. Therefore, the use of an *ozone scrubbing device* (*denuder*; see also chapter 6.4.5) in front of the filtering device and low-volume sampling is highly recommended (see Figure 11.4) [11].

Before the filter holder, a split air flow was passed through a diffusion denuder consisting of a glass jacket with a central tube of fine steel mesh surrounded by a dense packing of activated carbon. Experiments with the applied diffusion denuder have shown that it removes ozone, nitrogen dioxide and similarly reactive trace gases with an efficiency above 90%, while the diffusion losses of aerosol particles with

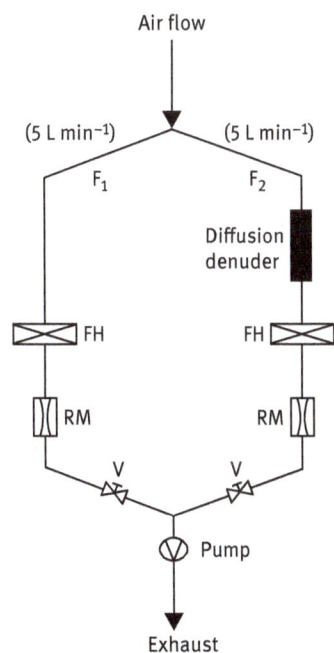

Figure 11.4: Setup of the applied low-volume aerosol filter sampling system: split flow F1 without denuder, split flow F2 with diffusion denuder filled with activated carbon; filter holders (FH), rotameters (RM) and valves (V). Reproduced with permission from the American Chemical Society.

(mobility equivalent) diameters above 50 nm are less than about 10%. The sampling intervals were 1–14 days corresponding to sampled air volumes of 7–100 m^3/filter. Throughout all sampling campaigns reported in this study, five- and six-ring PAHs and BaP were negatively correlated with ozone, which can be regarded as a tracer for the atmospheric oxidizing power which drives the chemical degradation of PAH.

Other experiments with artificially adsorbed PAHs on soot particles at high temperatures (25–250 °C), as in the after-treatment sections of cars or hot flue gas, revealed a considerable reactivity of pyrene and BaP with NO_2 leading to the formation of nitro-PAHs and other oxygenated PAHs [12].

PAH sampling from a stack needs different instrumentation. Most difficult is the separation of particle-bound PAH from gaseous PAH. One standardized sample train used is shown in Figure 11.5 [13].

Particulate matter is withdrawn isokinetically from a number of sampling points in an enclosed gas stream (stack, exhaust pipe). *Isokinetic sampling* (see also chapter 4.7.3) means that the linear velocity of the gas entering the sampling nozzle is equal to that of the undisturbed gas stream at the sampling point within the stack. Otherwise, the particle subsample will not be representative in mass to the main exhaust stream. The particulate sample is conducted through the nozzle, heated probe, cyclone (for pre-separation of large particles) and sampled on a quartz fiber filter, all maintained at a temperature of 120 ± 14 °C or at such a temperature as is necessary

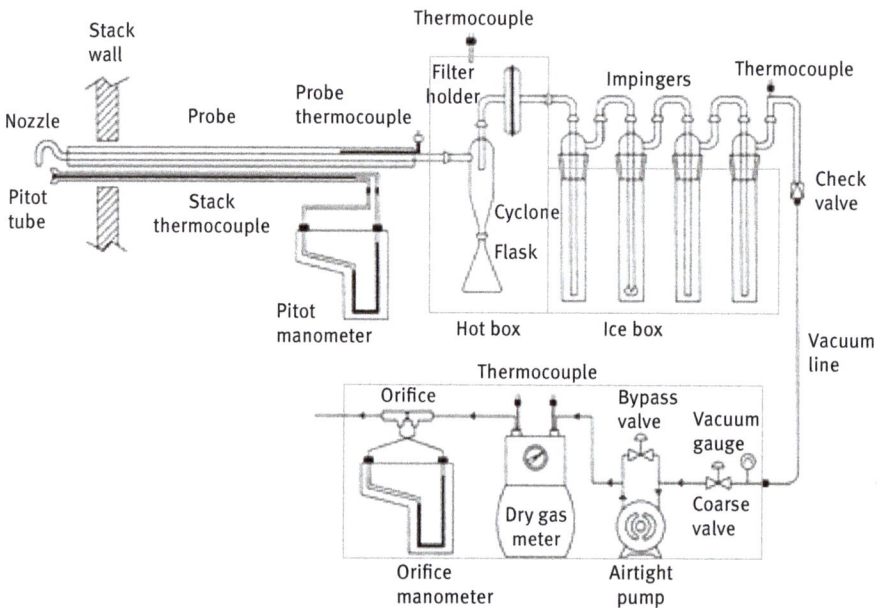

Figure 11.5: Stack sampling train for collection of particle and gases from combustion units.

to prevent blocking of the filter by condensation. The particulate matter on the filter is further processed after gravimetric total mass determination at constant humidity. Simultaneous determinations of the gas stream moisture content, velocity, pressure and temperature allow calculations of the particulate concentration and the particulate mass emission for standard temperature and pressure conditions (25 °C and 101.3 kPa).

11.1.2.3 PAHs in Soil

PAHs are frequently found as contaminations in soil [14]. Aside scarce natural sources (e.g., oilfield water aquifers connected to geothermal water wells sometimes contain huge amounts of light-weight PAHs), traffic accidents or fuel spillage at gas-filling stations, leaky oil tanks and residues from former coke oven plants in cities or industrial sites in use for wood-preservation are polluting soil. These impurities are severe permanent hazards to the drinking water supply through seepage water. Since removal and remediation of such polluted soils causes enormous costs, quick and reliable screening techniques are needed to localize the extent of such polluted areas. Moreover, success of a possible remediation needs to be monitored over a long period of time [15]. Soil sampling is the most critical step for it (see also chapter 4.5.2). Typically, information on all 16 US-EPA PAHs are required. Special strategies have been developed for a statistically sufficient 3D description of a contaminated site [16].

There have now been various individual directives published. Several environmental agencies ask for analysis of the fresh soil in its original state. The reason for this is the otherwise unavoidable loss of the light-weight PAH fraction by evaporation. The fresh soil becomes immediately extracted by acetone, water (with and without NaCl addition) and petrol ether in sequential order. All extracts are then subjected to a pre-chromatographic separation step and are afterward analyzed by GC–MS or liquid chromatography (LC), coupled with fluorescence detection (FLD) or an LC–MS combination.

Others apply air drying for 1 week, followed by sieving (mesh size: 154 µm). 1 g of the powdered sample is spiked with deuterated PAH standards. The next step is extraction in 10 mL hexane/dichloromethane (1:1) under ultrasonic agitation for 10 min. After breaking the suspension by centrifugation, a second extraction with the same organic solvent mixture follows. The combined extract is reduced in volume by evaporation until dryness. After reconstitution by adding 100 µL acetone and 15 mL water, *SPME* is applied. The analytical determination is performed by GC/FID or GC–MS [17].

The comparison of various extraction procedures ensures the superior extraction yield by using Soxhlet extraction versus other extraction techniques (accelerated pressure, supercritical CO_2 fluid, pressurized and ambient microwave assistance) [18].

11.1.2.4 PAHs in Body Fluids

Since PAHs are known as a possible threat to human health, many regulations exist for minimizing such risks. But finally, only monitoring the human body itself allows to estimate the toxicological risk after exposure. Usually, body fluids (urine for PAH metabolization studies or serum albumine for DNA-adducted studies) are the matter to assess (see also chapter 4.2.3).

A typical procedure for sample collection (urine) and sample preparation for PAH or PAH metabolite analysis is the following: spot urine samples from subjects are first collected in polyethylene containers. Volume and pH of each urine sample are measured, samples are encoded and 10-mL aliquots can be stored, frozen at −20 °C for no more than 14 days prior to analysis. To account for differences in dilution of the urine due to diuresis, urinary PAH metabolite concentrations are corrected for creatinine content. Sample preparation is performed according to Boos et al. [19]. After thawing and shaking of the frozen urine samples, an aliquot of 2.5 mL is placed in a glass flask and adjusted to pH 5.0 with 1 M hydrochloric acid. Then, the solution is filled up to a final volume of 5.0 mL with sodium acetate buffer (0.1 M, pH 5.0). To hydrolyze the conjugates of the metabolites, 7 μL of enzyme solution (4β-glucuronidase-arylsulfatase) are added and the solutions are incubated under continuous shaking for 3 h at 37 °C in a thermostated water bath. After that, the samples are centrifuged in glass tubes at 2,000g for 5 min, transferred into autosampler vials and frozen at −20 °C. After thawing, the solutions are injected into the analysis system without mixing.

In case of necessary pre-concentration, various techniques can be applied: SPE or *immunoextraction* [20, 21]. In the latter, sample enrichment and purification are performed with an immunoextraction cartridge containing 0.6 g of antibody-doped porous sol–gel glass. A 20-mL aliquot of the hydrolyzed urine is applied onto the column. To remove non-retained and weakly bound impurities, the immunosorbent is flushed successively with 3 mL of acetate buffer (0.1 M with 5% acetonitrile (ACN)) and 3 mL of water (with 5% ACN); both washing solutions contained 10 mg/L uric acid to inhibit oxidation of the OH−PAHs. The retained compounds are then desorbed from the immunoaffinity column with 3×1-mL fractions of ACN/water (1:1). The eluate is collected in a glass vial, which contained 15 μL of an aqueous ascorbate solution (1 g/L), and after filtration with a 0.45-μm syringe filter directly analyzed by the LC method or further concentrated by C_{18}−*SPE*.

11.1.3 Wet-chemical Analysis (HPLC, Thin-layer Chromatography)

11.1.3.1 HPLC Analysis of PAHs
HPLC has some advantages in PAHs analysis:
- separation of large PAHs and PAH isomers shows very good resolution;
- sufficient sensitivity and specificity of fluorescence detection (FLD);
- molecular sizes of PAHs can be estimated on the basis of the retention time using reverse-phase columns;
- possibility of determining compounds with high molecular mass (in contrast to GC, where sufficient vapor pressure is required);

- separations are usually carried out at ambient temperature, with no risk of thermal decomposition of analytes;
- by switching with a reducing precolumn, nitro-PAHs can be determined in parallel.

In many cases, PAH fluorescence analysis, in combination with a matrix-tolerant HPLC separation, is able to analyze the 16 priority US-EPA PAH pollutants, and this even in complicated matrices. The high symmetry and π-electron delocalization not only enables strong photon absorption but also yields a strong fluorescence. Only uneven symmetry of a PAH molecule, for example due to a five-ring system, shows deviations in this. Simple PAH mixtures may be analyzed by applying appropriate excitation and emission wavelengths, and in some tasks, synchronous FLD is sufficient. But more difficult matrices, such as for edible oil, need further assistance. The application of SPE as a cleanup procedure and HPLC/FLD for identification and quantification of PAHs is shown in Figure 11.6.

Programmable excitation and emission wavelengths enable the analysis of PAHs after SPE including washing the C_{18} sorbent with water, drying and eluting with pure hexane. Recoveries and selectivity using other sorbent materials (C_8, C_2, CH, PH and NH_2) have also been examined, with C_{18} being the best one. The recoveries ranged between 50% and 103% depending on the size of the PAH. The limits of quantitation are lower than 1 ng/g for most PAHs. The method was validated using certified reference materials from the Community Bureau of Reference (BCR)/Institute for Reference Materials and Measurement (Geel, Belgium).

11.1.3.2 *Thin-layer Chromatography* of PAHs

Thin-layer chromatography (TLC) belongs to the older analytical methods used for determination of PAHs in various matrices. TLC has the charm of being cheap, fast and easy in handling cleanup of difficult matrices, like plant and soil extracts, or tar.

One example for the latter is the determination of BaP in fresh cigarette smoke [23]. Cigarette smoke is immediately aging by air after release from a burning cigarette. A standardized procedure is applied for producing cigarette smoke aerosol samples. Cigarettes are smoked by a single pad smoking machine. The collection procedure for obtaining main stream smoke samples for the wet chemical analysis consists of a combination of denuder and filter sampling (see Figure 11.7).

Immediately after the filter tip, a borosilicate glass tube is mounted. The other end of the glass tube is connected to the inlet of a back-up-filter holder. A glass fiber filter is employed as collection medium for the smoke aerosol. By use of this arrangement, smoke from up to ten cigarettes could be sampled without clogging the filter. After having taken the sample, the glass fiber filters and denuder are immediately extracted with 3 mL of acetone under ultrasonic agitation. A volume of 1.7 mL of the extract is evaporated to dryness under nitrogen and the residue is resolved in

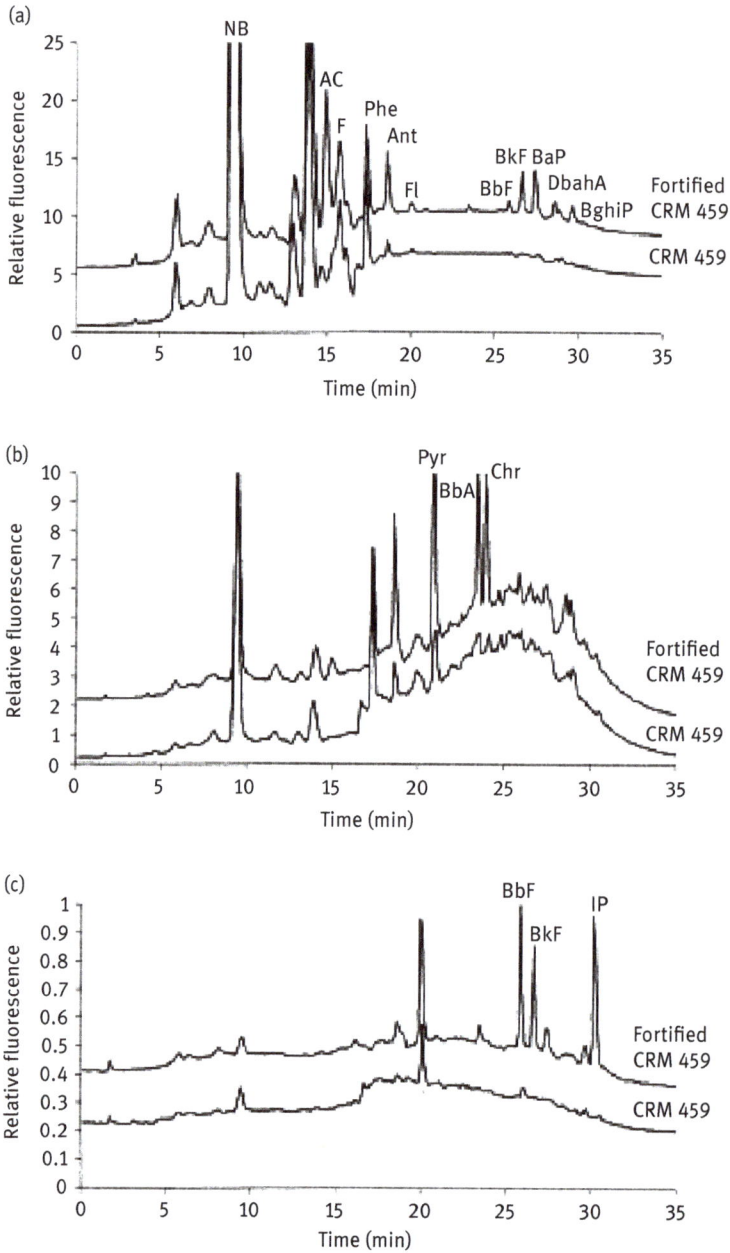

Figure 11.6: Chromatograms of a highly refined coconut oil (BCR CRM 459), fortified at a level corresponding to 2.5 ng/g of BaP. (a) Fluorescence program as in Table 11.2, (b) excitation wavelength as in Table 11.2, at emission wavelength 390 nm and (c) excitation wavelength as in Table 11.2 and emission wavelength at 500 nm [22]. Table 11.2 shows the excitation and emission wavelength program in HPLC/FLD analysis of PAHs. Reproduced with permission from Elsevier.

Table 11.2: Excitation and emission wavelength program in HPLC/FLD analysis of PAHs.

Time (min)	Excitation (nm)	Emission (nm)	PAH detected
0.0	275	330	Na, Ac, F
17.3	250	366	Phe
18.8	250	400	Ant
20.0	270	460	Fl
		390	Pyr, BaA, Chr
25.4	255	410	BbF, BkF, BaP
28.5	290	410	DbahA, BghiP
		500	IP

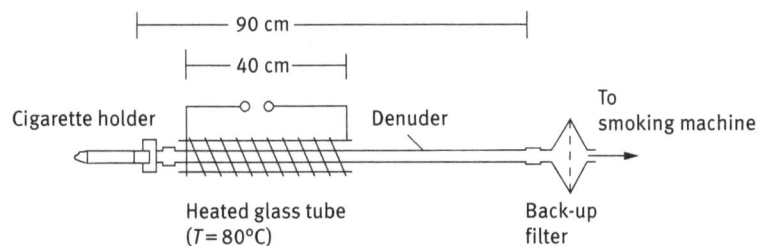

Figure 11.7: Device for sampling main stream smoke aerosol.

n-heptane (3 mL). A volume of 2.5 mL of the n-heptane solution is subjected to a further solid phase purification step. The elution is accomplished by 2.5 mL n-heptane and the whole extract re-evaporated again. The residual amounts are dissolved in 1.1 mL n-heptane. The content of 1 mL is concentrated to some microliters and the volume is determined by a syringe. A volume of 5 µL of each extract is used for the 2D chromatographic separation. TLC plates of 30% acetylated cellulose are used. The first run is performed with a ternary mixture of dichloromethane/ethanol/water (10:20:1). The second run is accomplished by application of pyridine, methanol and water (3:5:2). The chromatograms are developed in darkness to avoid the destruction of PAHs. Development time is 2 h. Between run 1 and 2, the plates are dried under nitrogen and darkness.

To determine the absolute amount of BaP on the TLC plate, the method of synchronous fluorimetry (SF) is directly applied. SF uses a constant $\Delta\lambda$ between excitation and emission wavelength and yields in favorable cases a very simple spectrum (see Figure 11.8). Small pieces of the developed TLC plate (Ø: 2 cm) are synchronously scanned by use of a spectrofluorimeter at $\Delta\lambda = 20$ nm. A strict linear relationship is observed within a range of 0.1–6 ng BaP (abs.). The detection limit is determined to approximately 50 pg BaP.

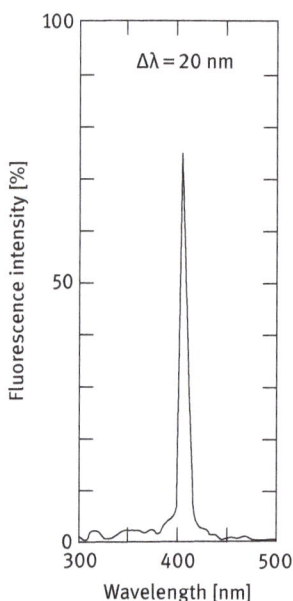

Figure 11.8: Synchronous fluorescence peak of 6.5 ng BaP obtained *in situ* on a TLC plate.

11.1.4 GC Analysis of PAH

Occurrence of PAHs in heavy oil, tar fractions or sediments from city harbors is an extremely complex analytical challenge. Not only a vast variety of PAHs is present, but also the concomitant other saturated and aromatic hydrocarbons, creosote, resins, siderophores, asphaltenes, and porphyrin, to name only a few fractions. Due to the history and genesis of these matrices, PAH determination needs the highest resolution techniques. Generally, GC offers the highest resolution, but with the limitation to high vapor pressure analytes in order to keep them in gas phase during separation.

The classical way for two–seven-ring PAH determination in heavy matrices is GC, coupled to FID or MS. Combined with appropriate sample pretreatment, robust and selective PAH analysis is possible, as demonstrated with GC–MS in Figure 11.9. In this study [24], different extraction methods have been applied to urban air matter. Supercritical CO_2 fluid extraction (SFE), Soxhlet extraction, with CH_2Cl_2–acetone, and extraction by subcritical water (known from water pressure cooking!) were performed. From the shown results, it becomes clear that each method has its own selectivity pattern. Soxhlet extraction yields the widest range of alkanes and PAHs, whereas SFE at low temperatures recovers an alkane-rich fraction. Subcritical water extraction provides the best solution but is not very common, due to technical expenditure.

GC–MS analysis of PAHs usually provides superior resolution capabilities, improved sensitivity and the possibility of further chemical structure confirmation. However, some isomers or structurally related compounds should be chromatographically separated before detection for proper identification and quantification. Neither

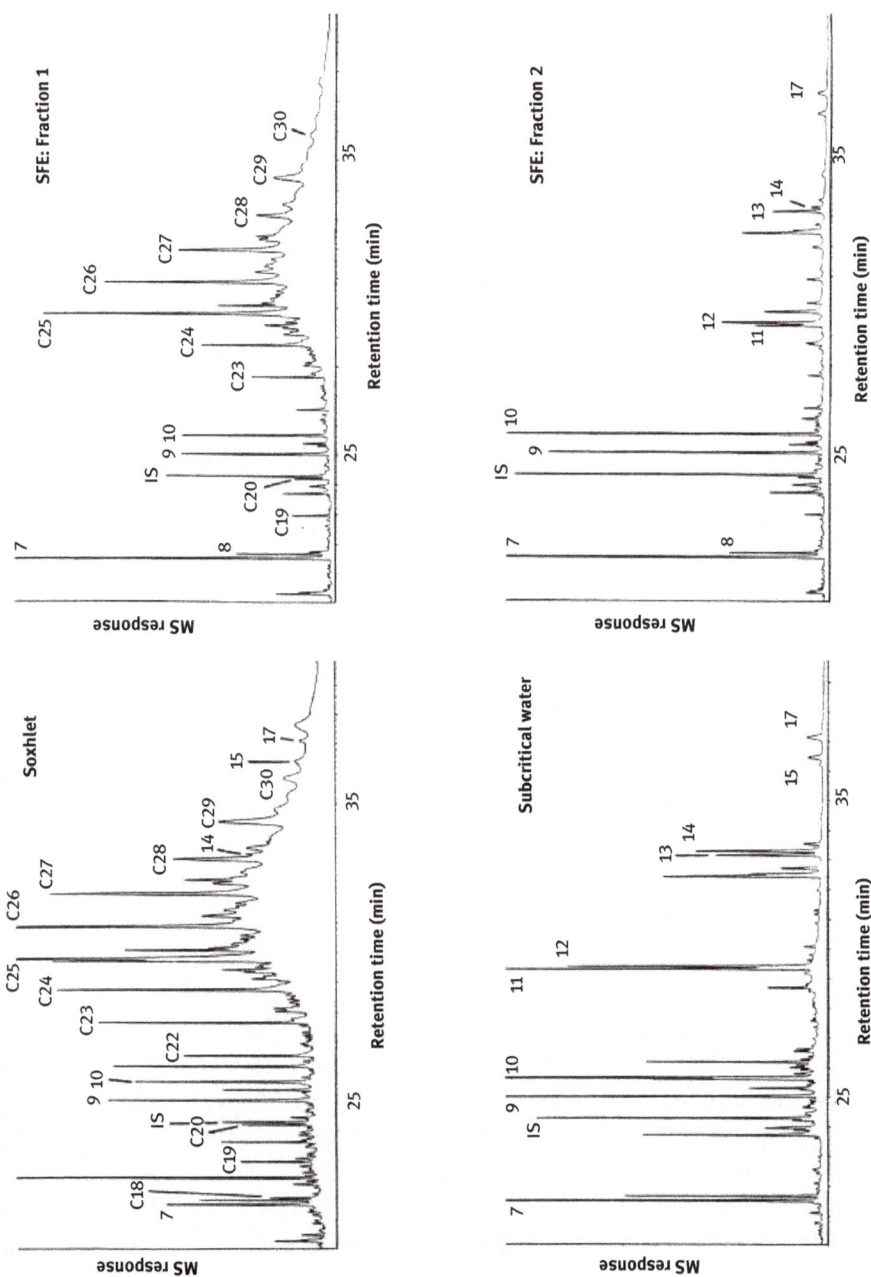

Figure 11.9: GC–MS chromatograms: selectivity of SFE with pure CO_2 and subcritical water compared to Soxhlet extraction for alkanes and PAHs from urban air particulate matter. All chromatograms are two composites of the GC-MS selected ions monitored for alkanes ($m/z = 57$) and PAHs (the molecular ion for each PAH). IS denotes the internal standard. The numbers above the peaks denote the EPA PAHs (e.g., 1 = naphthalene 17 = benzo[ghi]perylene). Major n-alkanes in the Soxhlet and first SFE fraction are designated by their chain length (e.g., C_8 denotes n-octadecane). Reproduced with permission from Elsevier.

GC phases applied for EPA PAH analysis nor any other commercially available station-ary phases provide a complete and simultaneous separation among the many com-pounds. Moreover, in recent years, some additional PAHs, including benzo[*c*]fluorene, cyclopenta[*cd*]pyrene (CCP), 5-methylchrysene, benzo[*j*]fluoranthene (BjF) and most of the dibenzopyrenes, have attracted special attention, especially by the European Union.

This insufficient separation capability often results in bias and overestimation of the reported concentrations. Under these circumstances, the use of *comprehens-ive two-dimensional GC* (GC × GC), that is, the combined use of two GC columns with distinct phases, can be advantageous (see also chapter 7.5.11.3).

In the following example [25], in a harbor sediment, not only EPA PAHs are sep-arated and determined, also a satisfactory separation of difficult pairs/group of PAHs, including BaA–CCP–Chr, BbF–BjF–BkF and IcdP–DahA, but also of other PAH pairs, which are usually undistinguishable even using MS detection, such as DacA and DahA, are shown.

Sample preparation consisted of the microwave-assisted extraction of 0.5 g of sediment sample with 30 mL of a 1:1 (v/v) mixture of *n*-hexane:acetone at 115 °C for 15 min. The cooled extract was filtered through a paper filter. A 1-mL aliquot of this filtered extract was purified by elution through a solid-phase extraction cartridge. The target analytes were eluted from the cartridge with 6 mL of *n*-hexane:isopropanol (1:0.034, v/v), concentrated under a gentle nitrogen stream and reconstituted in isooctane for final analysis. Identification of the studied PAHs in the purified extracts was performed using a GC–TOF-MS with EI ionization. Two nonpolar × mid-polar column sets were assayed: DB-5 (30 m) × BPX-50 (1.9 m), and HT-8 (30 m) × BPX-50 (1.6). The first dimension separation consisted of sample injection in the hot splitless mode in the DB-5 × BPX-50 column setup. The second dimension oven was programed to track the main oven with an off-set of 30 °C. A cryogenic modulator was used for sample focusing and injection in the second dimension column. The temperature of the modulator was set 30 °C above that of the main oven. The result of the ana-lysis of BCR CRM 535 (harbor sediment reference material) is shown in Figure 11.10. Limits of detection (LODs) are in the 6–60 µg/kg range.

11.1.5 Immunological Bioanalysis of PAHs

Immunoassays (IAs) are of importance for fast screening of organic traces of many samples in one run. IA are selective and sensitive methods that use biological binding proteins (antibodies) for target analyte detection. Rapidity, ease of use, low cost and the ability to process large numbers of samples at the same time make this technology of interest to the analytical laboratory. The often discussed bottleneck is the availabil-ity of acceptable antibodies for such bioanalyses. In clinical chemistry, many tests are in use since decades, whereas for food or environmental analysis, the rigid standards with possible legal consequences (e.g., closure of production sites), classical techniques like LC, GC or combinations with mass spectrometry, remain preferred.

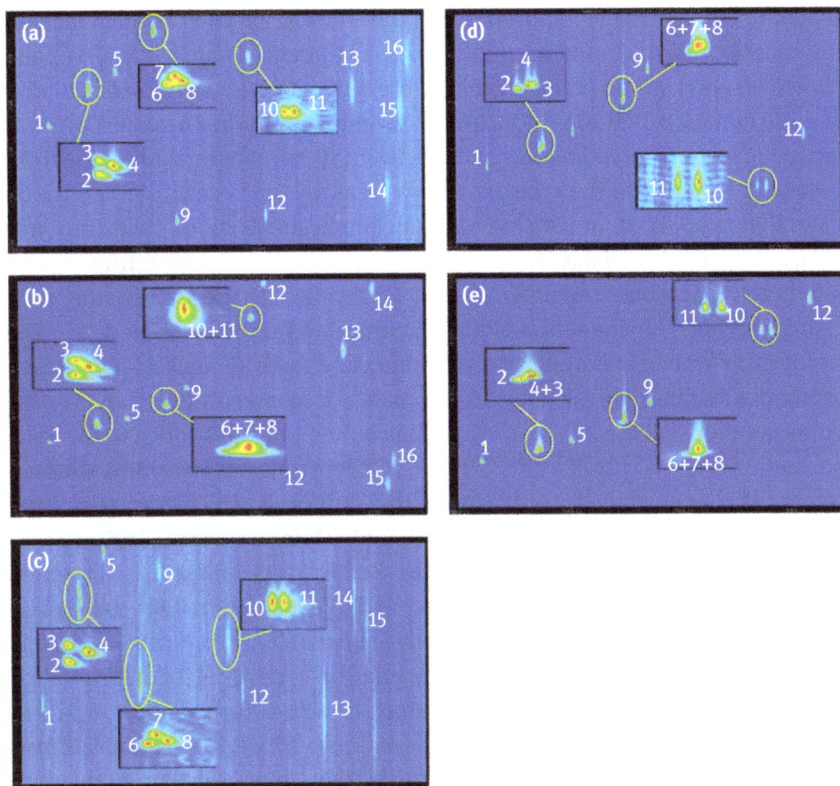

Figure 11.10: Optimization of the modulation period (P_M) for the 15 + 1 EU PAHs in the DB-5 × BPX-50 column set (left column) with a P_M of (a) 5, (b) 2 and (c) 10 s; and in the HT-8 × BPX-50 set (right column) with a P_M of (d) 5 and (e) 10 s. Selected m/z values: 216, 226, 228, 242, 252, 276, 278, 302 [25]. Reproduced with permission from Elsevier.

But exoneration of these priority measurement technologies by IA is possible, when antibodies are available and neither much sample pretreatment nor pre-concentration for subsequent analysis is needed, and sufficient selectivity and sensitivity is provided by IA to qualify a sample for further refined analysis.

11.1.5.1 Soil

PAH soil screening, as discussed in Chapter 10, needs tedious extraction steps, followed by spiking with internal standards and application of expensive equipment. Therefore, screening by a PAH IA is very helpful. In the following, one application is presented [26].

Soil samples were taken at different locations in the catchment area: forest soils, farmland, grassland, city ground and soil samples from a gas plant site. At each location, about 2–3 kg of soil was taken at a depth of 5–15 cm. If present, coarse plant material and flints were removed. Then, the soil was air dried at room temperature, thoroughly mixed, ground in a mortar, passed through <2-mm metal sieves and stored in brown glass bottles at room temperature until analysis. Additionally, a certified reference material (CRM 104–100, Resource Technology Corp., Laramie, USA) was included in the investigation. Soil was extracted with 10 mL of tetrahydrofurane in an ultrasonic bath for 1 h, and then the extract was removed and centrifuged at 5,000g for 15 min. Before analysis, in the case of tetrahydrofurane, 5 mL of the extract was concentrated to dryness under N_2 flow and redissolved in 5 mL of acetonitrile.

An indirect competitive microtiter plate ELISA format was used. As coating antigen, a 1-aminopyrene–bovine serum albumin conjugate (AP–BSA) was prepared. The ELISA assay is performed in polystyrene microtiter plates with 96 wells. Plate washing and absorbance readings are taken by an automatically operated washer and reader. The coating solution (0.25 µg AP–BSA/mL coating buffer, pH 9.6) is added to the wells (200 µL), and the plate is coated by drying at 55 °C overnight. After washing, sites not occupied by the coating antigen are blocked with 300 µL/well of blocking agent. After a further washing step, a standard (50 µL/well) and diluted antiserum (100 µL/well) are added and incubated for 30 min on a shaker. Soil extracts are applied in acetonitrile/water (10:90, v/v). Plates are again washed, and a goat anti-rabbit IgG-peroxidase conjugate (150 µL/well, dilution 1:25,000) is added and incubated for further 60 min as before. After a final washing step, the substrate solution (tetramethylbenzidine/H_2O_2, 150 µL/well) is added. Color development is stopped with 100 µL/well of 0.5 M sulfuric acid, and the absorbance is measured at 450 nm. All determinations are performed at least in triplicate. The sigmoidal calibration curve is set up by plotting the mean of the absorbance against the logarithm of the standard concentration and interpolating by Rodbard's logistic four-parameter equation (see Figure 11.11). Acetonitrile/Water (10:90, v/v) pyrene standard solutions covering the concentration range from 0.001 to 1,000 µg/L are daily prepared. The linear dynamic range of the method was 0.05–5 µg/L for aqueous standards. As can be seen, the presence of organic solvent has a distinct shift in calibration as consequence and needs to be considered.

The results are typical for ELISA tests: As often found with groundwater samples, ELISA determination of soil extracts resulted in an underestimation of PAH concentration when comparing the pyrene equivalents with the sum of the EPA PAHs from parallel HPLC measurements (see Figure 11.12). This is due to the presence of other PAHs and their metabolites, which will be recognized in part by the antibodies too. But classification of polluted sites of PAHs at environmentally significant levels in soil (<1, 1–10, 10–100, >100 mg/kg) is effectively possible.

Therefore, site-specific calibration is strongly recommended, that is, the target analyte partition of a few random samples from the contaminated site should be determined by a reference method as the base for the calibration of the ELISA with known cross-reactivities of the single compounds.

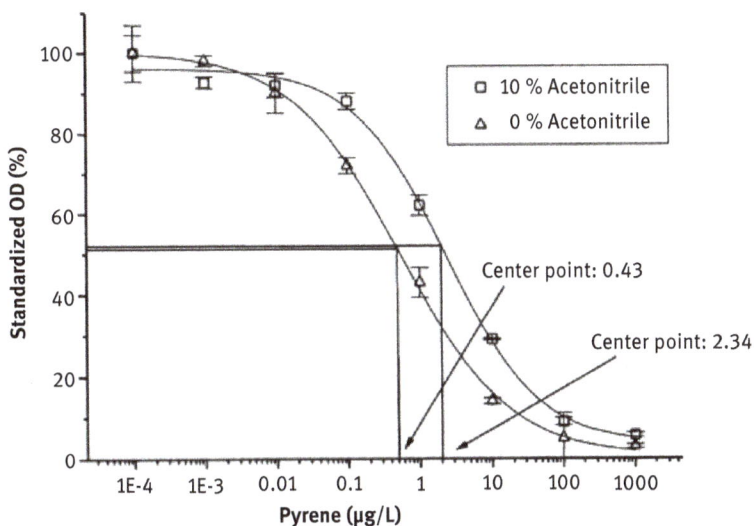

Figure 11.11: Calibration curves for pyrene obtained with spiked water samples and acetonitrile/water (10:90, v/v). Error bars represent ±1 standard deviation around the mean ($n = 3$). Standardized OD is the absorbance at 450 nm of a standard divided by the absorbance of the zero standard. Optical Density (OD). Reproduced with permission from the American Chemical Society.

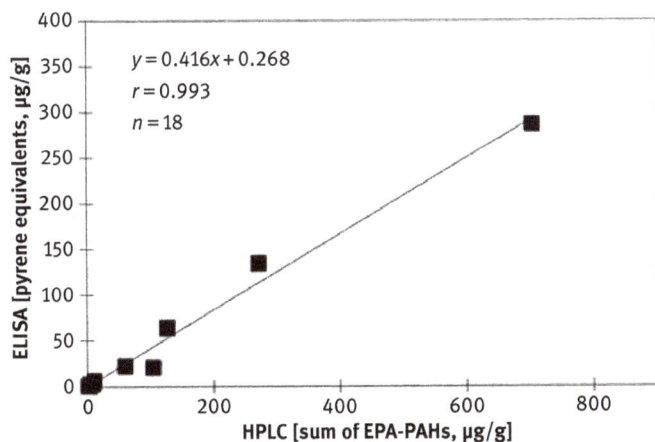

Figure 11.12: Comparison of HPLC- and ELISA–PAH determination in soil extracts. Individual values represent the sum of 15 EPA PAHs as measured by HPLC and were correlated to the amount of pyrene equivalents. Reproduced with permission from the American Chemical Society.

11.1.5.2 Water

Especially, BaP is of utmost concern as water pollutant. A trigger value of 10 ng/L is set for drinking water in Europe. This limit has been adopted by many agencies around the world. Currently, no standard technique is able to measure such low contamination without pre-concentration. However, antibodies with an extreme high

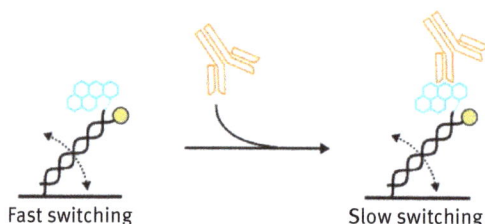

Fast switching   Slow switching

Figure 11.13: Electro-switching as sensor principle for BaP detection in water. Reproduced with permission from the American Chemical Society.

affinity constant, combined with an ultrasensitive detection principle, can handle this. Electro-switching as a sensor principle (see Figure 11.13) in a competitive format, together with a highly affine and selective monoclonal anti-BaP antibody, allows detection of far less than 10 ng/L [27]. DNA nanolevers of 48 bp are grafted onto gold microelectrodes via a sulfur linker. A square wave ac voltage is applied to the electrodes, which alternatingly repels and attracts the negatively charged DNA at 10 kHz frequency. As a consequence, the nanolevers switch their orientation from lying to standing and back. The upward and downward motions are measured by a time-resolved single-photon-counting module. The DNA strand that is covalently grafted to the electrode also features a fluorescent dye label. The complementary strand is modified with B[a]P and can be exchanged by DNA denaturation and re-hybridization with fresh ssDNA–B[a]P conjugates. In the competitive mode after mixing samples with limited amounts of antibodies, the remaining non-bound antibodies are attached to the BaP-labeled strand, with the result of changing nick frequency.

In practice, a violation of the permissible limit of 10 ng/L B[a]P in tap water could be detected [27] within less than an hour by monitoring the antibody association kinetics in real time, which makes the introduced approach the fastest and most sensitive detection method reported so far.

11.1.6 *In situ* PAH Analysis by Spectroscopic Techniques (Laser Fluorescence, Photoelectron Emission)

When continuous process monitoring or a highly unstable combustion situation control asks for an online analysis, a different strategy is demanded. Fortunately, very often, a stable PAH profile is observed, with varying total PAH concentration, but a stable relationship of the individual member PAHs to each other. In this case, also supported by the fixed PAH vapor pressure relationships of the PAH compounds, only one PAH needs to be observed instead of all individual 16 EPA PAHs.

11.1.6.1 Time-resolved Laser Fluorescence PAH Sensor
The long-term (for years) observation of a soil remediation measure, where a polluted catchment area has to be permanently flushed with groundwater, will serve as an

example [28]. The contaminated water of an old gas factory remediation site had to be examined. The contamination of the groundwater resulted from the PAH entry into the soil body due to a long-time coal pyrolysis producing gas for household heating over decades. In the beginning of 1990, building activities took place causing mobilization and wash-out of the PAHs with the lowered groundwater level. According to a government order, the water is permanently removed through groundwater wells and worked off by a cleaning station using activated carbon filtration. A continuous monitoring is necessary to realize a breakthrough through an exhausted adsorption bed in time.

Conventional analysis of PAHs in water is a time consuming process, based on frequent sampling, cleanup and chromatographic separation with subsequent UV or FLD detection. Due to this, a quasi-continuous surveillance of hazardous areas is not possible and in case of accidents, traditional methods have only a documentary character. To cope with this problem, the development of sensor systems gained interest. An obvious approach is the use of fluorescence spectroscopy for strongly fluorophoric PAHs. Unfortunately, attempts to detect PAHs in water by fluorescence spectroscopy are complicated due to their low concentrations ($\mu g/L$–pg/L), light scattering from hydrocolloids and the overlapping and featureless fluorescence spectra in combination with background fluorescence caused by humic compounds. Thus, an additional independent dimension – an excitation spectrum or decay time – is needed to increase the analytical information for fluorescence spectra.

Figure 11.14 shows a laser-induced PAH fluorescence sensor [28]. A pulsed N_2-laser (fixed excitation wavelength of 337 nm; pulse length: 5 ns) is focused on one distal end of a quartz glass fiber. The fiber-optic sensor head was designed as a dual bare fiber sensor, directly inserted into the flowing water. The useful laser intensity at the end of the excitation fiber was 20 μJ ($\pm$1 μJ), while the fluorescence was collected at an angle of 15° by a second 30-m fiber. The fluorescence light from the observation fiber is focused onto the entrance slit of a monochromator. The output is recorded by a photomultiplier and a digital storage oscilloscope.

From the time- and wavelength-resolved measurement, 3D spectral plots are achieved. Figure 11.15 shows this for 500 ppt BaP in water.

For a satisfactory analysis of time-resolved fluorescence emission spectra, several assumptions have to be made. First, reabsorption and reemission effects can be neglected so that the time decay of the PAHs can be described by a first-order kinetic. Thus, the observed intensity for a mixture of different fluorophores is at any time the sum of all single fluorescence decays. The algorithms applied for the detection of the compounds in mixtures are based on the KNORR–HARRIS algorithm [29]. The time-resolved emission matrix is thus decomposed in two submatrices containing the temporal and the spectral characteristics of the main components.

The best results for data evaluation have been achieved by placing the barriers for the decay times of the principle components in the following way: 0–2 ns representing straylight, 2–10 ns representing humic acids and short decay time PAH (BaP: 19 ns;

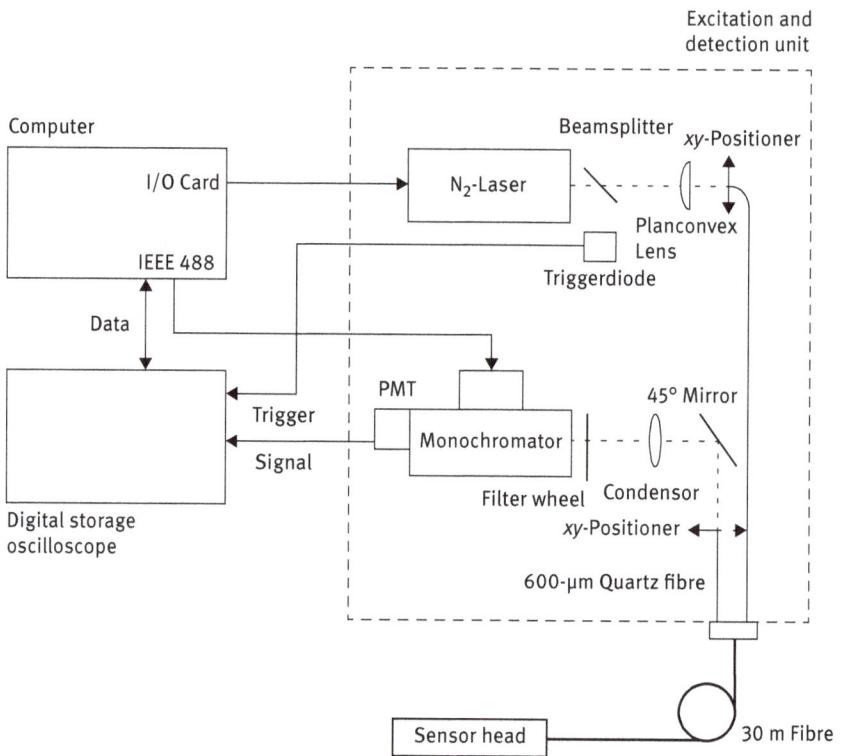

Figure 11.14: Experimental setup of the used PAH-sensor system. Reproduced with permission from Springer.

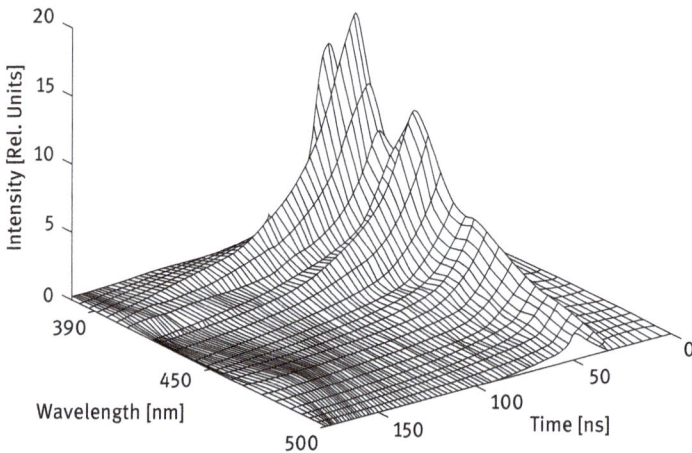

Figure 11.15: Time-resolved fluorescence spectrum of 500 ng/L BaP in water under excitation by pulsed laser light of 337 nm. Reproduced with permission from Springer.

pyrene: 123 ns) and 10–150 ns for the longer decay time PAH. Despite the laborious chemometric data treatment, sufficient good correlation ($R = 0.98$) to conventional HPLC analysis of 65 spot samples in range of 0.1–5 µg pyrene was found. Although selectivity and detection limits of the conventional instrumentary cannot be reached, the employment of pyrene as marker of a PAH presence in real samples results in a beneficial supplement to the common analytical instrumentation.

11.1.6.2 PAH Monitoring in Ambient Aerosol by Photoelectric Aerosol Charging

PAHs released from combustion units or engines are adsorbed onto carrier particles. The most thermodynamically stable inorganic compounds (carbon, metal oxides, silicates, etc.) form a condensation nucleus first. During cooling down, when leaving the combustion unit through stack or exhaust pipe, the accompanying volatile compounds (unburnt organic fuel constituents or synthesized PAHs and others) are condensed and/or adsorbed to them. Therefore, multilayered onion-like particles are carrying the PAH molecules. Depending on a changing fuel-to-oxygen ratio, fast change in PAH concentration is the consequence. For minimizing PAH emission, a rapid responding PAH sensor system would be extremely valuable.

Based on the photoelectric charging effect of PAH contaminated aerosol particles, a continuously sensing online and *in situ* sensor system could be developed [30]. Its basic working principle is shown in Figure 11.16. The model of a microstructured particle demonstrates the principle of charging a PAH-coated particle by photoelectron emission.

According to well-accepted models, the charging rate of particles under UV illumination is a function of the following quantities (particle size <100 nm):

$$\frac{dN^+}{dt} = f\left(\lambda_{uv},\ \pi r_p^2,\ F_{PAH},\ I_{UV},\ Y_{PAH},\ C_N,\ t_{irr}\right)$$

where N^+ is the positive-charged particles, λ_{UV} is the wavelength of the UV light, I_{UV}, is the photon flux of UV-light, F_{PAH} is the fraction of coverage of photo-emitting PAH

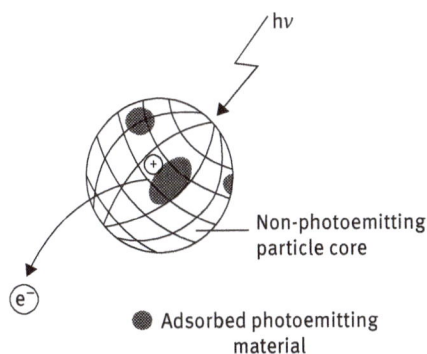

Figure 11.16: Principle of photoelectric aerosol charging. Reproduced with permission from Elsevier.

on a particle surface, Y_{PAH} is the material-dependent photoelectric yield, t_{irr} is the irradiation time, r_p is the particle radius, C_N is the number concentration of particles and t is the time.

If the variables l_{uv}, λ_{uv}, t_{irr}, C_N and r_p are kept constant, the observed fraction of charged particles is a direct measure of the total PAH-coated surface area and the PAH photoelectric yield itself. If the particle consists of pure PAH (100% covered by PAH, i.e., F is 1.0), the fraction of charged particles directly reflects the photoelectric yield of the PAH. Therefore, this fraction was measured (by observation of the current of an aerosol electrometer) after UV illumination, compared to a given number concentration of neutral particles at a fixed irradiation time, a given lamp spectrum and a fixed particle radius r_p. By calibration experiments with pure monodisperse PAH particles, the superior photoelectric yield of large five—seven-ring PAHs could be demonstrated [31]. For a fixed particle size spectrum, irradiation time and light intensity, the number concentration of charges produced is directly linearly related to particle concentration and PAH coverage, which is the total amount of adsorbed PAH. As long as only PAH sub-monolayers are prevailing, this reflects the extractable amount of PAHs and hence should correlate to classic PAH analysis.

The sensor system itself is depicted in Figure 11.17. The whole unit is similar to a laser cavity. The main advantage of this arrangement is that there is no contact between the particles and the UV lamp glass. Because of the geometry of the elliptical cavity, all UV light is reflected onto the particle surfaces, therefore requiring only a very low lamp power to obtain high UV intensity. Assisted by a sheath air flow, the light intensity cannot be influenced by deposition of particles on the quartz glass

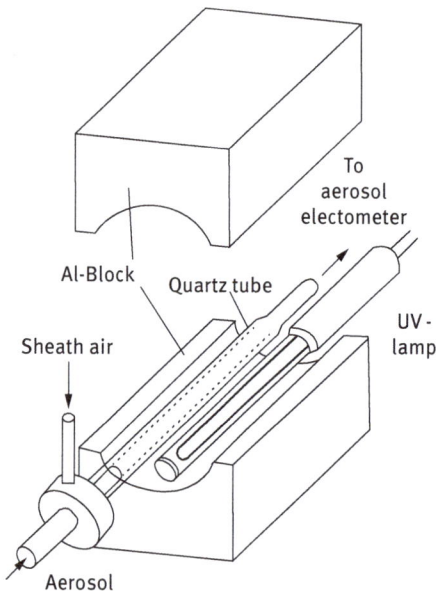

Figure 11.17: Photoelectric aerosol sensor, consisting of a UV lamp (λ_{uv} = 185 nm) and a quartz tube with sheath air ducted aerosol flow. UV lamp and quartz tube are aligned in the focal axes of an elliptically cut, UV-reflecting aluminum block [29]. Reproduced with permission from Elsevier.

surface. The glass surface itself offers a certain surface conductivity and serves as a sink for the photo-emitted electrons, preventing the formation of electrostatic negative "islands" on the glass surface, causing an unwanted deposition of already-charged particles. In order to excite photoelectrons from PAHs with a work function higher than 4.9 eV, a low-pressure mercury UV lamp was inserted as UV source. The flow rate through the quartz tube was set to 3.7 L/min, which resulted in an irradiation time of 125 ms. Due to this short irradiation time, only a fraction of the particles was charged. Additionally, the light intensity could be decreased by covering the UV lamp by a screen.

The number concentration of positive-charged particles is determined by the electric current produced by the singly charged particles and the constant flow rate through the aerosol electrometer. As long as the PAH profile is constant, and the aerosol size spectrum too, a variation in PAH coverage is directly observable by the electric current measured. The sensitivity for five—seven-ring PAHs is in the ng/m^3 range.

Many studies showed the applicability of the photoelectric aerosol sensor for ambient PAH monitoring. Not only rapidly changing emission situations (diesel engine, cigarette smoking, oil heating, etc.) but also fast changing ambient air quality like traffic near road crossings can be followed in diurnal profiles [32].

11.2 Polychlorinated Biphenyls, Dibenzodioxins and -Furans

11.2.1 General Remarks on Occurrence and Importance

Polychlorinated biphenyls (PCBs), polychlorinated dibenzodioxins (PCDDs) and polychlorinated furans (PCDFs) belong to those analytes which are tightly linked to the rise seen in instrumental analytical chemistry during the last 50 years, paired with increasing intensity of discussion about health impact of industrial technology on our present and future society. These substance classes also served as model compounds for studying the unexpected presence in biota, global distribution and accumulation in living organisms and food web, followed by allergenicity, inflammation, deterioration of reproduction or endocrine disruptive activities and finally abnormal cell proliferation. The name *xenobiotics* (formed from Greek words ξένος (*xenos*) = foreigner, stranger and βίος (*bios, vios*) = life) was coined for them, indicating undesirability. A common typical property of them is their inherent inertness and hydrophobicity, which is the reason for their presence worldwide, even in extremely remote areas like Antarctica or deep sea. They belong to the *persistent organic pollutants*.

These compounds occur in extremely low concentrations throughout. All matrices (water, air, soil, food, etc.) may be contaminated. Maximum residue levels set in Toxic Substances Control Acts responding to this show, for example, 750 fg WHO-PCDD/F-TEQ/g fat in vegetable oil, or 100 fg WHO-PCDD/F-TEQ per kg milk, are discussed for prevention. Toxic equivalents (TEQs) report the toxicity-weighted masses of mixtures of PCDDs, PCDFs and PCBs. The TEQ value provides toxicity information about the mixture of chemicals. For municipal waste combustion, a concentration of 100 pg PCDD-TEQ per cubic meter exhaust must not be exceeded.

The difficulty for analytical chemistry is the enormous variety of congeners, which means similar structural identities due to a common identical origin, for example biphenyl, acting as mother compound for 209 differently chlorinated derivatives of different halogen positions and numbers in case of PCBs. But only a few of them are toxicologically relevant, which demands a congener-specific analysis, and this in the ppq range.

Since the chlorine atom is part of the group of elements known as halogens (others are fluorine, bromine and iodine), polychlorinated aromatic compounds are only a part of a larger group of chemicals known as halogenated aromatic compounds.

For reasons of clarity, we refer in the following only to the chlorinated molecules. But the brominated and fluorinated compounds are entering the stage more and more, creating increasing concern and efforts.

11.2.1.1 PCBs

PCBs are organochlorinated compounds with the general formula $C_{12}H_{10-x}Cl_x$. The structural formula is depicted in Figure 11.18. Industrial PCBs are complex mixtures (e.g., Aroclor 1242, meaning 12 C atoms with 42% chlorination by mass) composed of up to 50 or 60 congeners (or individual chlorobiphenyls).

Despite a reduction in their usage, this group of substances is still found worldwide in many technical applications [33]. Their great inertness and physical properties (electrically nonconductive, high boiling point, high thermal conductivity, hydrophobic character) make them very appropriate as cooling, insulating or hydraulic fluids. PCBs were also often used as plasticizers (even considered as softener in chewing gum), impregnating agents and flame retardants.

Due to their great persistence, these compounds tend to accumulate in lipophilic sinks (milk, liver, fatty tissue, harbor sediment and microplastics). In the environment, there are only two not very effective degradation pathways known: aerobic oxidative processes and anaerobic reductive processes. Due to the observed accumulation effect in biota, in the 1970s, PCB production is banned in most countries and regulations concerning the presence of PCBs in the environment were tightened.

A number of 68 PCB congeners do not have chlorine atoms in the *ortho*-position or even only one, so the two phenyl rings may arrange in a planar configuration.

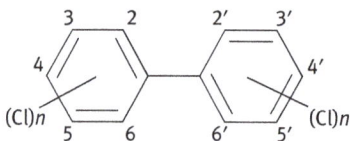

Figure 11.18: General structural formula of PCBs. The possible positions of chlorine atoms on the aromatic rings are assigned to the carbon atoms.

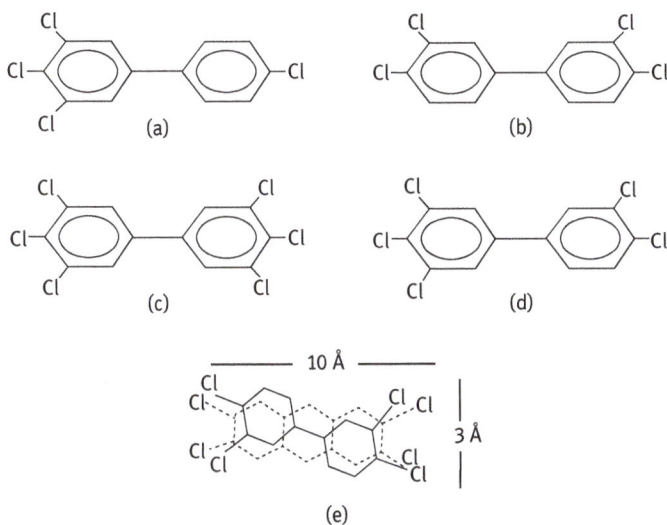

Figure 11.19: (a)–(d) Structure of four representative dl-PCBs; (e) molecular dimension of 2,3,7,8-tetrachloro-dibenzo-*p*-dioxin in comparison to 3′,4,4′,5-tetrachlorobiphenyl.

Such configurations are of toxicological relevance, since they may intercalate into the nucleic acid–base pairs (with the possible consequence of deteriorated replication of cell growth) [34]. Even more of interest are those PCB congeners where two or more of the *meta*-positions are chlorinated, the total number of chlorine is ≥4, and both *para*-positions are substituted by chlorine too. These 12 coplanar PCBs (examples seen in Figure 11.19) are considered as harmful and have to be identified individually. Substantial interaction with the *aryl hydrocarbon receptor (AhR)* has been found. As shown in Figure 11.19e, there is a similarity to the molecular dimensions of dioxins. This has led to the name "12 *d*(ioxin)*l*(ike) *coplanar PCBs*".

The acute PCB toxicity, expressed by the LD_{50} value, is in the range of 1 ppb (e.g., in water for shrimp), up to 2,000 ppm (in food for mice). According to the current knowledge, only the 12 dl-PCBs contribute to a health risk. Therefore, for risk assessment, these congeners are of prior-ranking concern and have to be analyzed individually and multiplied by *toxicity equivalence factor* (TEF) [35]. The TEF-corrected results are summarized and form a measure for the total human risk across the different species found by trace analysis.

11.2.1.2 Polychlorinated dibenzodioxins and –furans

The family of dioxins and dioxin-like compounds contain two chlorinated aromatic rings each, bridged either by one (PCDF) or by two (PCDD) oxygen atoms (see Figure 11.20).

There are 75 different congeners of PCDDs and 135 different congeners of PCDFs known. They are distinguished by the position and number of chlorine atoms attached to the two benzene rings. Dioxins and furans with the same number of chlorine atoms

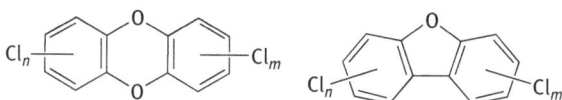

Figure 11.20: General structures of dibenzodioxins (left) and dibenzofurans (right).

constitute a homologous group of isomers. For toxicity purposes, only the homologues with four or more chlorine atoms are considered and these are called tetra (TCDD, TCDF or D4, F4), penta (PCDD, PCDF or D5, F5), hexa (HxCDD, HxCDF or D6, F6), hepta (HpCDD, HpCDF or D7, F7) and octa (OCDD, OCDF or D8, F8).

Physicochemically, these substances are hydrophobic, possess low boiling points and are relatively inert to decomposition [36]. Only high temperatures >900 °C destroy them completely.

These compounds are never intentionally produced for industrial use. They are formed as side products through unattended loss of control over chemical reactions, mainly by inappropriate temperature range (300–900 °C) during waste incineration of chlorinated polymers like PVC (leading to *de novo* synthesis; gas-phase initiated or solid-phase catalyzed) or during copper recycling [37, 38]. In the aqueous phase, partial dechlorination of octachlorinated dibenzodioxin/-furan or condensation of chlorophenols triggered by UV-light [39] or enzymatic degradation [40] leads to highly toxic PCCD/Fs (in the range of D/F4 to D/F8, i.e., dioxins and furanes with 4–8 chloro substituents). Other known sources are chlorination processes, for example pulp bleaching. Therefore, rigid emission control frames have been set over the last decades, with the consequence of continuously diminishing environmental concentrations observed.

Dibenzodioxins gained notoriety through the accident in Seveso (Italy) in the year 1976. At a chemical production site, where 2,4,5-trichlorophenol was produced by alkaline dechlorination of 1,2,4,5-tetrachlororobenzene, a sudden temperature rise overshot the critical temperature of 230 °C, opening the formation pathway to 2,3,7,8-tetrachloro-*p*-dibenzodioxin (TCDD). In the course of a reactor valve opening, an aerial release of 6 t of chemicals happened, which dispersed over 18 km^2 (6.9 sq mi) of the surrounding area. Among the substances released was 1 kg of TCDD. Within the next days, not only thousands of animals died within the affected area, but also more than 500 people developed serious chloracne and skin lesions. First findings on the long-term consequences showed an increase of cancer incidence and mortality, especially an increased occurrence of cancer of the gastrointestinal sites and of the lymphatic and hematopoietic tissue. Further studies confirmed in victims of the disaster that dioxin is carcinogenic to humans and corroborate its association with cardiovascular- and endocrine-related effects [41].

Toxicologically, the binding of these substances to the AhR is also responsible, like in the case of PCBs [42]. These findings suggest that some functions of the Ah receptor (AhR) in the toxic effects of dioxins are connected to teratogenicity, immunosuppression, tumor promotion, and carcinogenic and estrogenic consequences.

Since the toxicity of a dioxin congener depends on the level of chlorination and location of chlorine atoms on the benzene rings, different toxicity equivalence schemes, for example, denoted as NATO or CCMS (NATO Committee on the Challenges of Modern Society) have been developed. They are compiled in Table 11.3. They represent also the analytes of interest in a PCDD/DF trace analysis. Included are, because of similar health relevance, TEFs for the coplanar PCB congeners.

In consequence, extremely low threshold limits, for example, as maximum allowable concentration (MAK value), at a working place 100 pg/m^3 in air, are set. Since PCDD/Fs become enriched in the food chain, a variety of regulations similar to PCBs are in action and became tightened too. In the year 1998, WHO evaluated the health risk from PCDD/DFs and established a new *tolerable daily intake* between 1 and 4 pg I-TEQ/kg body weight. The *TEQ* is the toxicity-weighted sum of the found congener masses times their TEF value.

11.2.2 Sampling Strategies for Polyhalogenated Aromatic Hydrocarbons

The contamination level for polyhalogenated aromatic hydrocarbons searched in water, soil, sediment, combustion aerosol and food asks for detailed analysis of the coplanar congeners, roughly the 11 or 12 congeners in case of PCBs, PCDDs or DFs (see Table 11.3). The highest TEF for 2,3,7,8-TCDD/DF makes this compound the key substance. Since the given threshold limits are generally in the ppq range,

Table 11.3: *Toxic equivalency factors* (TEF) for PCDD/DFs and PCBs according to WHO (2005). Reproduced with permission from Elsevier.

PCDD congener	WHO-TEF	PCDF congener	WHO-TEF	PCB congener	WHO-TEF
2,3,7,8-TCDD	1	2,3,7,8-TCDF	0.1	*Non-ortho:*	
1,2,3,7,8-PeCDD	1	1,2,3,7,8-PeCDF	0.03	PCB No. 81	0.0003
1,2,3,4,7,8-HxCDD	0.1	2,3,4,7,8-PeCDF	0.3	PCB No. 77	0.0001
1,2,3,6,7,8-HxCDD	0.1	1,2,3,4,7,8-HxCDF	0.01	PCB No. 126	0.1
1,2,3,7,8,9-HxCDD	0.1	1,2,3,6,7,8-HxCDF	0.1	PCB No. 169	0.03
1,2,3,4,6,7,8-HpCDD	0.01	2,3,4,6,7,8-HxCDF	0.1	*Mono-ortho:*	
OCDD	0.0003	1,2,3,7,8,9-HxCDF	0.1	PCB No. 105	0.00003
		1,2,3,4,6,7,8-HpCDF	0.01	PCB No. 114	0.00003
		1,2,3,4,7,8,9-HpCDF	0.01	PCB No. 118	0.00003
		OCDF	0.003	PCB No. 123	0.00003
				PCB No. 156	0.00003
				PCB No. 157	0.00003
				PCB No. 167	0.00003
				PCB No. 189	0.00003

pre-concentration through large volume sampling is needed throughout. Moreover, common to all sampling steps is a spiking with trace amounts of stable isotope-labeled analytes (for validation purposes), which makes especially "dioxin" analysis quite expensive.

The applicable sampling and cleanup strategies are the same as discussed for PAHs (see Section 11.1.2). As these analytes are of lipophilic character, the removal of oil, fat or lipids prior to subsequent analysis is essential. For example, biomagnification of PCDD/PCDFs in fish presents a public health risk as approximately 90% of human exposure to PCDD/PCDFs comes from contaminated food.

In case of polluted soil, one first needs drying the sediment and homogenizing through a sieve or by grinding. Next, an aliquot of the dried sample is extracted in a Soxhlet apparatus with toluene. Following this, chromatographic cleanup of the extract involving several packed columns is performed. Finally, the eluted extracts are concentrated before each column and prior to instrumental analysis [43].

Such classical sampling preparation route can be performed more environmentally friendly by
– reducing or eliminating the use of solvents, reagents and preservatives,
– minimizing energy consumption,
– minimizing or simplifying management of analytical waste and
– resulting in a safer environment for the analyst.

Alternative schemes for sample treatment and cleanup have been developed. This concept became known under the acronym *QuEChERS* (Quick, Easy, Cheap, Effective, Rugged and Safe) [44]. QuEChERS utilizes solid–liquid extraction in acetonitrile (under ultrasonic agitation), followed by the partitioning of the acetonitrile from the solid and aqueous layers by salting out. An aliquot of the separated acetonitrile layer is then cleaned with dispersive sorbents and analyzed. This concept was first developed for pesticide analysis but is now extended to many other analytes in various matrices. This is schematically depicted in the following on occasion of PCDD/DF screening of sediments (see Table 11.4).

Fatty tissue samples are homogenized by use of dissecting and/or mortar equipment and frozen under liquid nitrogen before freeze drying [45]. The freeze-dried products are ground to obtain a fine powder. Pressurized-liquid extraction (PLE) is performed on dried powder using an accelerated solvent extractor with hexane as solvent. Fat extracts are dried over sodium sulfate before determination of their lipid content by gravimetric analysis.

Subsequent cleanup steps are then performed on these samples of fat. Briefly, the cleanup is based on the use of an open short chromatographic multilayer glass column freshly packed with acidic silica, deactivated alumina and sodium sulfate. The fat extract (0.5 g) diluted with 2 mL hexane is then applied on the top of the column and eluted with 20 mL hexane. After completion of the collection step, the unified

Table 11.4: Comparison of the classical Soxhlet extraction method with the modified QuEChERS method adapted for PCDD/PCDF screening in sediment samples [44]. Reproduced with permission from Elsevier.

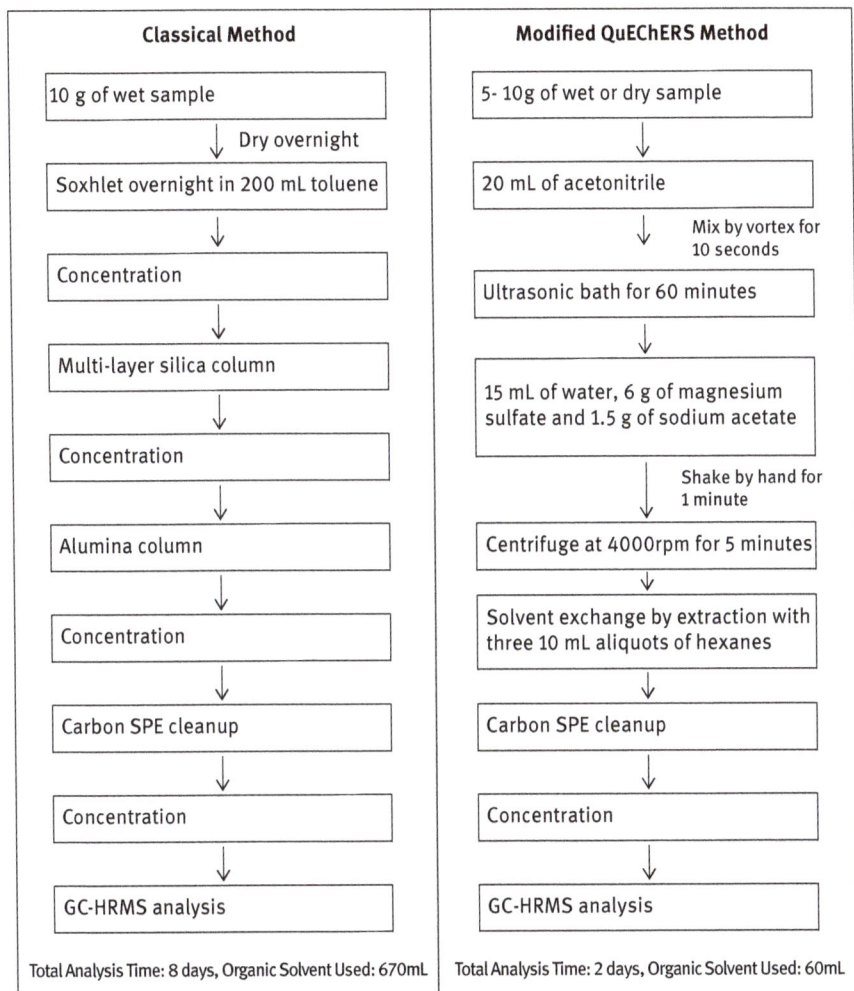

Classical Method	Modified QuEChERS Method
10 g of wet sample	5- 10g of wet or dry sample
↓ Dry overnight	↓
Soxhlet overnight in 200 mL toluene	20 mL of acetonitrile
↓	↓ Mix by vortex for 10 seconds
Concentration	Ultrasonic bath for 60 minutes
↓	↓
Multi-layer silica column	15 mL of water, 6 g of magnesium sulfate and 1.5 g of sodium acetate
↓	↓ Shake by hand for 1 minute
Concentration	Centrifuge at 4000rpm for 5 minutes
↓	↓
Alumina column	Solvent exchange by extraction with three 10 mL aliquots of hexanes
↓	↓
Concentration	Carbon SPE cleanup
↓	↓
Carbon SPE cleanup	Concentration
↓	↓
Concentration	GC-HRMS analysis
↓	
GC-HRMS analysis	
Total Analysis Time: 8 days, Organic Solvent Used: 670mL	Total Analysis Time: 2 days, Organic Solvent Used: 60mL

eluent is evaporated by rotary evaporation. Dodecane is added as keeper (to prevent evaporation losses) and the remaining solution is analyzed.

Parallel processing of suitable reference material is highly recommended.

11.2.3 GC Analysis of PCDDs/PCDFs and PCBs

As already outlined, the enormous diversification of congeners needs the highest separation resolution. Currently, most separation techniques don't provide the necessary resolution power, except GC with narrow-bore thin-film capillary columns. To meet the requested sensitivity in the absolute femtogram range (1 fg = 10^{-15} g), only the combination with an electron-capture detector (ECD) or MS detector can be used. Together with appropriate sampling and sample preparation, congener-specific analysis can be achieved [46].

11.2.3.1 GC-ECD

Historically, Ballschmiter was the pioneer for PCB analysis in complex matrices [47]. He also proposed a scheme for numbering the individual PCB congeners, nowadays known as BZ ("Ballschmiter Zahl") counting. Identification of PCB components that are often not available as reference compounds can be achieved by calculation of their retention indices. Retention indices of the PCB can be measured by ECD using n-alkyl trichloroacetates as reference homologues. This is helpful for identification of presence of technical PCB mixtures (e.g., Aroclor, Clophen, etc.). Figure 11.21 shows the GC profile of two widely used Aroclor mixtures. Detection limits are in the absolute pg range, so pre-concentration is needed.

11.2.3.2 GC × GC-ECD

Many attempts have been published to increase the achieved resolution by pre-separation into different polar fractions with LC. With the advent of multidimensional,

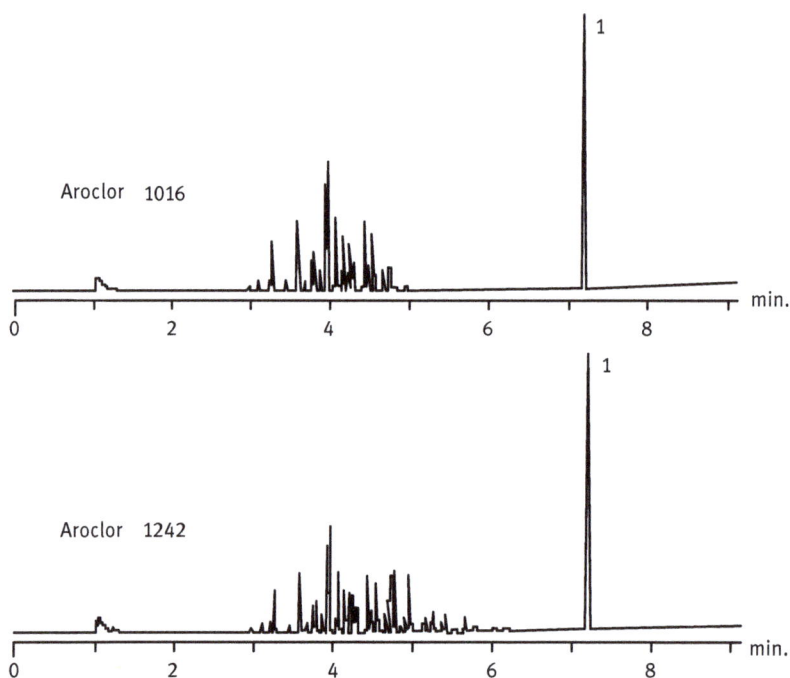

Figure 11.21: GC/ECD chromatogram of technical PCB mixtures. Aroclor 1242 = 42% chlorine; Aroclor 1016 = 41% chlorine (Monsanto, USA). Liquid phase for GC separation: methylpolysiloxan: SE 30. Numbering correlates to PCB structure [47].

orthogonal GC × GC separation, analyses even in difficult matrices like milk became feasible [48]. In the following, this is described in detail, since it shows exemplarily how many iterative pre-separation steps are needed to succeed.

For analysis of milk, internal standards (IS) (e.g., 1,2,3,4,6,7,9-HpCDD or 1,2,3,4-TCDD) had to be added, as the ECD does not differentiate between isotope-labeled and unlabeled congeners. All samples are fortified with 40 µL of IS before extraction (milk) or cleanup (oils and eel extracts). Milk aliquots of about 150 ml each are mixed with 50 mL of sodium oxalate-saturated ethanol and then liquid/liquid extracted three times with 225 mL diethyl ether/n-hexane (7:10). Ethanol (99.5%; 50 mL) is added and the solvents are removed by rotary evaporation at reduced pressure and a temperature of 30 °C. The fat content is determined gravimetrically. Next, the milk fat is dissolved in n-hexane and transferred to a multilayer silica column (Ø 35 mm) containing (from the bottom) glass wool, 6 g KOH-silica, 3 g silica, 17 g 40% H_2SO_4 on silica (w/w), 7 g 20% H_2SO_4 on silica (w/w), 3 g silica and 7 g Na_2SO_4. Prior to use, the silica columns are washed with n-hexane (2×100 mL). The samples are eluted with 200 mL n-hexane, and then the volumes are reduced to approx. 1 mL by rotary evaporation.

In the next step, an activated carbon column is used to fractionate the target compounds according to planarity. Activated carbon is mixed with Celite in the proportions 7.9/92.1. The carbon/Celite mixture (0.5 g) is packed in a glass pipette (10 mL, cut at both ends) with glass wool on either side. The extracts are transferred to the column with 3 × 1 mL n-hexane and eluted with 30 mL n-hexane followed by 40 mL n-hexane/DCM, 1/1 (v/v) and then 40 mL toluene. Before the elution of fraction 3, the column is turned upside down. This elution scheme results in three fractions. Most of the di- through tetra-*ortho* PCBs are recovered in fraction 1, the mono-*ortho* PCBs in fraction 2 and the non-*ortho* PCBs and dioxins in fraction 3. After reducing the solvent volume to approx. 1 mL, the samples are transferred to washed, miniature, multilayer silica columns (Ø 5 mm) containing KOH–silica, silica, 40% H_2SO_4–silica and Na_2SO_4, and then eluted with 8 mL n-hexane. Prior to injection, 40 µL of the solution containing 1,2,3,4-tetrachloronaphthalene (16 pg/µL) and octachloro-naphthalene (8 pg/µL) is added to serve as both syringe spike and retention reference standards. Finally, the samples are evaporated to ca. 30 µL under a gentle stream of nitrogen. With this, a full congener analysis in the low pg TEQ/g range is achieved. Figure 11.22 depicts the separation visually in a contour plot.

Comparison of the GC–ECD and GC–MS congener quantitations suggested that the ECD detectors were measuring many peaks in a nonlinear response range rendering the single point calibration inadequate for accurate quantitation [49]. Hence, nowadays, GC–MS is the preferred instrumental setup for PCB, PCDD and PCDF determination.

11.2.3.3 GC/MS

Today, in majority of applications, GC/MS combination is the preferred tool for trace analysis of polyhalogenated aromatic hydrocarbons. Not only the superior resolution

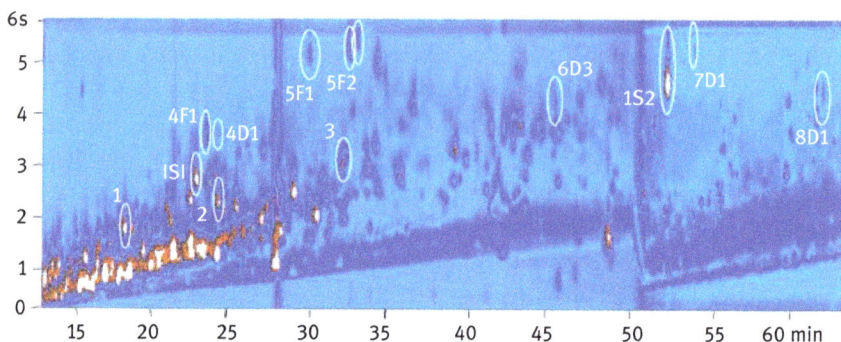

Figure 11.22: GC × GC-ECD contour plot (PCDD/F and non-*ortho* PCB fraction) with DB-XLB × LC-50 column combination. Assignment (CB, chlorinated biphenyls): 1, CB 77; 2, CB 126; 3, CB 169; IS1, 1,2,3,4-TCDD; 4F1, 2,3,7,8-TCDF; 4D1, 2,3,7,8-TCDD; 5F1, 1,2,3,7,8-PeCDF; 5F2, 2,3,4,7,8-PeCDF; 5D1, 1,2,3,7,8-PeCDD; 6D3, 1,2,3,7,8,9-HxCDD; IS2, 1,2,3,4,6,7,9-HpCDD; 7D1, 1,2,3,4,6,7,8-HpCDD; 8D1, OCDD. Numbering of the congeners has been published, for example, Ref. [47]. Reproduced with permission from Elsevier.

power of latest GC instrumentation (hyphenation; high boiling stationary phases for narrow-bore separation columns), but also latest high-resolution MS technology (multi-stage MS, selective ionization) allows not only fast screening with acceptable sample pretreatment before injection in GC/MS but also confirmation of found congeners in the ppq range. Especially, usage of stable isotope-labeled analytes makes these techniques extremely reliable and juridically robust. Especially, the latter aspect is of importance to meet possible consequences like banning of food marketing. The most recent European Union Regulations (589/2014 and 709/2014) refer to the use of an appropriate confirmatory method for checking compliance with the maximum level set for food and feed control, respectively.

In the 1980s and 1990s, throughout the sector field MS became popular for detecting TEF-afflicted analytes. Nowadays, less expensive MS combinations do the same job [50]. To understand the challenge in Figure 11.23, it is exemplarily demonstrated that at least a resolution of 50,000 is needed to separate important congeners of PCBs and PCDDs in one run, for example, a hexa-CB (m/z 357.8444) from the monitored mass of PeCDD (m/z 357.8518). So, pre-fractionation of PCBs and PCDD/F is frequently used to surpass this difficulty.

Aside from HRGC–HRMS (sector field), which is used for confirmation analysis, Orbitrap-MS, HRGC–MS/MS or GC × GC–MS configurations seem to become favored more and more [52].

A new dimension has been opened by a 3D hyphenated technique, which combines GC, MS and high resolution UV ionization [53]. In principle, the 2D laser mass spectrometric REMPI (*resonance enhanced multiphoton ionization*)–TOF-MS method is used as a highly compound-selective GC detector (see also chapter 9.3.2).

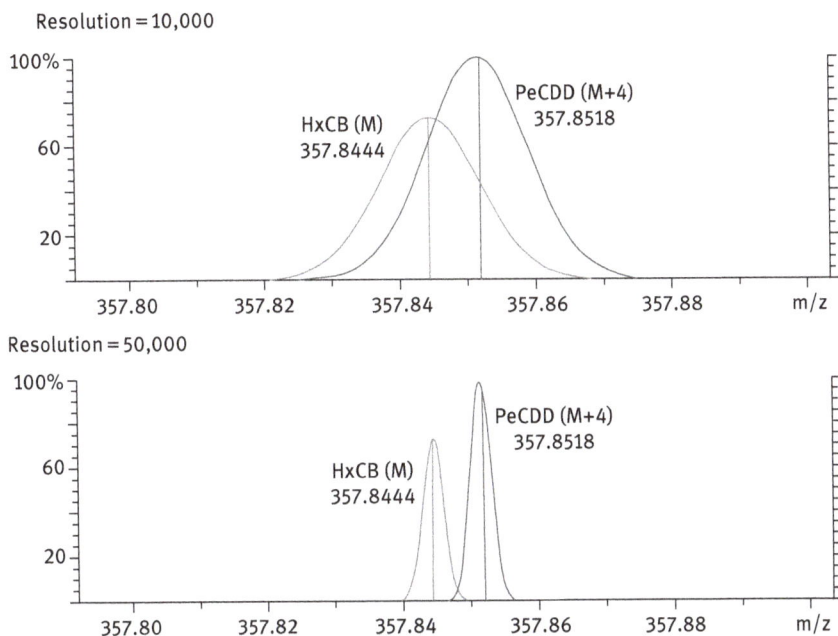

Figure 11.23: Mass spectra of hexa-CBs and PeCDDs obtained by EI-MS. A resolution power of 50,000 is required to resolve the interfering mass of the hexa-CB (*m/z* 357.8444, corresponding to the [M] fragment) from the monitored mass of PeCDD (*m/z* 357.8518, corresponding to the [M + 4] fragment) [51].

REMPI ionization means that an intermediate state of the analyte of interest is se-lectively excited by absorption of a laser photon (the wavelength of a tunable laser is set in resonance with the transition). The excited molecules are subsequently ionized by absorption of an additional laser photon. Thereby, the ionization selectivity is introduced by the resonance absorption of the first photon. However, conventional UV spectra of polyatomic molecules exhibit relatively broad and continuous spectral features, allowing only a medium selectivity. By application of a pulsed supersonic beam system, molecules can be prepared cold (5–50 K) and interaction free in the gas phase. Under these conditions, well-pronounced spectral features (often as sharp as those found in IR) are observed in the molecular UV spectra. The optical selectivity is dramatically increased by this jet approach, even isomeric compounds can be differ-entiated by jet REMPI. A remaining problem is still to gain sufficient ionized molecules within the laser focal volume.

Pulsed *two-photon ionization* is currently the latest trend in order to achieve a se-lective ionization in combination with GC–MS [21]. Figure 11.24 depicts the way to generate femtosecond UV photons by stimulated Raman emission. In combination with supersonic jet technology, this may someday open the possibility for an *on-line PCDD/DF process monitoring* [54].

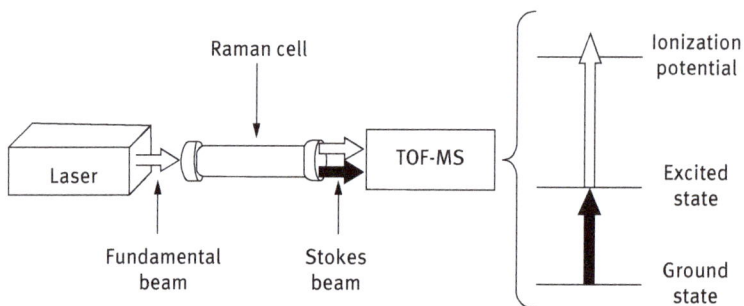

11.2.4 Bioanalytical Methods for PCDD/PCDF and PCB Screening

In view of the cumbersome and lengthy procedure to assess the toxicologically relevant representatives of polyhalogenated aromatic compounds, a fast and inexpensive prescreening would be very helpful in order to exonerate the tremendous lab workload. If such tests would not confirm the presence, further refined analysis would become abdicable. Only a positive result would invoke further detailed analysis. Of course, such prescreening assays should allow high-throughput selection.

Currently, two approaches are available: antibody-based IAs or effect-based screening.

11.2.4.1 CALUX

An established effect-based bioassay is the *Chemical Activated LUciferase gene eXpression* (CALUX) test [55]. It is based on gene-modified cells, whose AhRs become activated through (mainly) dioxin-like molecules. The consequence of such binding and activation of signaling pathways is a gene expression. The recombinant cells used in the CALUX bioassay contain a transfected AhR-responsive firefly luciferase reporter gene. This gene responds to dioxin-like analytes that can bind to and activate the AhR, leading to the induction of luciferase gene expression, which is then measured through its fluorescence. For assaying the congeners, rat or mouse hepatoma cell lines, stably transfected with the luciferase reporter gene plasmid pGudLuc1.1, are used.

The assay is usually performed in 96-well microtiter plates with a fast fluorescence readout. Typically, fractionation of samples to be screened is necessary as explained already for chromatography.

The produced signal depends on the dose, which means it is time and concentration related (see Figure 11.25).

This approach assumes that the extracts analyzed behave like a diluted or concentrated solution of the standard. This implies that the dose–response curves of the sample and of the standard are parallel and that the maximal achievable response for the standard and the sample is identical. From Figure 11.25, it is obvious that this isn't

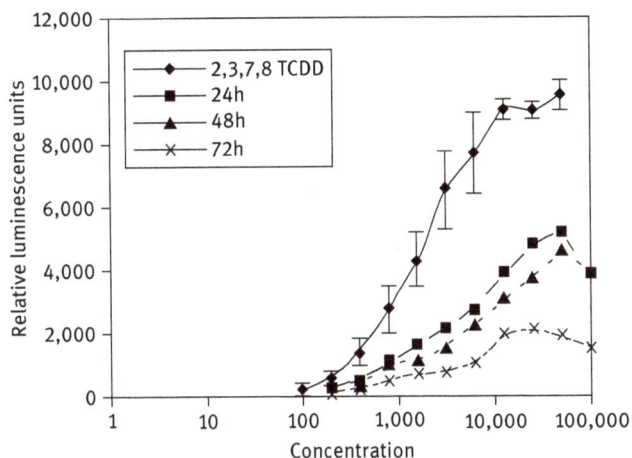

Figure 11.25: Dose–response curves measured for sediment extract after different reaction times, compared to the dose–response curve of the reference 2,3,7,8-TCDD. Concentrations expressed in fg/well for TCDD [56]. Reproduced with permission from the American Chemical Society.

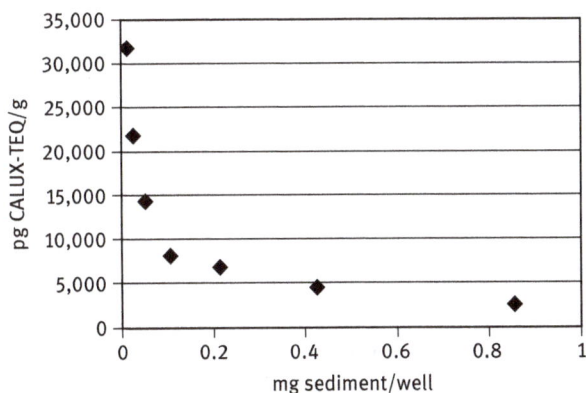

Figure 11.26: Variation of the concentration measured by CALUX as a function of the amount of sediment analyzed. Reproduced with permission from the American Chemical Society.

the case. The measured concentrations become dose dependent. This requires a dilution of the sample and repeated measurements in parallel. It has been proposed to use differently diluted samples to retrieve a range, where at least concentrations can be deduced approximately (see Figure 11.26).

The result for the highest amount of sample presented in the graph is neglected, since the extract may be too toxic for the cells (as it is for higher amounts). Results for the other measurements vary between 4,500 and 31,800 pg CALUX-TEQ/g. Consequently, it has been decided to quantify only the samples giving results in the lower half of the calibration curve (range of concentration between 781 and 3,125 fg/well). If only results obtained with in these limits are considered, the concentrations measured for the sediment samples vary between 4,500 and 14,400 pg CALUX-TEQ/g.

Of course, as with most biological receptor-based assays, there is a multitude of possible interferences with substances acting as activators (agonists) or antagonists to the AhR. An individual response is not gained, rather it is the sum of all AhR-binding chemicals and their potency for binding. The measured fluorescence resulting from exposure to a chemical or chemical mixture is converted into a bioassay toxic equivalency (CALUX-TEQ) value by the direct comparison of the response for a given sample to a dose–response curve obtained with 2,3,7,8-tetrachlorodibenzo-p-dioxin [56].

Interlaboratory comparisons (ring tests) of dioxin-like compounds in food samples have led to similar results. The obtained values are on the same order of magnitude when a similar or quite similar set of parameters is applied for the analysis of cod liver oil (1.8–26.9 pg TEQ/g) or fly ash extract (446–7,361 pg TEQ/g), but the range of results is still broad and the relative standard deviation is high for some laboratories (>44%). Comparison with instrumental analyses sometimes showed satisfying accordance, while other samples not. The effect of pre-fractionation obviously determines the success considerably by removing agonists or antagonists.

In essence, CALUX and instrumental analysis are different tools and they are better seen as complementary. The results of the CALUX bioassay may be used as a method for screening and prioritization of samples for subsequent instrumental analysis.

11.2.4.2 Immunoassay (IA) for PCDD/PCDF Analysis

IAs are well accepted as fast, sensitive and capable high-throughput tools, especially in clinical and food chemistry. They need only a few microliters of aqueous sample to assess contamination by fg-to-pg of xenobiotics. Restrictions are the need for highly selective and affine antibodies, and the presence of the antigen within an aqueous medium. The latter poses a certain constraint for analyzing lipophilic polyhalogenated aromatic hydrocarbons. Addition of organic solvents and/or surfactants may change the tertiary structure of the optimal antigen–antibody complex. As known from numerous studies on nonpolar analytes (e.g., ELISA for PAHs [57] or nitro-PAHs [58]), response curves produced in water are no longer applicable in a diluted organic solvent for quantitative determinations. This fact also biases the determination of the predominantly hydrophobic PCDD/DFs and PCBs.

To meet the required specificity for distinct high-impact congeners, like 2,3,7,8-TCDD, is an enormous challenge for an antibody. One has to be aware that the immunization step with a hapten (=small molecules that create an immune response only when attached to a large carrier protein), linked to *Keyhole Limpet Hemocyanin* (KLH) or BSA, stimulates immune response against this construct with its covalent linkage and not only to the non-linked "naked" hapten molecule itself. The consequence in case of PCDD/DF or PCB molecules without functional groups is a certain cross-reactivity toward other similar molecules. In essence, an antibody raised against hydrophobic analytes without functional groups always exhibits a tolerable cross-reactivity toward isomeric molecules. Sometimes this is wanted, if a multitude of analytes comprise a group of toxic substances. Again, such tests would allow enormous savings in terms of workload, expenditure and time. Only those samples recognized as contaminated would become processed further.

The group around Hammock [59, 60] used for immunization haptens like the following, in order to acquire antibodies specific for 2,3,7,8-TCDD (see Figure 11.27).

Polyclonal antibodies were raised in rabbits according to a standard protocol. For development of the direct competitive ELISA and optimization instead of the expensive and toxic 2,3,7,8-TCDD molecule, the nontoxic 2,3,7-trichloro-8-methyldibenzo-*p*-dioxin (TMDD) was used as surrogate standard. This format has the big advantage that the coating of the microtiter plate does not need handling of the very toxic dioxin congeners. Figure 11.28 shows the final calibration curve for TCMD within a mixture of DMSO/PBS buffer (50:50).

The best antibody Ab7598 yielded a LOD of 4 ppt TCMD in a microtiter plate well volume of 100 µL. This is equal to a mass of 400 fg TCMD. Table 11.5 lists the determined cross-reactivity for the optimized test. The compounds listed in Table 11.5 were tested for cross-reactivity by preparing each compound in 50% DMSO-PBS and determining the I_{50} in the ELISA. Cross-reactivity values were calculated as follows: CR% = (I_{50} of TMDD/I_{50} of tested compound) × 100.

For assay validation, extracts from fish and egg samples were analyzed by both GC–MS and ELISA in a blind fashion, applying the already discussed sample preparation technique by a pre-chromatographic step (cleanup with Florisil according to US EPA Method 1613). A good agreement between GC–MS and ELISA measured TMDD equivalent was obtained from linear regression analysis. No matrix effects were found for these extracts. A fairly good correlation between ELISA and TEF values was also observed. The slope value of the linear regression equation showed an overestimation by

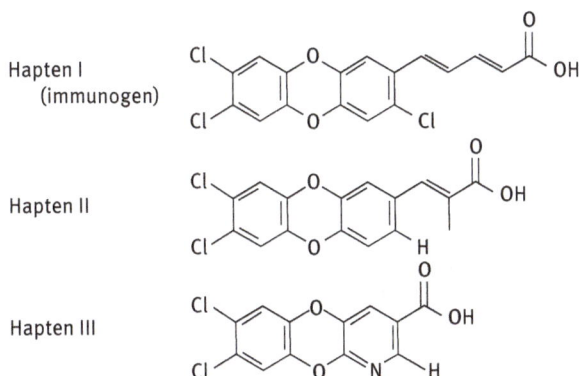

Figure 11.27: Hapten molecules designed for 2,3,7,8-TCDD immunoassay development.

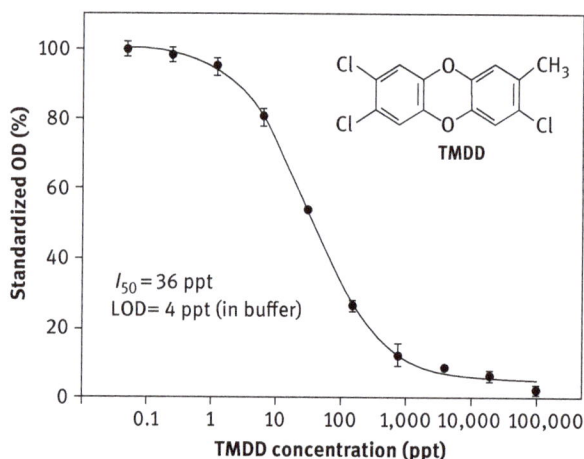

Figure 11.28: ELISA inhibition curve for TMDD. Bars represent standard error. The standard curve represents the average of 14 curves [60]. Optical Density (OD). Reproduced with permission from Elsevier.

Table 11.5: Comparison of cross-reactivities for dioxins, PCBs and dibenzofurans of antibody Ab7598 [28]. Reproduced with permission from Elsevier.

Surrogate standard	Congener	Cross-reactivity (%)	TEF value
	TMDD	100	
PCDDs	1-CDD	< 0.01	< 0.001
	2,7-DiCDD	0.19	< 0.001
	2,3,7-TriDD	6.7	< 0.001
	1,3,7,8-TCDD	43	0.1
	1,2,3,4,-TCDD	0.01	< 0.001
	2,3,7,8-TCDD	129	1.0
	1,2,3,7,8-PentaCDD	72.9	1.0
	1,2,3,4,7,8-HexaCDD	1	0.1
	1,2,3,4,6,7,8-HeptaCDD	0.3	0.01
	OCDD	< 0.01	0.001
	2-Br,3,7,8-TriCDD	110	1.0
	2,3-DiBr,7,8-DiCDD	115	1.0
PCBs	3,3'4,4'-TCB	0.10	0.0001
	3,3',4,4',5-PCB	< 0.01	0.1
	3,3',4,4',5,5'-HCB	< 0.01	0.01
PCDFs	2,3,7,8-TCDF	26	0.1
	2,3,4,7,8-PentaCDF	9.0	0.5
	1,2,3,7,8-PentaCDF	0.1	0.05
	1,2,3,6,7,8/1,2,3,7,8,9-HCDF	5.4	0.1
	1,2,3,4,7,8-HCDF	< 0.01	0.1
	1,2,3,4,6,7,8-HeptaCDF	0.06	0.01
	OCDF	< 0.01	0.0001

ELISA in comparison to TEF values. However, a strong correlation ($R^2 = 0.90$) between ELISA and TEF values suggests that this ELISA is useful for TEF screening of dioxins in these samples.

The mentioned overestimation by ELISA is a frequently found characteristic of immunological techniques as similar molecular structures (e.g., other congeners or metabolites) can't be discriminated by the antibodies raised by nonideal immunogens. Ideal would mean that the non-modified pure analyte molecule can act as an immunogen.

11.2.4.3 Immunoassay (IA) for PCB Sum Analysis

For PCB screening by IA techniques, the issue is similarly difficult. Not only the dioxin-like 12 PCB congeners are of interest. Often the contamination of a matrix consists of a variety of differently lipophilic PCBs, stemming from industrial applications like Aroclor/Clophen mixtures.

Up to now, the development of a satisfying group selective PCB assay has failed, since a priori the composition of a sample with cross-reacting interfering compounds is not known. Hence, a correction is not possible. To overcome this situation, a different concept was proposed: sum determination of chlorinated PCBs after a reductive dechlorination step, followed by biphenyl determination by an ELISA.

To develop anti-biphenyl antibodies, biphenylhexanoic acid (see Figure 11.29) was coupled to KLH or BSA by the NHS-ester technique and used for immunization in rabbits [61].

Again, the direct competitive ELISA protocol was applied. Biphenylacetic acid–BSA conjugate diluted in coating buffer was used for coating the microtiter plates.

The schematic of the dechlorination procedure is depicted in Figure 11.30. Dechlorination is complete after the addition of 150 mg Pd/BaSO$_4$ and 50 mg NH$_4$HCOO in 3 mL methanolic extract after 1 h reaction time, followed by a centrifugation step. The assay is performed within 200 µL per well. LOD was 100 µg biphenyl/l, which is equivalent to an absolute detection limit of 20 ng biphenyl.

First, validation experiments with artificially Clophen A50 contaminated soil showed a fair correlation between the detected biphenyl by ELISA and the added

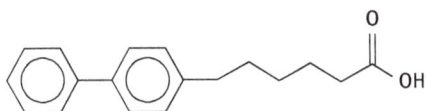

Figure 11.29: Biphenylhexanoic acid used as hapten for immunization against biphenyl.

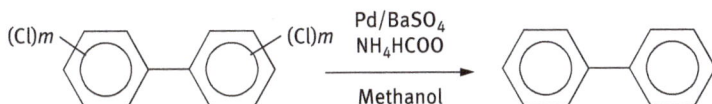

Figure 11.30: Reductive dechlorination of PCBs.

amount of Clophen A50. Especially for low contamination levels an overestimation was observed. It was concluded that the interaction of humic substances with the antibodies impairs the test.

Based on the experience so far for bioassays and IAs as well, only the classical sample treatment with fractionated cleanup and chromatographic multidimensional determination can tackle this analytical problem in a satisfying manner. Nevertheless, the pressure to develop screening tools will remain.

11.3 Organophosphorus Compounds

11.3.1 General Remarks on Occurrence and Importance

Organophosphates (OPs) have been used for many decades as warfare agents, pesticides, flame retardants (replacing the older persistent PCBs and polybrominated diphenyl ethers), solvents, antifoaming agents, plasticizer and additives in further applications. They represent synthetic esters, amides or thiol derivatives of phosphoric, phosphonic, phosphorothioic or phosphonothioic acids (Figure 11.31).

Many of the OPs are used as additives in common products, that is, furniture, textile coatings, upholstery, electronics, paints, polyvinyl chloride plastics, polyurethane foams, lubricants and hydraulic fluids, resulting in an easy release to the environment via volatilization, leaching and abrasion. Organophosphorus pesticides have been used as insecticides against agricultural, household and structural pests since decades. Advantages of these pesticides are their rapid degradation in the environment (if not used at elevated concentrations much above good agricultural practice), for example, by hydrolysis and photolysis, the rapid formation of often less toxic degradation products and their low potential for bioaccumulation in organisms. However, they are itself highly toxic due the inhibition of acetylcholinesterase (AChE), essential for the functioning of the central nervous system of mammals and insects, by blocking the serine in the active site through nucleophilic attack to produce a serine phosphoester. The phosphorylated enzyme regenerates very slowly (half-life hours to days) and the enzyme is quasi-irreversibly inhibited. This results in the accumulation

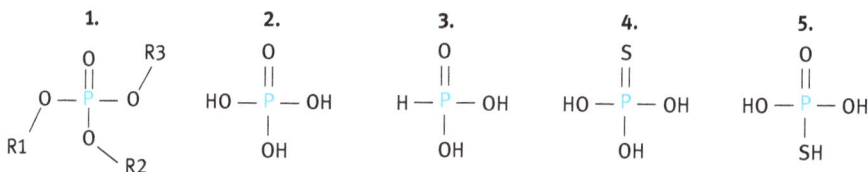

Figure 11.31: (1) Organophosphorus compounds as esters from (2) phosphoric acid. Corresponding derivatives can be formed with (3) phosphonic acid, (4) phosphorothioic acid and (5) phosphonothioic acid. In these structures, the phosphorus atom has the oxidation state +5.

of the neurotransmitter acetylcholine, which would be cleaved to acetate and cholin by active AChE, causing muscular and myocardial injury and, dose dependent, finally to death by excessive cholinergic stimulation and respiratory failure. Many poisonings during application of OPs and due to suicides have been reported. Therapy for nerve agent poisoning is by dosage of a nucleophilic oxime that can reactivate the enzyme by displacing the OP residue. Sublethal exposure to OPs can also lead to neurobehavioral and neuropsychological disorders such as decreased mental alertness, anxiety and depression [62].

Some examples for more than 40 available OP insecticides are methylparathion, malathion, chlorpyrifos, diazinon, dichlorvos and fenitrothion (Figure 11.32), some of which are banned in many countries but some still in use in others.

As an example, chlorpyrifos is, because of its high efficiency in pest control and despite of its high toxicity, registered in most of the countries worldwide in ever increasing amounts. Although its use in certain applications has been restricted in several countries, markets still do exist in most of the developing and non-developing countries. The extent of toxicity increases not only for the first metabolite chlorpyrifos oxon, but also the further degradation products 3,5,6-trichloro-2-pyridinol and 3,5,6-trichloro-2-pyridinol (Figure 11.33) which are more hazardous pollutants than the parent [63].

Sarin, tabun and soman are examples of highly toxic OPs that have been used as military nerve gases acting in the same way as described above (Figure 11.34). Sarin and soman represent fluorinated OPs. Such terrible weapons are still in use in recent

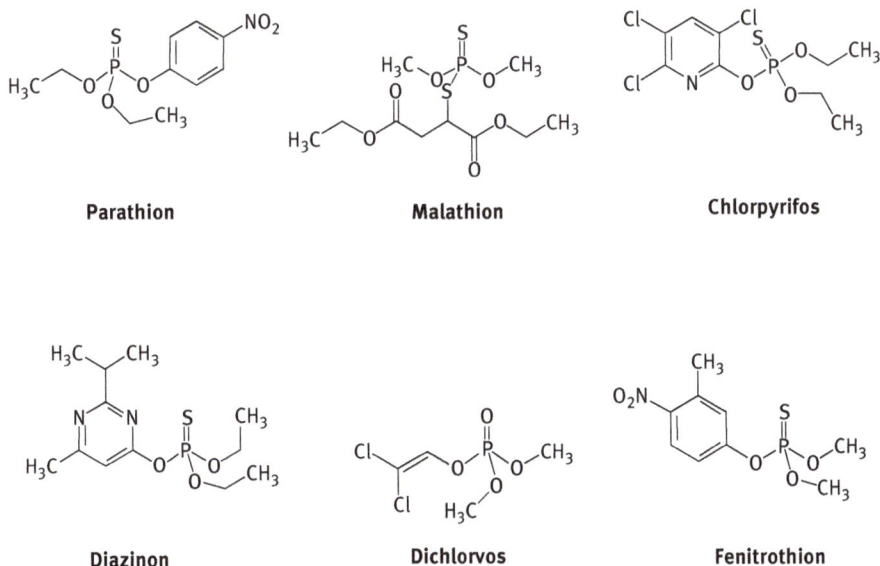

| Parathion | Malathion | Chlorpyrifos |

| Diazinon | Dichlorvos | Fenitrothion |

Figure 11.32: Some organophosphate insecticides.

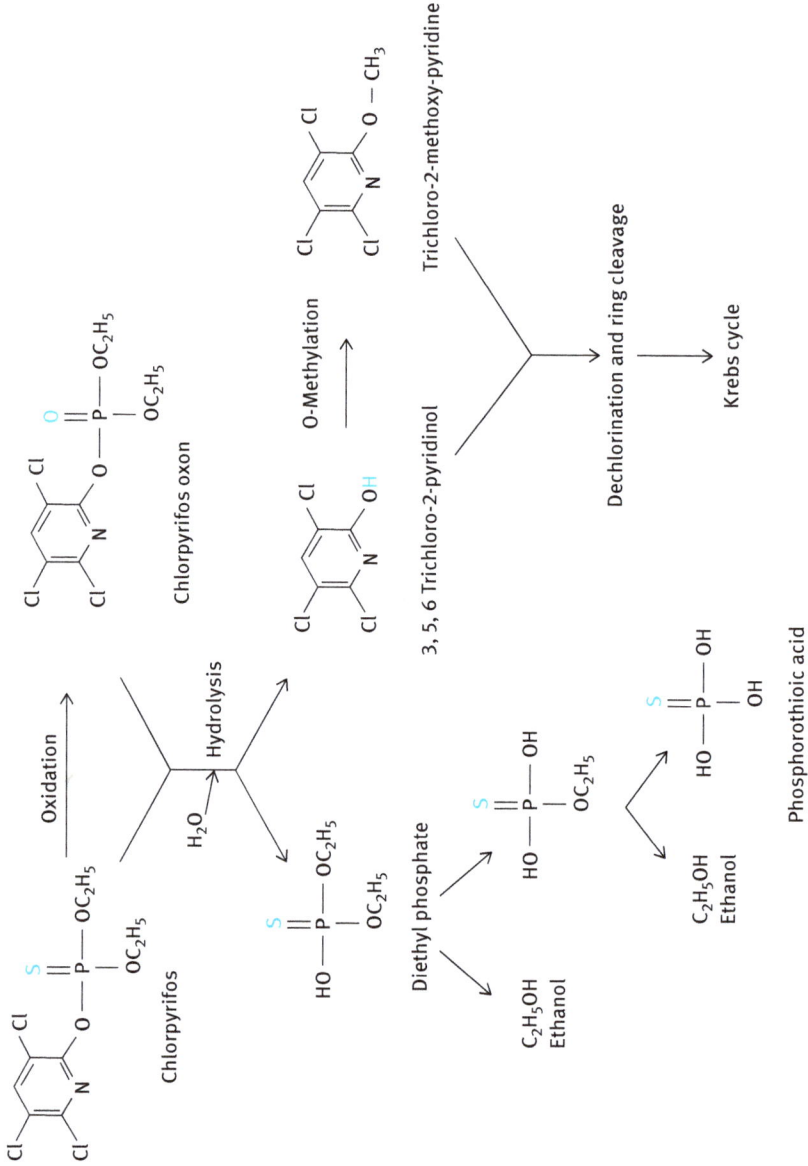

Figure 11.33: Biodegradation of chlorpyrifos in the environment (modified according to Singh et al. [64]).

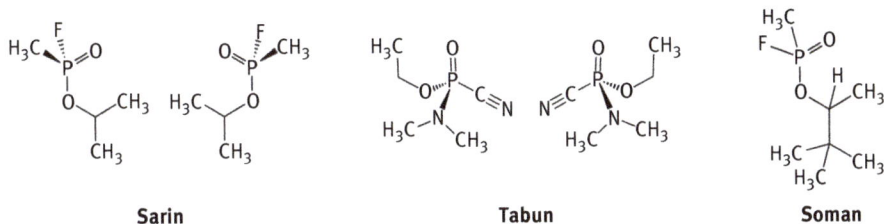

Figure 11.34: Organophosphate nerve gases. Sarin and Soman represent fluorinated compounds, Tabun is non-fluorinated.

military wars and terrorist attacks like in the Tokyo subway in 1995. Symptoms of exposed humans have been described as eyesight difficulties, tightness in the chest and then, after gradual failing of other body functions, falling into a coma and death. One possibility to screen for poisoning is the detection of urinary metabolites of such nerve agents which are cleaved to alkylated methylphosphonic acids; this and further methods to detect OP metabolites in human matrices as biomarkers of poisoning have been reviewed [65–67].

Not covered in this overview are OPs used as herbicides, like glyphosate and glufosinate, not acting as AChE inhibitors (Figure 11.35) but inhibiting a plant enzyme involved in the synthesis of aromatic amino acids (glyphosate) or an enzyme necessary for the production of glutamine and for ammonia detoxification (glufosinate) resulting in plant death.

Other uses of OP are coordination complexes with Zn (Figure 11.36) as additives to lubricants, for instance as greases, gear oils and motor oils in amounts of a few hundred to a few thousand ppm in the products. The market share of such compounds is huge (production of several thousand tons/year).

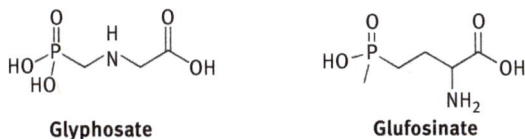

Figure 11.35: Two organophosphate herbicides.

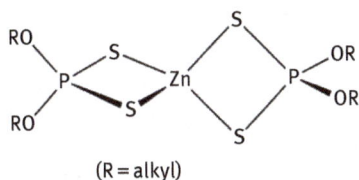

Figure 11.36: Zn containing organophosphates.

Other OP examples are the industrial chemicals triphenyl phosphate and bisphenol A-bis(diphenylphosphate), both used a plasticizers and flame retardants.

Organophosphorus compounds, due to their broad usage patterns, have been detected in the environment, that is, in dust (mg/kg concentrations ranges), air (ng/m^3), water (µg/L), sewage sludge (µg/kg), sediment (ng/kg-µg/kg), soil (µg/kg) and biota samples (µg/kg lipid) [68, 69] as well as in food and nutraceuticals [70]. Various analytical methods (GC, HPLC, CE, MS) are at hand but there is an expanding trend to develop easy and rapid, sensitive and selective, low cost but robust detection methods at trace levels by use of biosensors. In the following, classical and emerging sampling analytical methods for OPs will be described. Deuterated triethyl phosphate, tri-*n*-butyl phosphate and triphenyl phosphate, are commercially available and can be used as internal standards when using GC–MS or HPLC–MS analytical methods.

11.3.2 Sampling Strategies for OPs (Water, Soil, Air, Body Fluid, Food and Nutraceuticals, Living Tissue)

11.3.2.1 Water

A rapid and sensitive method using dispersive liquid-liquid microextraction (DLLME) (acetonitrile as disperser solvent and dichloromethane as extraction solvent) was developed for organophosphorus pesticides from water [71]. A volume of 2 mL of acetonitrile containing 0.3 mL CH_2Cl_2 was added to 10 mL of a water sample. The mixture was shaken by hand and centrifuged to sediment the CH_2Cl_2 phase to the bottom of the centrifuge tube. After removing the upper phase, the CH_2Cl_2 was evaporated to dryness under a nitrogen stream. The residue was dissolved with 250 mL of 15 mM borate buffer and analyzed by capillary electrophoresis (see Section 11.3.4). Parameters that affect the efficiency of the extraction, such as the kind and volume of the extraction and disperser solvents, or salt addition were optimized. Liquid-phase microextraction can also be modified by solidifying the extractant after the OP has partitioned into it, if a suitable solvent is used such as 1-undecanol. The solidified solvent can be transferred into a small vial by a spatula and melted at room temperature to be injected into a gas chromatograph [72]. As an alternative, *cloud point (micellar) extraction* (details see Chapter 6) with the solvents polyoxyethylene lauryl ether and oligoethylene glycol monoalkyl ether has proven successful for OP detection in water [73]. Reviews on analytical techniques for detection of OPs in water have been published [74].

Functionalized magnetic nanomaterials are solid-phase extraction adsorbents (magnetic *SPE* materials) which due to their paramagnetic property, large surface area and selective adsorption capacity are able to extract and enrich various analytes incl. OPs [75]. The analytes sorbed on the magnetic particles are separated from the solution using an external magnetic field and the analytes are consequently eluted from the adsorbents for further analysis. Most often, functionalized Fe_3O_4 nanomaterials have been used for this purpose; coating with organic polymers can prevent oxidation and agglomeration of the particles which is a problem with bare iron nanoparticles. C_{18}-functionalized magnetic silica Fe_3O_4 nanoparticles have been successfully used for extracting OPs from environmental water samples [76]. Similarly, the iron particles can be coated by carbon (the particles are mixed with a glucose solution and treated by a hydrothermal reaction) which also have a high sorption potential for OPs [77].

11.3.2.2 Plants and Food

Dimethoate, methyl parathion, ethyl parathion, chlorpyrifos, chlorpyrifos-methyl and trichlorfon have been extracted by C_{18} SPE cartridges with mean recoveries between 80% and 90% from vegetable samples and from 80% to 100% from fruit samples [78]. In another plant study, OPs were extracted by the *QuEChERS* method: The plant material was dried, 1 g mixed with 10 mL acetonitrile, vortexed, centrifuged in a tube containing an amine and carbon black. A volume of 1.5 mL of the acetonitrile phase was mixed with 0.3 mL chloroform and the mixture rapidly added to 5 mL of water containing 4% of sodium chloride. The cloudy solution was vortexed, centrifuged and the sedimented phase after removal of the supernatant evaporated and dissolved in 5 mM borate buffer for capillary electrophoresis analysis [69]. Moreover, SPME has been used for OP extraction from water samples using 1-cm long fused-silica fibers coated with a 0.1-mm layer of PDMS; analysis was performed with GC–NPD and GC–MS after thermal desorption [79]. Sample preparation techniques and analytical parameters for OP detection by GC in various vegetables have been summarized by Farina et al. [80] and Sharma [81], the latter also including LC techniques.

Stir bar sorptive extraction (*SBSE*) is similar to *SPME* (see Chapter 6) but has the advantage of a higher sorption capacity due to the larger volume of the coating material. In SBSE, a 10–40-mm long magnetic stir bar coated with a layer (about 0.3–1.0 mm, and thus 50–250 times thicker than in SPME) of the extracting phase which has most often been PDMS for unipolar analytes but phases for other analytes are available like carbon adsorbents, molecularly imprinted polymers (MIP), ionic liquids, microporous monoliths, sol–gel prepared coatings and dual phase material [82]. The magnetic bar is stirred for a certain period of time in the sample solution to increase the diffusion of analytes into the extracting matrix by decreasing the thickness of the boundary layer around the coating. After removal, the bar is either extracted with an organic solvent or the analyte is thermally desorbed, the latter enabling an online extraction method for GC analysis. Besides the volume of the polymer coating the stir bar, the extraction time, pH, the potential addition of salt, the stirring rate, the extraction temperature, the sample volume and the possible addition of organic solvents (<5 vol%) determine the extraction efficiency and kinetics. SBSE is an equilibrium technique [83] but to reduce extraction and analysis time, it can be conducted under non-equilibrium conditions [84]. SBSE bars coated with poly(phthalazine ether sulfonic ketone), prepared by immersion precipitation, were used for the extraction of OPs in grape and peach samples. The fruit juice was diluted with water (1:20, v/v) and spiked with the OPs; extraction was performed at 40 °C for 30 min under stirring at 600 rpm. The extracted analytes were desorbed into a GC-ECD system. The calibration curve gave linearity between 20 and 1,000 ng/L, LODs were in the range of 0.2–2 and 2–10 ng/L in grape and peach juices, respectively [85]. Other examples for OP analyses by SBSE are available in the literature [86–88].

A variety of extraction and cleanup procedures for pesticides incl. various OPs in honey [89], juice and wine [90] has been summarized including LLE, *solid-phase extraction*, QuEChERS, dispersive liquid–liquid and *solid-phase microextraction* and so on as well as subsequent GC and HPLC methods.

Since nutraceuticals represent a concentrated form of a food or plant, OPs and other pesticides may be detected in such products which can be extracted by the original or modified QuEChERS methods sometimes accompanied by further *SPE* cleanup; also PLE and gel permeation chromatography for cleanup have been reported [70].

11.3.2.3 OPs in Soil

OPs have been extracted from soils by microwave extraction using a water–methanol mixture for desorption and simultaneous partitioning on *n*-hexane. A volume of 1 g soil samples was transferred to microwave extraction vessels. A volume of 1–3 mL of the extractant was added and after shaking, 5 mL of hexane was added for partitioning. The vessels were preheated for 2 min at 250 W and then for 3–15 min at 300–600 W in a microwave oven. Afterward, the vessels were cooled, and the hexane evaporated to dryness under nitrogen. The residue was dissolved in 1 mL hexane for GC analysis with a flame photometric detector (FPD). Under optimized conditions, recoveries of OP pesticides from different soils were above 73%, but lower for methyl parathion [91].

Molecular imprinted polymers (MIP) (see Chapter 6) have been used for extracting and enrichment of OPs [92]. As an example, a parathion sensor based on MIP sol–gel films has been developed with a high specific binding affinity for parathion [93]. If combined with *SPE*, the method is called molecularly imprinted solid-phase extraction. Monocrotophos, mevinphos, phosphamidon and omethoate have been analyzed by this technique in water and soil samples. Binding studies demonstrated that the polymer showed excellent affinity and high selectivity to monocrotophos. The recovery of the four polar organophosphorus pesticides from 1 L of river water at a 100-ng/L spike level was in the range of 77.5–99.1% and from a 5-g soil sample at 100 µg/kg 79.3–93.5%. The limit of detection varied from 10 to 32 ng/L in water and from 12 to 34 µg/kg in soil samples [94].

However, MIP extractions have some limitations, such as limited and slow binding of the analyte because much of the imprinted binding sites are embedded in the interior of a highly rigid polymer matrix water. For the same reason, MIP have a tendency towards bleeding, that is, after recovery, still some residual analyte can be released interfering with subsequent analysis.

11.3.2.4 OPs in Animal and Human Tissues

Matrix solid-phase dispersion by use of a C_{18} sorbent combined with silica gel cleanup and acetonitrile elution was used to extract chlorpyrifos, chlorfenvinphos. diazinon, fenitrothion and parathion-methyl from bovine samples. The method had recovery values above 94%, except for chlorfenvinphos in liver (55%) [95].

Analysis of OP pesticides in human matrices like blood, urine, breast milk, meconium and hair as well as the analysis of relevant OP metabolites – which may be also used as target analytes for monitoring the exposure to OPs – has been reviewed by Yusa et al. [96, 97]. A short review on the detection of acephate and its toxic metabolite metamidophos, both organophosphorus compounds, summarizes the sampling methods and detection techniques in human blood and serum, urine, in vegetables, fruits, environmental water, soil and plants: mainly LLE and *liquid–solid phase extraction* (alumina, Florisil, ion exchange, silica gel and silica-based reverse phase sorbents and graphitized black carbon) have been applied and various GC- and HPLC-mass spectrometry methods [98].

11.3.2.5 Air

For OP detection in air, active sampling is most often used. The air is pumped through a device comprising a glass-fiber or quartz-fiber filter and subsequently a solid sorbent. The filter retains the fraction of OPs bound to any particulate matter and sorbs those in the gaseous state on the solid matrix. The sorbing matrix typically consists of polyurethane or XAD resins, but Florisil, Chromosorb, Tenax^{TM}, Porasil, C_{18} membranes, SepPakC$_{18}$, PDMS and octadecylsiloxane have also been used as trapping matrix. *SPE cartridges* packed with aminopropyl silica gel have been used for isolation and preconcentration of organophosphorus flame retardants [99]. Further trapping sorbents, desorption solvents and air-flow rates for monitoring OPs have been published [74]. Since OPs are mainly associated to particulate matter, the fiber filters are able to retain them more or less quantitatively. If solid sorbents are used without a filter, both OPs associated with particulate matter as well as those in gas phase are sampled (Figure 11.37).

Van der Veen and de Boer reviewed sampling methods for OP flame retardants in air and other environmental matrices [68]. As an example for air sampling, passive flux samplers or polyurethane foam plugs at a sampling rate of 4 L/min for 8 h have been applied. The plugs were extracted with methylene chloride under ultrasonication and the extracts transferred into hexane followed by rotary evaporation. Alternatively, *SPE membranes* and *SPE cartridges* can be used as sampling matrix, although the latter reveal a relatively high resistance against the air pumping flow.

Another air sampling technique for OP analysis is *solid-phase microextraction*. In quantitative *SPME*, concentrations on the fiber are normally measured after the analyte has reached partitioning equilibrium between the fiber and the sample matrix. However, equilibrium settling of semi-volatile compounds in air with *SPME* often takes several hours. Time-weighted average sampling using *SPME* under non-equilibrium

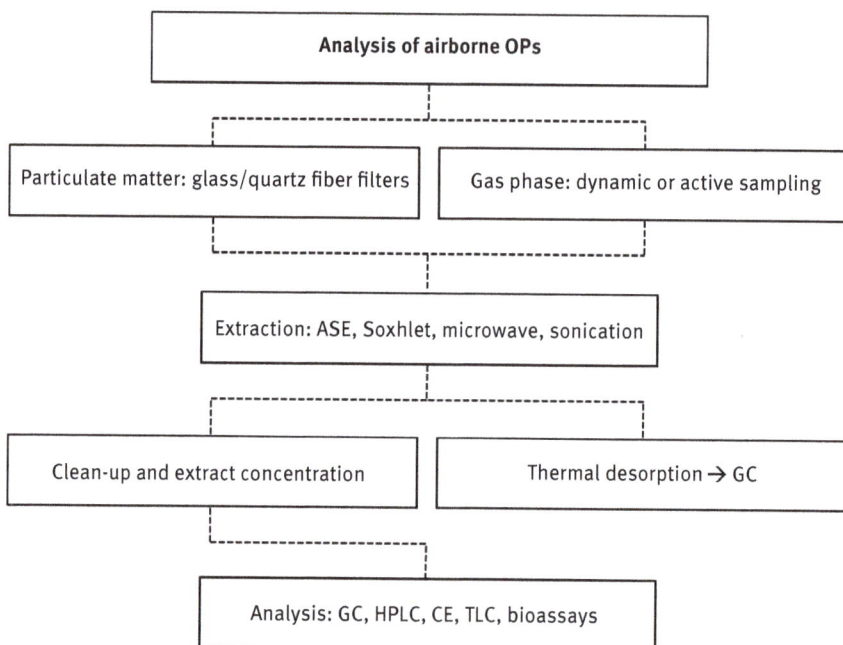

Figure 11.37: Experimental steps for analysis of airborne OPs in the gaseous state and adsorbed to particulate matter.

conditions can be applied as alternative. Also passive sampling, for example, by a C_{18} membrane in a glass cylinder, has been applied, for instance to study the release of OP flame retardants from materials in-house, for example, ceilings, walls, computer screens and TV sets [100, 101].

11.3.3 GC Analysis of OPs

GC is most often used for OP determination since they are sufficiently volatile and since an NPD detector or a MS detector provides good selectivity and sensitivity.

By GC in combination with a *FPD* (see Chapter 7, Section 7.5.9.2), various OPs in water were analyzed with a linearity of signal between 10 and 100,000 ng/L and LODs of 3–10 ng/L; the pesticides were enriched by *DLLME* [102]. GC–FPD analysis of OP in water after liquid-phase microextraction resulted in a LOD of 10–40 ng/L [72]. For OP detection in soil extracts, the addition of olive oil at 0.3% (v/v) as a matrix mimic significantly improved the peak shape and intensity in FPD detection, particularly for methyl parathion and parathion. LODs were in the range of 6–12 µg/kg [91]. Also for the analysis of OP in tea, the principle of dispersive LLE was successful; subsequent analysis by GC–FPD resulted in LOD of 0.03–1 µg/kg [103].

Quantification of OP pesticides in surface water was possible by LLE with petroleum ether and dichloromethane (70:30, v/v) and GC–MS analysis. The analysis method had a linear range between 0.02–ca. 1 µg/L and a LOD below ca. 50 ng/L [104]. The combination of *SPE* fractionation and enrichment with GC–MS–MS analysis resulted in low- to sub-ppt levels' LODs for OPs [105]. Optimal conditions for detection of OP pesticides by GC–MS detection by negative chemical ionization was evaluated comparing isobutane, methane, ammonia in methane and pure ammonia as ionizing gases. Pure ammonia improved the signal-to-noise ratio best as well as the overall sensitivity of the method [106].

One of the most often used GC column is fused silica with a stationary phase of 95% dimethyl–5% diphenylpolysiloxane; other more or less polar stationary phases have been applied. The majority of reported studies used the splitless injection mode to achieve a high sensitivity. Reviews on GC separations of OPs and other pesticides are available [70].

OP residues in bees (as bio-collectors of pesticides in agricultural landscapes) can be analyzed to monitor the exposure: Several grams of freeze-dried bees were crushed in a mortar with diatomaceous earth. The pesticide residues were extracted with dichloromethane and the extract was purified by gel-permeation chromatography with Biobeads SX resin. The fraction containing the pesticides, dried and resuspended in acetone, was analyzed by GC using a nitrogen–phosphorus detector. Subsequent GC–MS analysis confirmed the presence of OP residues at µg/kg dry weight quantities [107]. Alternatively, OP residues in bees have been analyzed with LC–APCI–MS (LOD 1–15 µg/kg dry weight) [108].

11.3.4 Wet-chemical Analysis of OPs (*HPLC–MS, TLC, Capillary Electrophoresis*)

11.3.4.1 HPLC for OP Analysis

As in HPLC analysis of other pesticides, for OPs, the most used column phase is C_{18}-bonded silica.

HPLC coupled to a UV detector with preconcentration by in-tube *SPME* in an open capillary column has been used for OP detection in drinking water. *SPME* columns were coated with 95% dimethylpolysiloxane–5% diphenylpolysiloxane. LODs were 0.1–10 µg/L, and thus at least 100 times lower than that obtained by direct injection of the samples to the LC system [109]. Moreover, cloud-point extraction leads to an efficient up-concentration of OP in water which were analyzed by HPLC coupled to a diode array detector [73]. OP in samples from bovine tissues after C_{18} *SPE* cleanup and enrichment were detected by HPLC–DAD with LOD below 1 mg/kg [95].

Acephate and other OPs were analyzed in water samples using *hydrophilic interaction liquid chromatography (HILIC)*. Sample preparation was carried out with GL-Pak-activated carbon cartridges. A volume of 50 mL water were loaded on the cartridges and eluted with 5 mL of 0.2% (v/v) formic acid in acetonitrile/isopropanol (95/5, v/v). D_6-acephate was added as internal standard. A HILIC silica column was used for the separation of the OPs using a mobile phase consisting of acetonitrile/isopropanol/200 mM ammonium formate in water (pH 3) (92:5:3, v/v) under isocratic conditions. Acephate was detected in river water samples at low concentrations (<0.1 µg/L) [110].

A method based on LC–MS–MS was developed for determination of rather polar OPs, that is, acephate, methamidophos, monocrotophos, omethoate, oxydemeton-methyl and vamidothion in cabbage and grapes. Ethyl acetate was the most favorable solvent for extraction, although for subsequent HPLC analysis, a solvent switch but without further cleanup was required (switch to 0.1% acetic acid/water). Extracts were analyzed on a C_{18} column with polar endcapping. The pesticides were ionized using atmospheric pressure chemical ionization on a tandem mass spectrometer in multiple reaction monitoring mode. The method was validated at the 0.01 and 0.5 mg/kg level, for both cabbage and grapes. Recoveries were between 80% and 101% and the limits of detection were between 0.001 and 0.004 mg/kg [111].

Simultaneously, dimethoate, fenthion, diazinon and chlorpyrifos were analyzed in human blood by HPLC–tandem mass spectrometry. The pesticides were extracted by a simple one-step protein precipitation procedure. Chromatography was performed on a C_{18} column. The assay was linear from 0.5 to 100 µg/L and the lower limit of quantification was 0.5 µg/L [112].

11.3.4.2 TLC for OP Analysis

TLC (see Chapter 7, Section 7.2) – though being an "old" technique for analysis of all kind of environmental pollutants since many decades – has still remained an important method to study the environmental fate (distribution and degradation) of contaminants. Excellent reviews are available summarizing TLC data of pesticides, including many examples of OPs, in forensic science, water, animal, fish, food and beverages [113–116]. Not to forget the ready-to-apply preparative TLC with its various surface properties allowing the enrichment of contaminants with different properties for further analysis like MS and NMR spectroscopy. Further advantages of TLC comprise

– the possibility for automated analysis,
– low cost, simplicity, rapidity and high sample throughput,
– minimal sample cleanup because TLC plates are used only once,
– high separation power (development with a single sorbent or coupled layers and 2D separation),
– multiple detection methods: chemical and biological visualization, radioactivity analysis, densitometry with UV or visible light,
– repeated detection and quantification because chromatograms are stored on the plate,
– documentation and quantitative analysis by videoscans or photographs,
– spectral analysis of separated zones by multiwavelength scanning (e.g., using a diode array detector),
– hyphenation with MS for structural elucidation of separated compounds.

Many determinations are performed on commercial silica gel TLC or HPTLC glass plates, aluminum plates or plastic sheets. The layers often contain a fluorescence dye to easily detect the compound spots that absorb 254 nm light as dark zones on a green or blue background. Modified silica layers such as C_{18}, *amino* or *cyano* can also be used.

Dichlorvos in grain was detected by silica TLC, cyclohexane–acetone–methanol (8:3:0.5) as mobile phase and 2% 2-thiobarbituric acid as detection reagent. Upon alkaline hydrolysis by spraying 2% NaOH, dichlorvos is split to dimethylphosphoric acid and dichloroacetaldehyde, the latter reacting with the detection reagent and forming a pink zone (LOD on the TLC plate was 18 μg) [117]. OPs in rice were extracted with ethyl acetate and extracts cleaned by gel permeation chromatography and subsequently analyzed by TLC with cholinesterase inhibition detection and GC with a FPD [118]. Monocrotophos was analyzed on TLC silica plates by an *enzyme inhibition* technique: succinate dehydrogenase from egg albumin, the substrate sodium succinate, and the chromogenic reagent [2-(4-iodo-phenyl)-3-(4-nitrophenyl)-5 phenyl tetrazolium chloride] and *N*-methyl phenazonium methosulfate were used revealing white spots on a pink background with an LOD of 6 μg on the plate [119]. OP and carbamate pesticides were determined in apple juice and water samples after extraction of separated zones from silica TLC plates using methanol–0.1% formic acid (95:5, v:v). TLC analysis used a *multi-enzyme inhibition* assay (HPTLC–EI) (Chapter 10) based on rabbit liver esterase and *Bacillus subtilis* esterase. Because choline esterase inhibition is more effective after conversion of thiophosphate thions into their corresponding oxons, a pre-oxidation step was added to the HPTLC–EI assay using bromine vapor. The eluates were subsequently analyzed by quadrupole ESI mass spectrometry [120].

Polar (methamidophos and acephate) and nonpolar organophosphorus insecticides (fenitrothion, diazinon and *O*-ethyl *O*-4-nitrophenylphenyl phosphonothioate [EPN]) have been analyzed by TLC directly coupled to a TOF mass spectrometer (*DART–TOF-MS*, see Chapter 9, equipped with a TLC auto slider) in fat containing precooked food (dumpling with >10% fat). LODs were 23 ng for methamidophos, 9 ng for acephate, 5 ng for fenitrothion, 0.3 ng for diazinon and 0.2 ng for EPN, respectively. HPTLC silica plates with a fluorescent indicator were run in *n*-hexane/acetone (3:1, v:v) as developing solvent. The method shows good linearity (Figure 11.38), and therefor seems applicable to residues at levels associated with food product inspections. In addition, for the simultaneous analysis of polar and nonpolar OPs, the method is more applicable than GC–MS and GC–MS–MS analyses [121].

11.3.4.3 Capillary Electrophoresis for OP Analysis

Soisungnoen et al. have developed a rapid and sensitive method using DLLME (acetonitrile as disperser solvent and dichloromethane as extraction solvent) coupled with *micellar CE* (see Chapter 8) for the analysis of five organophosphorus pesticides. The linearity of the method for parathion, azinphos and fenitrithion was in the range of

20–1,000 µg/L, and for malathion and diazinon in the range of 50–1,000 µg/L. The limits of detection were 3–15 µg/L [71].

A rapid CE–MS method using an amino group-modified capillary for the determination of glyphosate, glufosinate, bialaphos, aminomethylphosphonic acid and

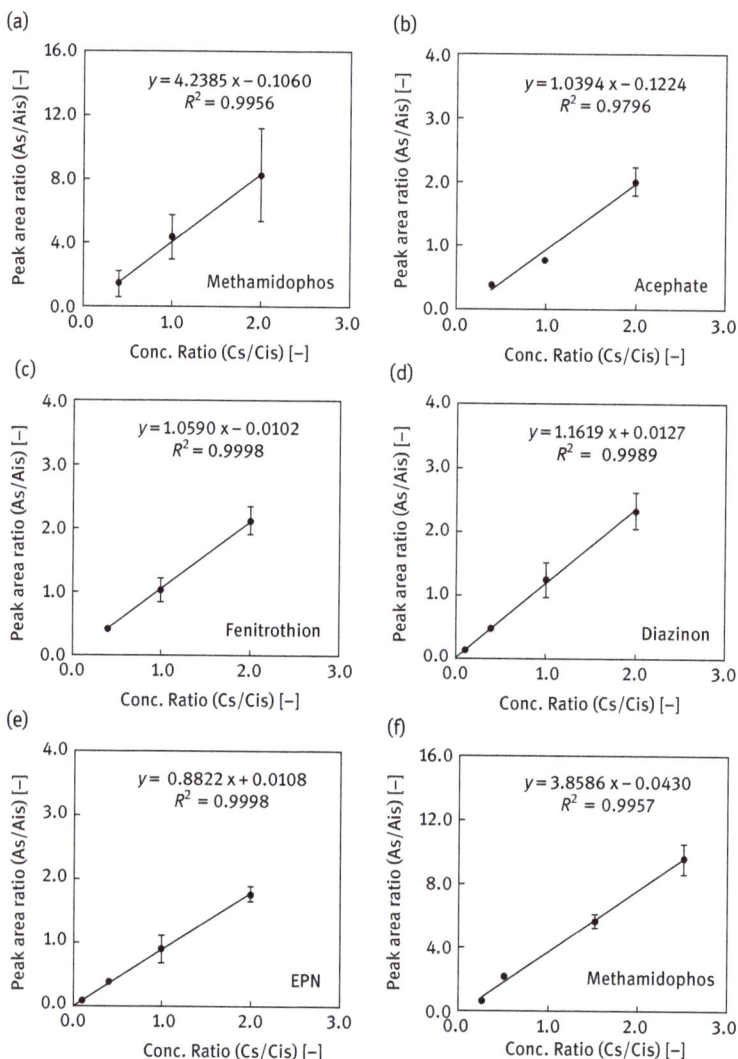

Figure 11.38: Linearity of target pesticides by TLC/DART–TOF-MS. Lower calibration standard concentration range: 0.05–5 µg (labeled internal standards 2.5 µg), (a–e); higher calibration standard concentration range: 2.5–25 µg (labeled internal standards 10 µg), (f–j); methamidophos was calculated using acephate-d6; x-axis: concentration ratio of calibration standard, y-axis: peak area ratio of calibration standard; Cs, charge amount of sample; Cis, charge amount of labeled internal standard; As, peak area of sample; Ais, peak area of labeled internal standard [121]. Reproduced with permission from Elsevier.

(g)

$y = 0.8112x + 0.1233$
$R^2 = 0.9967$

Peak area ratio (As/Ais) [–]

Conc. Ratio (Cs/Cis) [–]

Acephate

(h)

$y = 1.2087x + 0.0223$
$R^2 = 0.9991$

Peak area ratio (As/Ais) [–]

Conc. Ratio (Cs/Cis) [–]

Fenitrothion

(i)

$y = 1.1197x - 0.0233$
$R^2 = 0.9953$

Peak area ratio (As/Ais) [–]

Conc. Ratio (Cs/Cis) [–]

Diazinon

(j)

$y = 0.9677x - 0.0004$
$R^2 = 0.9988$

Peak area ratio (As/Ais) [–]

Conc. Ratio (Cs/Cis) [–]

EPN

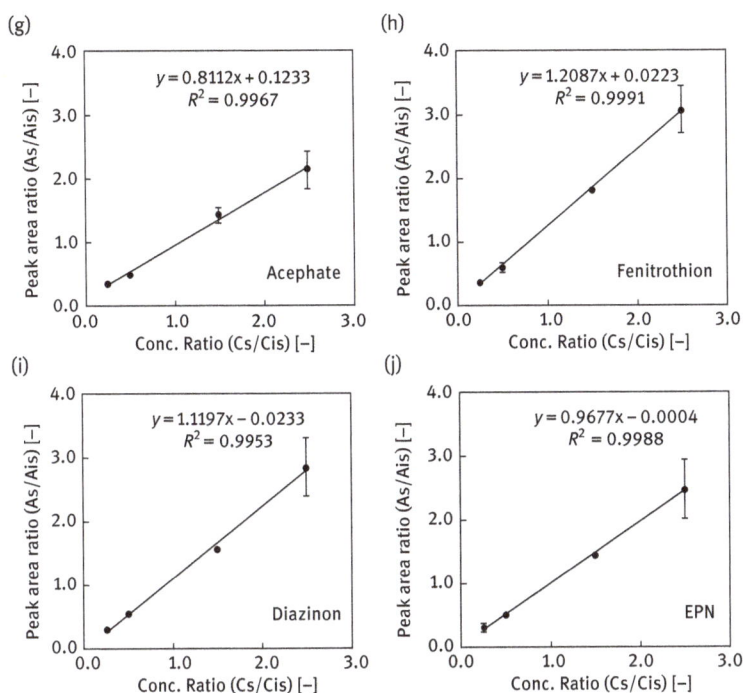

Figure 11.38: (continued)

3-methylphosphinicopropionic acid in soil and tea beverages has been developed [122]. CE coupled to laser-induced FLD using 5-(4,6-dichlorotriazinyl)amino fluorescein as derivatization agent has been used for analysis of glyphosate, glufosinate and aminomethylphosphonic acid in soil and water samples with LODs of 3, 6, and 2 ng/kg, respectively [123].

Sung et al. determined glufosinate-ammonium, aminomethylphosphonic acid and glyphosate residues on the external surface of apples by in-line coupled liquid extraction and CE analysis. The analytes sprayed on the apple surface were directly extracted into a liquid microjunction formed by dispensing the extractant from the inlet tip of a separation capillary. After extraction, the analytes were derivatized in capillary with the fluorophore 4-fluoro-7-nitro-2,1,3-benzoxadiazole and analyzed for subsequent CE-laser-induced fluorescence analysis. The limits of detection for glufosinate-ammonium, aminomethylphosphonic acid and glyphosate were 2.5, 1 and 10 ppb, respectively [124].

Iron oxide nanoparticles interface-chelated with Ti^{4+} using polydopamine as bridging molecules were employed to develop a magnetic *solid-phase extraction* method for glyphosate and its main degradation product AMPA. After extraction and elution, the purified and enriched analytes were derivatized prior to CE with diode array UV detection. The detection limits were 0.4 µg/L for both compounds in river water [125].

11.3.5 Bioanalysis of OPs

The principles and variants of receptor-based biosensors have been introduced in Chapter 10. Reviews on this approach for detection of pesticides incl. OPs are available [126–128].

11.3.5.1 Detection in Biosensing of OPs

The development of sensitive, cheap and small-sized biosensors has been improved by the emergence of suitable nanomaterials. Particularly, *gold nanoparticles* are ideal candidates for bio-based nanosensors for detection of OPs: (a) they are biocompatible for appropriate ligands as recognition elements that can be immobilized on the gold surface; (b) they possess optical and electronic properties for detection of organic molecules or metal ions, for example, by surface-enhanced Raman scattering (SERS), colorimetric, fluorimetric and electrochemical methods [129].

When gold nanoparticles (*AuNPs*) are irradiated by light at wavelengths much larger than the *AuNP* size, the electrons at the surface are displaced with respect to the *AuNP* nuclei generating a restoring force by Coulomb attraction between electrons and nuclei, which leads to an oscillation of these electrons. At a specific frequency, the surface electrons oscillate in coherence with the incident light, resulting in localized *surface plasmonic resonance* (*SPR*) (Figure 11.39). The resonance condition is established when the frequency of incident photons matches the frequency of surface electrons oscillating against the restoring force of positive nuclei. With nonspherical particles (such as *gold nanowires*), however, one dimension of the particles equals the light wavelength. In this case, the nanoparticles propagate surface plasmons, a phenomenon which is sensitive to the presence of substrates close to the particles surface [129]. Thus, gold nanoparticles can be used for very sensitive detection of organic molecules like OPs by colorimetric and fluorescence methods, as described below.

Dispersions of gold nanoparticles sized between 3 and 10 nm are red due to their strong SPR near 520 nm. Agglomeration of the particles leads to interparticle surface plasmon coupling and a corresponding color change from red to blue and, upon redispersion, back to the red. Since the agglomeration behavior of the particles is sensitive to properties of the solution, like the pH, ionic strength and ligands and substrates at the surface, gold nanoparticles provide a platform for *absorption-based* colorimetric sensors (Figure 11.40). Lanthanum (La^{3+})-functionalized gold nanoparticles have been used for analysis of methylparathion: Binding induces aggregation of the nanoparticles and changes the color of the solution from red to blue. Such assay can have detection limits up to 0.1 nM [130]. Dichlorvos binding to ascorbic acid capped *gold NP* leads to the same aggregation effect and this method was applied in river water, apple and wheat residue analysis [131].

Mostly, fluorescence-based detection is based on *fluorescence resonance energy transfer* (*FRET*) which is the resonant energy transfer occurring between an excited

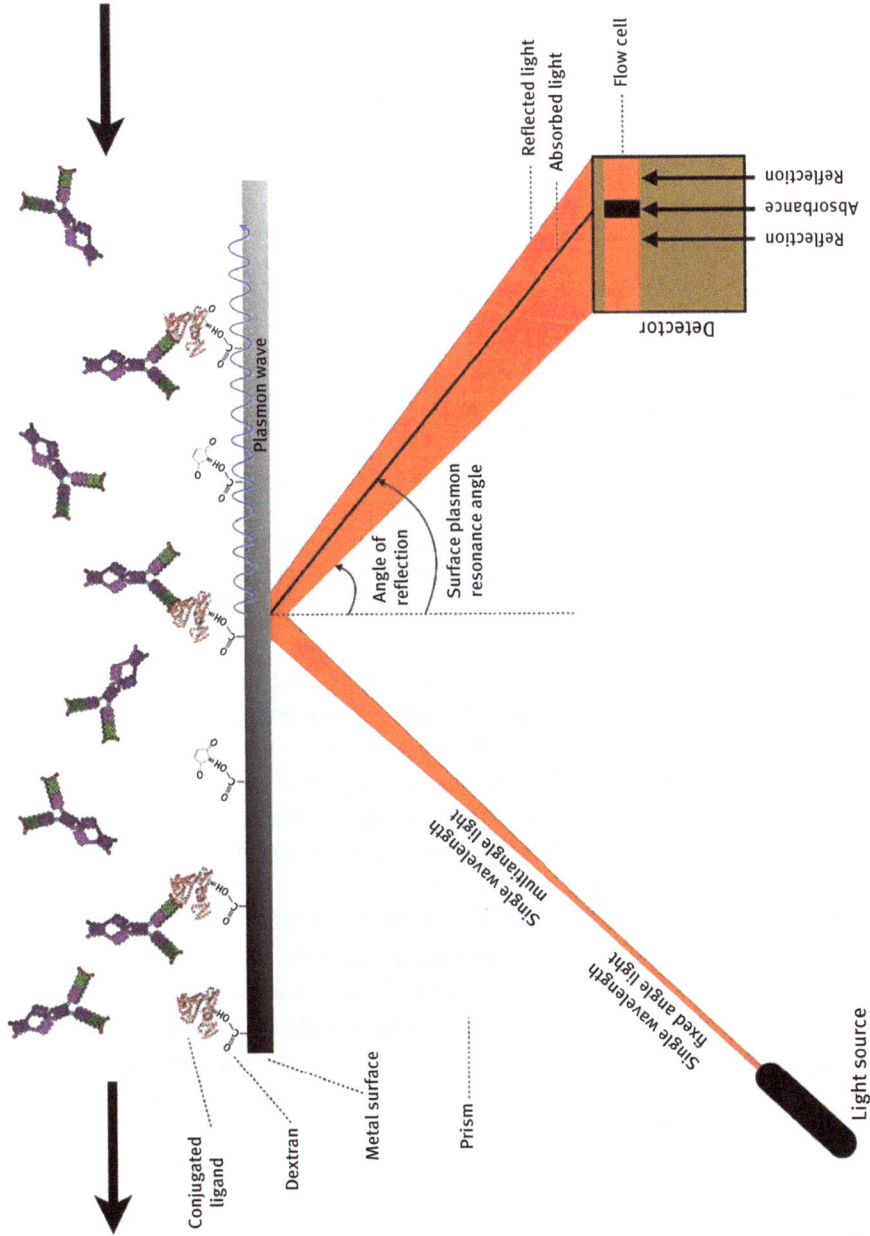

Figure 11.39: Surface plasmonic resonance (SPR) and propagation of plasmon waves if the frequency of incident photons matches the frequency of surface electrons oscillating against the restoring force of positive nuclei. The surface plasmons are sensitive to the presence of analytes binding to attached ligands on the metal surface. Copyright: Wikimedia Commons. https://en.wikipedia.org/wiki/Surface_plasmon_resonance.

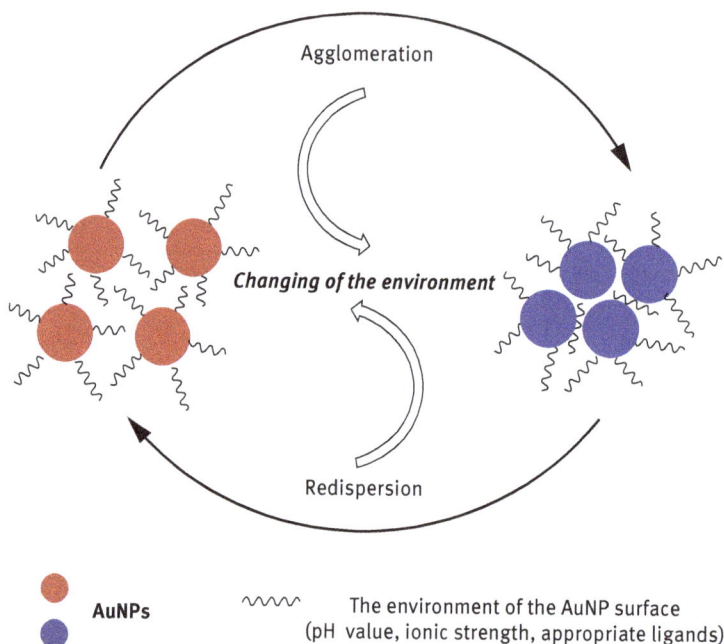

Figure 11.40: The agglomeration behavior of AuNPs is sensitive to the pH value and the ionic strength of the solution and – important for bioanalyzing pollutants – to the presence of analytes that bind to appropriate ligands at the metal surface [129]. Since the color changes upon agglomeration or re-dispersion of AuNPs, this principle can be used for optical measurements. Reproduced with permission from Elsevier.

donor fluorophore and an acceptor fluorophore via induced dipole–dipole inter-actions (Figure 11.41). The acceptor must of course absorb energy at the emission wavelength of the donor. The rate and efficiency of energy transfer depends on the distance between the donor and acceptor, usually, at distances between 1 to 10 nm (Figure 11.41). A FRET-based sensor for OP pesticides based on coumarin was developed which uses the quenching of fluorescence in presence of p-nitrophenol-substituted OPs like methyl parathion and fenitrothion [132].

Platinum 1,2-enedithiolate complexes with an attached primary alcohol were syn-thesized to construct a fluorescent chemosensor. In presence of an OP and triazole as activation agent in dichloromethane, the alcohol forms a phosphate ester with the OP and intramolecularly a fluorescent cyclic product. Several nerve agents have been analyzed by this method very effectively [133]. Sensor systems with thienylpyridyl and phenylpyridyl which also undergo intramolecular cyclization reactions upon OP exposure are based on bathochromic shifts in the absorption and fluorescence. Diiso-propylfluorophosphate has been analyzed selectively with a sensitivity of 10 ppm [134]. Another chemosensor type is based on the suppression of a photo-induced elec-tron transfer triggering a fluorescence signal (pyrene as fluorophore) using a spacer

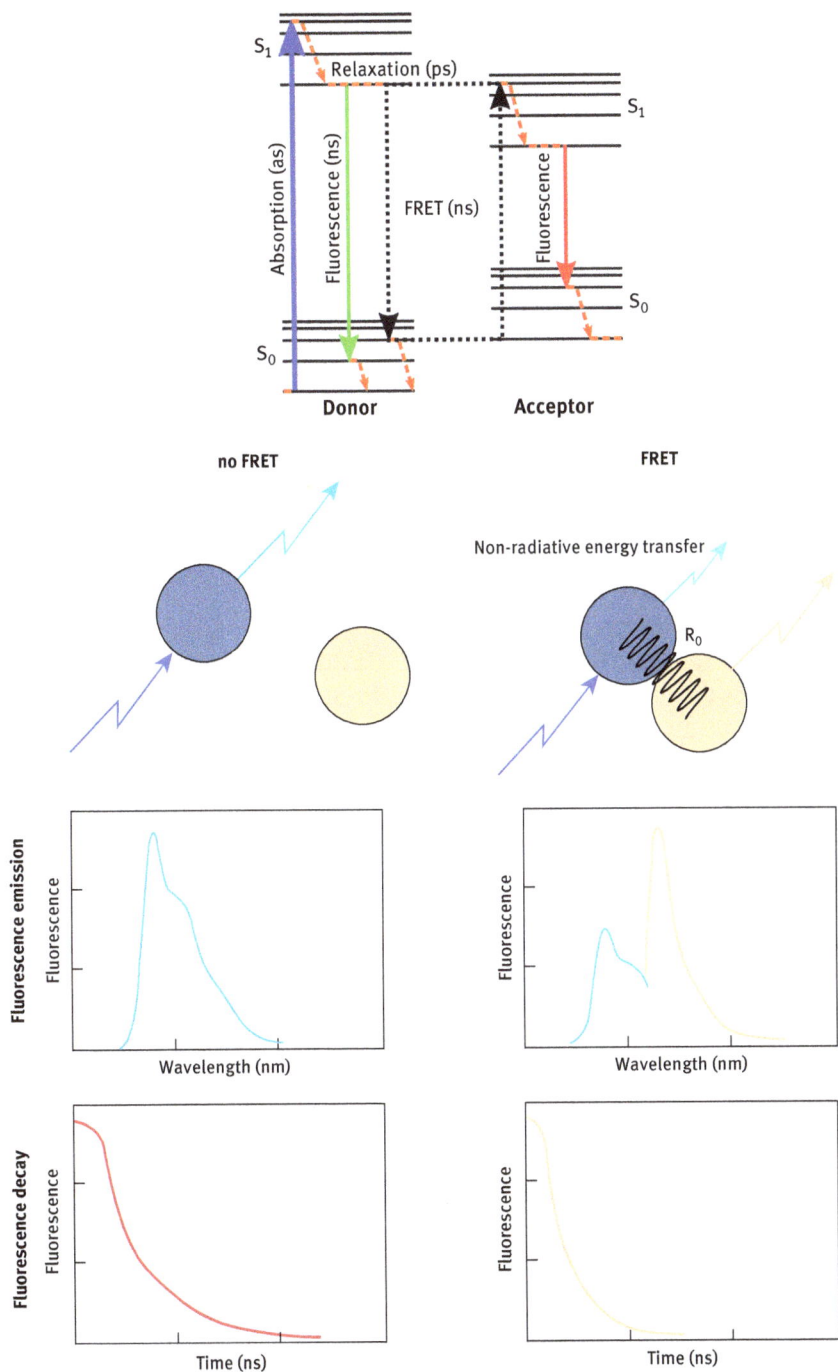

Figure 11.41: (Top) A donor chromophore, initially in its electronic excited state (S_0 ground state, S_1 excited state), may transfer energy to an acceptor chromophore through non-radiative dipole–dipole coupling. The efficiency of this energy transfer is inversely proportional to the sixth power of the distance between donor and acceptor, making FRET sensitive to small changes in distance. (Bottom) Fluorescence changes occur if the distance between the donor and the acceptor becomes short (1–10 nm). AuNPs are excellent fluorescence absorbers in FRET-based assays. Reproduced with permission from Elsevier.

that links the fluorophore and an amine. Upon binding of diethylchlorophosphate showed a significant increase in fluorescence intensity on exposure to 10 ppm of OP in vapor [135].

Obare et al. utilized the electrophilic reactivity of the pentavalent phosphorus atom of OPs to a nucleophilic fluorophore. Signal transduction is accomplished via the π-electronic system of azastilbene upon complexation of the electrophilic phosphorus to the nucleophilic binding site. The azastilbene sensor responds to the OPs ethion, malathion, parathion and fenthion and can be used for detection based on changes in UV–visible absorption (Figure 11.42), fluorescence and cyclic voltammogram signals. Fluorescence was significantly reduced when μM concentrations of the OPs were tested [136].

11.3.5.2 Enzymatic Assays for Analysis of OPs

During the last 25 years, about 1,500 journal articles, book chapters and patents on pesticide sensors have been published, of which 40–50% are based on *enzyme inhibition* mechanisms. In principle, *enzyme inhibition*-based biosensoring includes first the determination of initial enzymatic activity, and second measurement of the residual activity after exposure of the biosensor to the pesticide.

AChE biotests are simple, rapid and sensitive analytical tools for detection of organophosphorus pesticides and are a welcome alternative to classical analytical methods since the sample preparation is easier and they can be even used in the field [137]. Many papers on AChE-based electrochemical biosensors for the detection of OPs have been published, and LODs range from μM to pM levels depending on the source of the enzyme and the transduction mechanism employed [138]. The basic principle has been described in Chapter 10. Many techniques have been applied to immobilize the enzyme on the working electrode such as physical entrapment, adsorption and covalent binding. Recently, nanomaterials have been used for this purpose, especially single or multiwalled carbon nanotubes, due to their chemical stability, mechanical strength and improved electron transfer properties [139].

Although a *AChE biotest* cannot selectively detect and quantify different AChE inhibitors such as carbamate pesticides, natural toxins, heavy metals and some drugs, it can be used for rapid screening of samples to discriminate between samples that do or do not contain such inhibitors. Since the selectivity to discriminate between different AChE-inhibiting substrates is poor, arrays of different AChEs can be used which each have a more specific sensitivity for a certain OP [140–142].

Recombinant AChEs can be used in biosensing sometimes with significantly better LOD by increasing the affinity of the target analyte and the accessibility of the active site [143]. The use of a highly sensitive double mutant of AChE increased the sensitivity to detect dichlorvos down to 10^{-17} mol/L, that is, several orders of magnitude lower compared to the wild type [144]. Flow-through biosensors with rapid analysis times have been developed for quantification of OPs in milk [145]. Many other examples of biosensors featuring recombinant AChE have been summarized by Songa et al. [138].

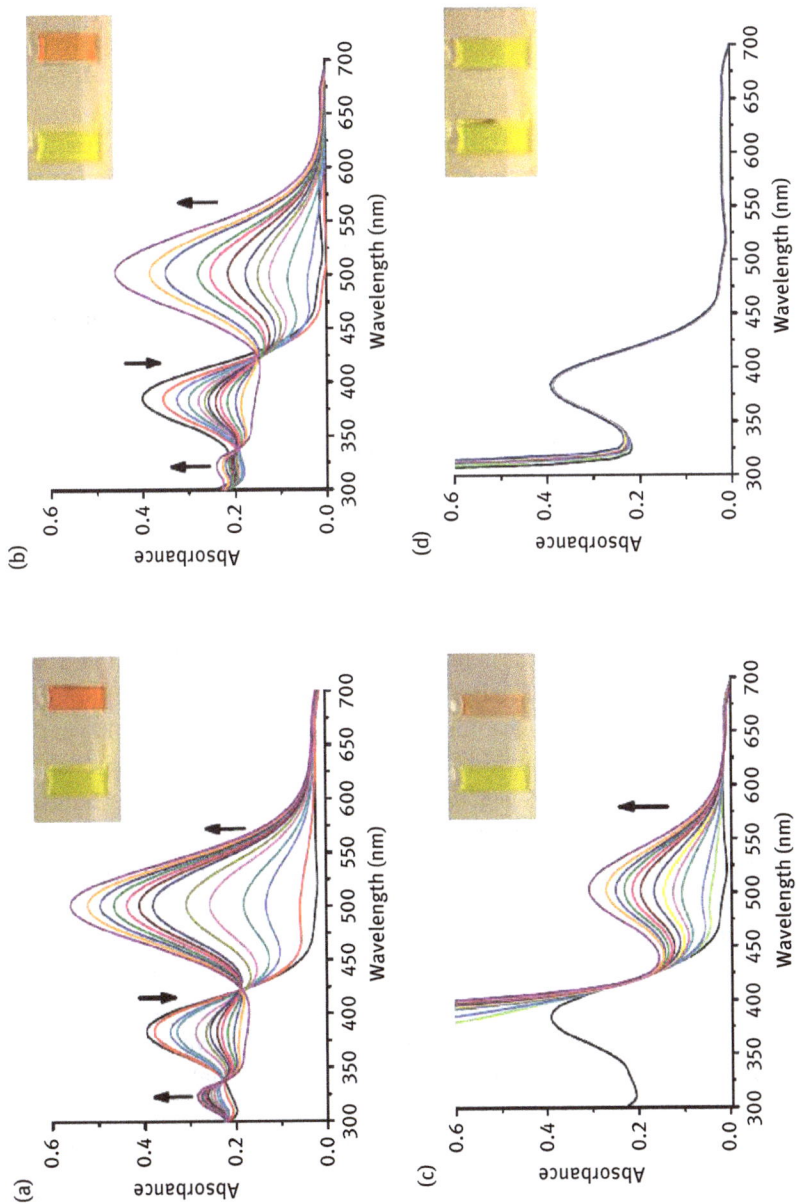

Figure 11.42: Changes in UV–visible absorbance of an azastilbene upon binding to OP pesticides: (a) titration with ethion, (b) titration with malathion, (c) titration with parathion and (d) titration with fenthion. In each case, the direction of the arrow indicates concentrations of 0, 2, 4, 6, 8, 10, 12, 14, 16, 18, 20, 22 and 24 µM [136].

The electrochemical transduction mode of AChE biosensors is commonly based on potentiometric or amperometric analyses. Potentiometry relies on the measurement of the electromotive force of a galvanic element, consisting of an analytical and a reference electrode. The potential of the analytical electrode should depend on the analyte concentration (Nernst equation), the potential of the reference electrode being constant. The dynamic range of such systems is large (3–4 orders of magnitude), but the accuracy and precision is limited. Amperometry on the other hand involves the measurement of the current response of an indicator electrode, as a function of the concentration of the analyte at a constant potential. The amperometric detection in comparison to potentiometric analyses has a higher sensitivity, good precision and better linearity.

Analysis by potentiometric AChE biosensors is based on the pH shift resulting from the hydrolysis of the choline esters. Amperometric AChE biosensors measure the change in concentration of thiocholine produced by hydrolysis of acetylthiocholine which is used as a substrate. Thiocholine will produce an irreversible oxidation peak. In presence of OPs, AChE is inhibited resulting in less thiocholine production. Monitoring of the oxidation peak current of thiocholine without and with OP present allows the OP concentration to be quantified. Thiocholin can be measured directly using a variety of electrode materials, such as gold and platinum, graphite, immobilized carbon nanotubes and recently also boron-doped diamond electrodes. Chlorpyrifos detection was based on the inhibition of AChE bound to magnetic beads allowing detection limits in the sub-nM range [146].

The hydrolysis of acetylthiocholine by AChE can be used for a spectrophotometric method to detect OPs. Acetylthiocholin is strongly sorbed to the surface of gold nanoparticles which affects their aggregation behavior and correspondingly the color of the dispersion of the nanoparticles as explained above. In presence of OPs, the hydrolytic formation of acetylthiocholin is inhibited and the color change diminishes [147]. Thiocholin can also be quantified by thermal lens spectrometry, an absorbance measurement using laser irradiance leading to very high sensitivity with LODs in the sub-micromolar range [148]. *SERS* provides a promising alternative for highly sensitive detection of OPs: Paraoxon analysis was based on the AChE-inhibition mechanism. The reaction product thiocholine sorbed on silver nanoparticles and the variation in the SERS signal correlated well with the reduced concentration of thiocholine in presence of paraoxon (LOD 5 ppb) [149].

The most commonly used enzyme for the development of OP *biosensors* is AChE [150] but organophosphorus hydrolase (OPH) and organophosphorus acid anhydrolase are alternative recognition enzymes for the use in biosensors. OPHs catalyze the hydrolysis of OP compounds and generate hydrogen ions by cleavage of P–O, P–F, P–CN or P–S bonds, while P–O or P–S bond hydrolysis is minimal. Thus, such enzymes provide a highly discriminatory potential for selective detection of OP [151]. Examples of products that can be cleaved by such enzymes comprise pesticides, for example, paraoxon, (methyl) parathion and nerve agents, for example, sarin, soman and tabun. Since the hydrolysis produces a change in pH (release of protons) and generates electroactive products (for instance, *p*-nitrophenol by hydrolysis of parathion, methyl parathion, paraoxon, EPN and fenitrothion), direct detection in a single step is possible, using potentiometric (pH sensitive) or amperometric transducers [152]. If the reaction products are chromophoric, for example, *p*-nitrophenol formed by hydrolysis of parathion and other OPs, optical detection is also possible. Moreover, bi-enzymatic biosensors using AChE and OP-hydrolase to discriminate OPs and carbamates have been developed with LOD of 0.5 µmol/L for paraoxon and a wide linear range [153]. Enzyme-based fluorescence sensors for detection of OP comprise two classes on the basis of the enzyme employed, that is, AChE and organophosphorous hydrolase. Since in both hydrolytic reactions, protons are released, most of such biosensors rely on pH-responsive fluorophores like pyranine, fluorescein isothiocyanate, 2-butyl-6-(4-methylpiperazin-1-yl)benzo[de]isoquinolin-1,3-dione, carboxyseminaphthofluorescein or carboxynaphthofluorescein.

The decrease in fluorescence of fluorescein isothiocyanate covalently immobilized to OP hydrolase that was adsorbed to poly(methyl methacrylate) beads is based on the substrate-dependent change in pH at the local vicinity of the enzyme: parathion, methylparathion, fensulfothion, diazinon and dichlorvos have been analyzed by this method [154]. Moreover, other detection techniques (potentiometric, optical, amperometric) based on OP hydrolases have been described [155].

Despite some disadvantages of enzymatic assays, they are very useful for screening environmental samples for the presence of OPs which in case of positive results can be accompanied by other technologies such as traditional chromatography-MS measurements. These have usually lower LODs (in the ng/L range) than that obtained by enzymatic assays (LOD in the µg/L range). Other advantages and disadvantages of enzymatic tests – for instance, the sensitivity of enzymes against environmental constituents in environmental samples like humic substances or heavy metals ions – have been thoroughly discussed by Van Dyk and Pletschke [128]. A higher selectivity can be achieved by observation of inhibition kinetics and the influence of inhibition modulating additives as well [156].

A flow injection calorimetric biosensor has been developed for detection of dichlorvos with chicken liver esterase as enzyme and acetyl-1-naphthol as substrate. The enzyme was immobilized on an ion exchange resin in the reaction cell (the reference cell contained the same resin and the inactivated enzyme). The detection was based on the inhibition of the enzyme by the analyte, measuring the difference of the signal comparing the difference in temperatures at the two cells. The assays were however not used for analysis of dichlorvos in real environmental samples but rather in samples spiked with the pure reference compound; the sensitivity was only in the range of mg/L [157].

A piezoelectric biosensor was developed for the detection of OPs [158]. A derivative of cocaine, a reversible inhibitor of cholinesterase, was immobilized on the surface of the sensor. The binding of AChE to the inhibitor was monitored with a mass sensitive piezoelectric quartz crystal. In presence of an inhibiting substance in the sample, the binding of the enzyme to the immobilized cocaine derivative was reduced, and the decrease of mass change was proportional to the concentration of the analyte in the sample. This sensor was applied to the determination of the pesticides in river water samples [LODs for paraoxon, diisopropylfluorophosphate, chlorpyriphos and chlorfenvinphos down to 10^{-10} mol/L (0.02 µg/L)].

11.3.5.3 Immunoassays for Analysis of OPs

Most *enzyme inhibition*-based biosensors give a sum parameter of the inhibition efficiency of various analytes, without *qualitative or quantitative information* about individual OPs. However, the sensitivity of such tests needed to detect important pollutants like pesticide residues in ground water (legal trigger value 0.1 µg/L) is not sufficient. Another source of uncertainty in quantitative determinations is the fact that, for example, AChEs differ substantially when comparing different sources with respect to the detection of OPs. Especially, well suited in terms of sensitivity are AChEs from the brain of insects like the house fly [159].

Unlike enzymatic assays, immunoassays (IAs) can reach detection limits in the lower ppt range. *IAs* are based on a bioactive receptor reacting to a specific recognition between an antibody and a corresponding analyte, and a physicochemical transducer converting the binding event into a quantifiable optical or electrical signal (see also chapter 10.4). Such assays are more prone to identify a single OP or a group of very similar OPs.

Parathion-methyl was detected by a fluorescence polarization IA by use of a monoclonal antibody. Fluorescein-labeled parathion was synthesized and purified by TLC. Methanol extracts of vegetable, fruit and soil samples were diluted 1/10 for the analysis and the detection limit for this compound was 15 ppb. Recovery in spiked samples averaged between 85% and 110% [160].

Bi-specific antibodies (Figure 11.43) that are able to specifically bind two different antigens (in our case OPs) have already been used in medical sciences in the field of immunochemical diagnosis for different disorders. For specific pesticides like OPs, this technology is still in its infancy and so far seems rather limited to general structural types [161], although this method was successfully applied to detect parathion and imidacloprid simultaneously [162]. Alternatively, different antibodies may be combined in an immunological sensor that specifically bind different pesticides: Three or even four antibodies were used in this way to detect chlorpyrifos-methyl, fenitrothion and pirimiphos-methyl [163] or diazinon, fenthion, malathion and chlorpyrifos [164], respectively.

Time-resolved fluorescence IAs using lanthanide fluorescent chelates and a monoclonal antibody with broad specificity to a class of OPs have been developed. Thirteen OPs have been analyzed with a detection limit below 10 µg/L [165]. With a similar sensitivity, a fluorescent polarization IA enabled the simultaneous

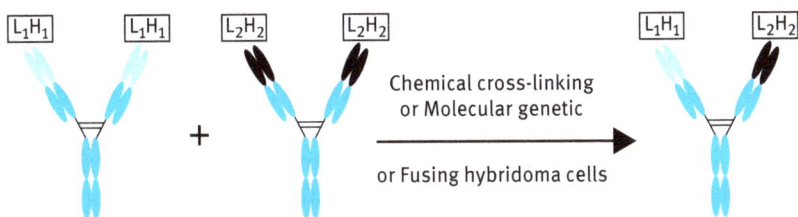

Figure 11.43: Sketch of bispecific antibodies (according to [161]). Reproduced with permission from the Royal Society of Chemistry.

determination of parathion, phoxim, coumaphos, quinalphos and triazophos using a broad-specificity monoclonal antibody and fluorescein isothiocyanate isomer-labeled haptens [166].

Honeybees and bee products can be used to monitor the exposure with pesticides in agricultural landscapes, both by following the bees' mortality and quantifying the residues inside or adsorbed on their bodies [167] or in the hive products either by, for example, chromatography–mass spectrometry or by immunochemical methods. Azinphos-methyl was analyzed in bee extracts by an enzymatic IA with electrochemical detection, by exploiting horse radish peroxidase as antibody label. The detection was based on competition for binding to monoclonal antibodies with an ovalbumin conjugate, followed by the incubation with anti-mouse IgG labeled with horse radish peroxidase. The enzyme activity was measured amperometrically with hydrochinone as substrate. The sensitivity, estimated as IC_{50} value, for the amperometric assay was 1.2 nmol/L [168].

The insecticide chlorpyrifos was quantitatively determined with a selective ELISA in extracts from honeybees. The sensitivity (IC_{50}) of the assay was 3.5 µg/L with a very high selectivity (LOD 1–1.8 µg/L [108]).

11.3.5.4 Whole Cell Bioassays

Biosensors with whole native or genetically engineered microorganisms such as bacteria, viruses and fungi are utilized in environmental monitoring, food safety, military defense and medicine, mostly single-celled and applied either fresh or freeze-dried. Such biosensors have several advantages: microorganisms are often easy to culture, that is, the costs of purification of specific enzymes are omitted. They are also more tolerant against environmental conditions that could inactivate an isolated enzyme. They are multipurpose catalysts containing a number of enzymes that can metabolize suitable substrates. The immobilized cells can act as both sensing components and generators of the recognition signals [169].

Microbial biosensors can be classified according to different signal transducers: electrochemical, fuel cells and optical. Amperometric sensors function at a fixed potential with respect to a reference electrode, and a current is obtained due to the oxidation or reduction of electroactive species at the electrode surface. Primarily, oxygen electrodes are used because oxygen is often consumed during enzymatic reactions. Most microbial sensors work on that principle. The less sensitive conductometric biosensors measure changes in conductivity in the medium caused by target analytes. Microbial fuel cell biosensors generate electricity as a signal when biodegrading organic molecules; however, the use of such sensors in environmental monitoring has not been yet established. Optical microbial biosensors utilize bacteria cells for producing signals in light absorption, fluorescence, luminescence or refractive index due to the interaction with analytes.

Recombinant bacteria that express OPH on the cell surface are able to detect OP pesticides [170] and nerve agents [171]. Alternatively, the degradation of OP insecticides by wild-type *Flavobacterium* sp. has been used to develop a potentiometric biosensor by applying either immobilized whole cells or cytoplasmic membrane fractions of wild-type *Flavobacterium* sp. on the surface of a glass pH electrode. The sensor with cytoplasmic membrane fractions was superior to the one with whole cells and showed a linear range for paraoxon between 0.01 to 0.47 mM and 3 weeks working stability [172].

Whole-cell biosensors based on OPH expressed in periplasm and cytoplasmic membranes have been constructed with *Pseudomonas diminuta*, *Flavactenium* sp. and *Sphingomonas* sp. that contain the OP degrading gene. Moreover, recombinant organisms with surface-expressed OP hydrolases have been used, such as *Moraxella* sp., *Pseudomonas putida* or *Escherichia coli* [126]. The detection principle is identical to that for enzymatic OP hydrolase-based biosensors. The enzyme catalyzes the hydrolysis of several OPs to release *p*-nitrophenol, which can be detected by electrochemical or colorimetric methods. Reviews describing the application of microbial and hybrid biosensors (co-immobilization of OP hydrolases and bacteria) for OP detection have been published [173–177].

Many of the published results of biosensor analyses so far refer to "clean" standard solutions of OP which makes it difficult to evaluate their usefulness for real sample monitoring in the environment. Few examples have been given above; a valuable summary of further examples of "real" environmental analyses is given by Liu et al. [126]

Further Reading: Polycyclic Aromatic Hydrocarbons

Bjoerseth A, Ramdahl T. Handbook of polycyclic aromatic hydrocarbons vol. 2: emission sources and recent progress in analytical chemistry. New York: Marcel Dekker, 1985

Clar E. Polycyclic hydrocarbons. New York: Academic Press, 1964

Fetzer J. The chemistry and analysis of the large polycyclic aromatic hydrocarbons. Polycyclic Aromat Compd. 2000;27:143–62

Mastral A, Callen A. A review on polycyclic aromatic hydrocarbon (PAH) emissions from energy generation. Environ Sci Technol 2000;34:3051–7

Ravindra K, Sokhi R, Van Grieken R. Atmospheric polycyclic aromatic hydrocarbons: source attribution, emission factors and regulation. Atmos Environ 2008;42:2895–921

Further Reading: Polychlorinated Biphenyls (PCBs), Dibenzodioxins (PCDDs) and –Furans (PCDFs)

Erickson M. Analytical chemistry of PCBs. Boca Raton: CRC, 1997.

Safe S, Hutzinger O, Hill T. Environmental toxin series, 3: polychlorinated dibenzo-p-dioxins and -furans (PCDDs/PCDFs): sources and environmental impact, epidemiology mechanisms of action, health risks. Berlin: Springer-Verlag, 1990.

Schecter A. Dioxins and health including other persistent organic pollutants and endocrine disruptors. Hoboken: John Wiley & Sons, 2012.

Van den Berg M, Birnbaum L, Bosveld A, Brunstrom B, Cook P, Feeley M, Giesy J, Hanberg A, Hasegawa R, Kennedy S, Kubiak T, Larsen JC, van Leeuwen FX, Liem AK, Nolt C, Peterson RE, Poellinger L, Safe S, Schrenk D, Tillitt D, Tysklind M, Younes M, Waern F, Zacharewski T. Toxic equivalency factors (TEFs) for PCBs, PCDDs, PCDFs for humans and wildlife. Environ Health Perspect 1998;106:775–92.

Further Reading: Organophosphates

Chambers JE, Levi PE. Organophosphates chemistry, fate, and effects. Oakville, ON: Academic Press, 1992.

Karalliedde L, Henry JA. Organophosphates and health. London: Imperial College Press, 2001.

Gupta R. Toxicology of organophosphate and carbamate compounds. Amsterdam, The Netherlands: Elsevier, 2005.

Sundkvist AM. Organophosphorus flame retardants and plasticizers: levels and sources in indoor and outdoor environments. Saarbrücken: VDM Verlag Dr. Müller, 2008.

Bibliography

[1]   Boffetta P, Jourenkova N, Gustavsson P. Cancer risk from occupational and environmental exposure to polycyclic aromatic hydrocarbons. Cancer Causes Control 1997;8(3): 444–72.

[2]   Blumer M. Curtisite, idrialite and pendletonite, polycyclic aromatic hydrocarbon minerals – their composition and origin. Chem Geol 1975;16(4):245–56.

[3]   Keith L. The Source of U.S. EPA's Sixteen PAH Priority Pollutants. Polycylic Aromat Compd 2015;35:147–60.

[4]   Achten C, Andersson J. Overview on polycyclic compounds (PAC). Polycyclic Aromat Compd 2015;35:177–86.

[5]   Elordui-Zapatarietxe S, Fettig I, Richter J, Philipp R, Gantois F, Lalere B, Swart C, Emteborg H. Interaction of 15 priority substances for water monitoring at ng L^{-1} levels with glass, aluminium and fluorinated polyethylene bottles for the containment of water reference materials. Accredit Qual Assur 2015;20(6):447–55.

[6]   Węgrzyn E, Grześkiewicz S, Popławska W, Głód BK. Modified analytical method for polycyclic aromatic hydrocarbons, using SEC for sample preparation and RP-HPL with fluorescence detection. Application to different food samples. Acta Chromatogr 2006;17:233–49.

[7]   Johnsen S, Krane J, Carlberg GE, Aamot E, Schou L. PAH in water systems, effect of aquatic humus and chlorination. In Organic micropollutants in the aquatic environment. Amsterdam: Springer Netherlands, 1986:440–8.

[8]   May WE, Chesler SN, Hertz HS, Wise SA. Analytical standards and methods for the determination of polynuclear aromatic hydrocarbons in environmental samples. Int J Environ Anal Chem 1982;12(3-4):259–75.

[9]   Thrane KE, Mikalsen A, Stray H. Monitoring method for airborne polycyclic aromatic hydrocarbons. Int J Environ Anal Chem 1985;23(1-2):111–34.

[10]   Miguel AH, Deandrade JB, Hering SV. Desorptivity versus chemical reactivity of polycyclic aromatic hydrocarbons (PAHs) in atmospheric aerosols collected on quartz fiber filters. Int J Environ Anal Chem 1986;26(3-4):265–78.

[11]   Schauer C, Niessner R, Poschl U. Polycyclic aromatic hydrocarbons in urban air particulate matter: decadal and seasonal trends, chemical degradation, and sampling artifacts. Environ Sci Technol 2003;37(13):2861–8.

[12]   Carrara M, Wolf JC, Niessner R. Nitro-PAH formation studied by interacting artificially PAH-coated soot aerosol with NO_2 in the temperature range of 295–523 K. Atmos Environ 2010;44(32):3878–85.

[13]   Environment Canada. Reference method for source testing: measurement of releases of fine particulate matter from stationary sources. In Reference Method EPS 1/RM/55: 2013,1–50.

[14]   Wilcke W. Polycyclic aromatic hydrocarbons (PAHs) in soil – a review. J Plant Nutr Soil Sci-Z Pflanzenernahr Bodenkd 2000;163(3):229–48.

[15]   Wilson SC, Jones KC. Bioremediation of soil contaminated with polynuclear aromatic hydrocarbons (PAHs): a review. Environ Poll 1993;81(3):229–49.

[16]   Carlon C, Nathanail CP, Critto C, Marcomini A. Bayesian statistics-based procedure for sampling of contaminated sites. Soil Sediment Contam 2004;13(4):329–45.

[17]   Wang YH, Zhang J, Ding YC, Zhou J, Ni LX, Sun C. Quantitative determination of 16 polycyclic aromatic hydrocarbons in soil samples using solid-phase microextraction. J Sep Sci 2009;32(22):3951–7.

[18]   Saim N, Dean JR, Abdullah MP, Zakaria Z. Extraction of polycyclic aromatic hydrocarbons from contaminated soil using Soxhlet extraction, pressurised and atmospheric microwave-assisted extraction, supercritical fluid extraction and accelerated solvent extraction. J Chromatogr A 1997;791(1-2):361–6.

[19]   Boos KS, Lintelmann J, Kettrup A. Coupled-column high-performance liquid chromatographic method for the determination of 1-hydroxypyrene in urine of subjects exposed to polycyclic aromatic hydrocarbons. J Chromatogr 1992;600(2):189–94.

[20]   Knopp D, Schedl M, Achatz S, Kettrup A, Niessner R. Immunochemical test to monitor human exposure to polycyclic aromatic hydrocarbons: urine as sample source. Anal Chim Acta 1999;399(1-2):115–26.

[21]   Schedl M, Wilharm G, Achatz S, Kettrup A, Niessner R, Knopp D. Monitoring polycyclic aromatic hydrocarbon metabolites in human urine: extraction and purification with a sol-gel glass immunosorbent. Anal Chem 2001;73(23):5669–76.

[22]   Barranco A, Alonso-Salces RM, Bakkali A, Berrueta LA, Gallo B, Vicente F, Sarobe M. Solid-phase clean-up in the liquid chromatographic determination of polycyclic aromatic hydrocarbons in edible oils. J Chromatogr A 2003;988(1):33–40.

[23]   Niessner R, Walendzik G. The photoelectric aerosol sensor as a fast-responding and sensitive detection system for cigarette smoke analysis. F Z Anal Chem 1989;333(2):129–33.

[24]   Hawthorne SB, Grabanski CB, Martin E, Miller DJ. Comparisons of Soxhlet extraction, pressurized liquid extraction, supercritical fluid extraction and subcritical water extraction for environmental solids: recovery, selectivity and effects on sample matrix. J Chromatogr A 2000;892(1-2):421–33.

[25]   Pena-Abaurrea M, Ye F, Blasco J, Ramos L. Evaluation of comprehensive two-dimensional gas chromatography time-of-flight-mass spectrometry for the analysis of polycyclic aromatic hydrocarbons in sediments. J Chromatogr A 2012;1256:222–31.

[26]   Knopp D, Seifert M, Vaananen V, Niessner R. Determination of polycyclic aromatic hydrocarbons in contaminated water and soil samples by immunological and chromatographic methods. Environ Sci Technol 2000;34(10):2035–41.

[27]   Lux G, Langer A, Pschenitza M, Karsunke X, Strasser R, Niessner R, Knopp D, Rant U. Detection of the carcinogenic water pollutant benzo[a]pyrene with an electro-switchable biosurface. Anal Chem 2015;87(8):4538–45.

[28]   Kotzick R, Niessner R. Application of time-resolved, laser-induced and fiber-optically guided fluorescence for monitoring of a PAH-contaminated remediation site. Fresenius J Anal Chem 1996;354(1):72–6.

[29]   Lewitzka F, Niessner R. Application of time-resolved fluorescence spectroscopy on the analysis of PAH-coated aerosols. Aerosol Sci Technol 1995;23(3):454–64.

[30]   Niessner R. The chemical response of the photoelectric aerosol sensor (PAS) to different aerosol systems. J Aerosol Sci 1986;17(4):705–14.

[31]   Niessner R, Robers W, Wilbring P. Laboratory experiments on the determination of polycyclic aromatic hydrocarbon coverage of submicrometer particles by laser-induced aerosol photoemission. Anal Chem 1989;61(4):320–5.

[32]   Cheng Y, Ho KF, Wu WJ, Ho SS, Lee SC, Huang Y, Zhang YW, Yau PS, Gao Y, Chan CS. Real-time characterization of particle-bound polycyclic aromatic hydrocarbons at a heavily trafficked roadside site. Aerosol Air Qual Res 2012;12(6):1181–8.

[33]   Rossberg M, Lendle W, Pfleiderer G, Tögel A, Dreher E, Langer E, Rassaerts H, Kleinschmidt P, Strack H, Cook R, Beck U, Lipper K, Torkelson T, Löser EK, Mann T. Chlorinated hydrocarbons, Ullmann's Encyclopedia of Industrial Chemistry. Weinheim: VCH, 2006:186.

[34]   Abtouche S, Very T, Monari A, Brahimi M, Assfeld X. Insight on the interaction of polychlorobiphenyl with nucleic acid-base. J Mol Model 2013;19(2):581–8.

[35]   US-EPA. Recommended toxicity equivalence factors (TEFs) for human health risk assessments of 2,3,7,8-tetrachlorodibenzo-p-dioxin and dioxin-like compounds. 2010, Vol. Document EPA/100/R 10/05.

[36]   Shiu WY, Doucette W, Gobas F, Andren A, Mackay D. Physical-chemical properties of chlorinated dibenzo-p –dioxins. Environ Sci Technol 1988;22(6):651–8.

[37]   Environment Australia. Incineration and dioxins: review of formation processes. In 1999; Vol. Consultancy report prepared by environmental and safety services for environment, Australia, Commonwealth, Department of the Environment and Heritage, Canberra.

[38]   Everaert K, Baeyens J. The formation and emission of dioxins in large scale thermal processes. Chemosphere 2002;46(3):439–48.

[39]   Vollmuth S, Zajc A, Niessner R. Formation of polychlorinated dibenzo-p-dioxins and polychlorinated dibenzofurans during the photolysis of pentachlorophenol-containing water. Environ Sci Technol 1994;28(6):1145–9.

[40]   Muñoz-Fernandez M, Gomez-Rico MF, Font R, Garrido M. Study of the enzymatic formation of PCDD/F during composting of sewage sludge. Organohalogen Compd 2014;76:792–5.

[41]   Pesatori AC, Consonni D, Bachetti S, Zocchetti C, Bonzini M, Baccarelli A, Bertazzi PA. Short- and long-term morbidity and mortality in the population exposed to dioxin after the "Seveso accident". Ind Health 2003;41(3):127–38.

[42]   Birnbaum LS. The mechanism of dioxin toxicity: relationship to risk assessment. Environ Health Perspect 1994;102:157–67.

[43]   Van Loco J, Van Leeuwen SP, Roos P, Carbonnelle S, de Boer J, Goeyens L, Beernaert H. The international validation of bio- and chemical- analytical screening methods for dioxins and dioxin-like PCBs: the DIFFERENCE project rounds 1 and 2. Talanta 2004;63(5):1169–82.

[44]   Haimovici L, Reiner EJ, Besevic S, Jobst KJ, Robson M, Kolic T, MacPherson K. A modified QuEChERS approach for the screening of dioxins and furans in sediments. Anal Bioanal Chem 2016;408(15):4043–54.

[45] Pirard C, Focant JF, De Pauw E. An improved clean-up strategy for simultaneous analysis of polychlorinated dibenzo-p-dioxins (PCDD), polychlorinated dibenzofurans (PCDF), and polychlorinated biphenyls (PCB) in fatty food samples. Anal Bioanal Chem 2002;372(2):373–81.

[46] Hess P, Deboer J, Cofino WP, Leonards PE, Wells DE. Critical review of the analysis of non- and mono-ortho-chlorobiphenyls. J Chromatogr A 1995;703(1-2):417–65.

[47] Ballschmiter K, Zell M. Analysis of polychlorinated biphenyls (PCB) by glass capillary gas chromatography. Composition of technical Aroclor-PCB and Clophen-PCB mixtures. Fresenius Z Anal Chem 1980;302(1):20–31.

[48] Danielsson C, Wiberg K, Korytar P, Bergek S, Brinkman UA, Haglund P. Trace analysis of polychlorinated dibenzo-p-dioxins, dibenzofurans and WHO polychlorinated biphenyls in food using comprehensive two-dimensional gas chromatography with electron-capture detection. J Chromatogr A 2005;1086(1-2):61–70.

[49] Frame GM, Cochran JW, Bowadt SS. Complete PCB congener distributions for 17 Aroclor mixtures determined by 3 HRGC systems optimized for comprehensive, quantitative, congener-specific analysis. HRC J High Res Chromatogr 1996;19(12):657–68.

[50] Abalos M, Cojocariu C, Silcock P, Roberts D, Pemberthy D, Saulo J, Abad E. Meeting the European Commission performance criteria for the use of triple quadrupole GC-MS/MS as a confirmatory method for PCDD/Fs and dl-PCBs in food and feed samples. Anal Bioanal Chem 2016;408(13):3511–25.

[51] Eljarrat E, Barcelo D. Congener-specific determination of dioxins and related compounds by gas chromatography coupled to LRMS, HRMS, MS/MS and TOFMS. J Mass Spectrom 2002;37(11):1105–17.

[52] Hernandez F, Sancho JV, Ibanez M, Abad E, Portoles T, Mattioli L. Current use of high-resolution mass spectrometry in the environmental sciences. Anal Bioanal Chem 2012;403(5):1251–64.

[53] Zimmermann R, Boesl U, Heger HJ, Rohwer ER, Ortner EK, Schlag EW, Kettrup A. Hyphenation of gas chromatography and resonance-enhanced laser mass spectrometry (REMPI-TOFMS): a multidimensional analytical technique. HRC J High Res Chromatogr 1997;20(9):461–70.

[54] Uchimura T. Sensitive and selective analysis of polychlorinated dibenzo-p-dioxins/dibenzofurans and their precursors by supersonic jet/resonance-enhanced multiphoton ionization/time-of-flight mass spectrometry. Anal Sci 2005;21(12):1395–400.

[55] Murk AJ, Legler J, Denison MS, Giesy JP, vandeGuchte C, Brouwer A. Chemical-activated luciferase gene expression (CALUX): a novel in vitro bioassay for Ah receptor active compounds in sediments and pore water. Fundam Appl Toxicol 1996;33(1):149–60.

[56] Windal I, Denison MS, Birnbaum LS, Van Wouwe N, Baeyens W, Goeyens L. Chemically activated luciferase gene expression (CALUX) cell bioassay analysis for the estimation of dioxin-like activity: critical parameters of the CALUX procedure that impact assay results. Environ Sci Technol 2005;39(19):7357–64.

[57] Meisenecker, K, Knopp, D, Niessner R. Development of an enzyme-linked immunosorbent assay (ELISA) for pyrene. Anal Methods Instrum 1993;1:114–18.

[58] Knopp, D, Vaananen, V, Zuhlke, J, Niessner R. Development of an enzyme-linked immunosorbent assay for 1-nitropyrene – a possible marker compound for diesel exhaust emissions. In Aga DS, Thurman EM, editors. Immunochemical technology for environmental applications. 1997;657:61–76.

[59] Sugawara, Y, Gee SJ, Sanborn JR, Gilman SD, Hammock BD. Development of a highly sensitive enzyme-linked immunosorbent assay based on polyclonal antibodies for the detection of polychlorinated dibenzo-p-dioxins. Anal Chem 1998;70(6):1092–9.

[60] Shan G, Leeman WR, Gee SJ, Sanborn JR, Jones AD, Chang DP, Hammock BD. Highly sensitive dioxin immunoassay and its application to soil and biota samples. Anal Chim Acta 2001;444(1):169–78.

[61] Schuetz AJ, Weller MG, Niessner R. A novel method for the determination of a PCB sum value by enzyme immunoassay to overcome the cross-reactivity problem. Fresenius J Anal Chem 1999;363(8):777–82.

[62] Stallones L, Beseler CL. Assessing the connection between organophosphate pesticide poisoning and mental health: a comparison of neuropsychological symptoms from clinical observations, animal models and epidemiological studies. Cortex 2016;74:405–16.

[63] John EM, Shaike JM. Chlorpyrifos: pollution and remediation. Environ Chem Letters 2015;13(3):269–91.

[64] Singh BK, Walker A, Morgan JA, Wright DJ. Biodegradation of chlorpyrifos by Enterobacter strain B-14 and its use in bioremediation of contaminated soils. Appl Environ Microbiol 2004;70(8):4855–63.

[65] Black RM, Reed RW. Biological markers of exposure to organophosphorus nerve agents. Arch Toxicol 2013;87:421–37.

[66] Dirtu AC, Van den Eede N, Malarvannan G, Ionas AC, Covaci A. Analytical methods for selected emerging contaminants in human matrices – a review. Anal Bioanal Chem 2012;404(9):2555–81.

[67] Kapka-Skrzypczak L, Cyranka M, Skrzypczak M, Kruszewski M. Biomonitoring and biomarkers of organophosphate pesticides exposure – state of the art. Ann Agric Environ Med 2011;18(2):294–303.

[68] van der Veen I, de Boer J. Phosphorus flame retardants: properties, production, environmental occurrence, toxicity and analysis. Chemosphere 2012;88(10):1119–53.

[69] Wei JC, Cao JL, Tian K, Hu YJ, Su HX, Wan JB, Li P. Trace determination of five organophosphorus pesticides by using QuEChERS coupled with dispersive liquid-liquid microextraction and stacking before micellar electrokinetic chromatography. Anal Methods 2015;7(14):5801–7.

[70] Martinez-Dominguez G, Plaza-Bolanos P, Romero-Gonzalez R, Garrido-Frenich A. Analytical approaches for the determination of pesticide residues in nutraceutical products and related matrices by chromatographic techniques coupled to mass spectrometry. Talanta 2014;118:277–91.

[71] Soisungnoen P, Burakham R, Srijaranai S. Determination of organophosphorus pesticides using dispersive liquid-liquid microextraction combined with reversed electrode polarity stacking mode-micellar electrokinetic chromatography. Talanta 2012;98:62–8.

[72] Khalili-Zanjani MR, Yamini Y, Yazdanfar N, Shariati S. Extraction and determination of organophosphorus pesticides in water samples by a new liquid phase microextraction-gas chromatography-flame photometric detection. Anal Chim Acta 2008;606(2):202–8.

[73] Sanz CP, Halko R, Ferrera ZS, Rodriguez JJS. Micellar extraction of organophosphorus pesticides and their determination by liquid chromatography. Anal Chim Acta 2004;524(1-2):265–70.

[74] Quintana JB, Rodil R, Reemtsma T, Garcia-Lopez M, Rodriguez I. Organophosphorus flame retardants and plasticizers in water and air II. Analytical methodology. TRAC Trends Anal Chem 2008;27(10):904–15.

[75] Huang DN, Deng CH, Zhang XM. Functionalized magnetic nanomaterials as solid-phase extraction adsorbents for organic pollutants in environmental analysis. Anal Methods 2014;6(18):7130–41.

[76] Xiong ZC, Zhang LY, Zhang RS, Zhang YR, Chen JH, Zhang WB. Solid-phase extraction based on magnetic core-shell silica nanoparticles coupled with gas chromatography-mass spectrometry for the determination of low concentration pesticides in aqueous samples. J Sep Sci 2012;35(18):2430–7.

[77] Heidari H, Razmi H. Multi-response optimization of magnetic solid phase extraction based on carbon coated Fe_3O_4 nanoparticles using desirability function approach for the determination of the organophosphorus pesticides in aquatic samples by HPLC-UV. Talanta 2012;99:13–21.

[78] Wu WM, Wu YM, Zheng MM, Yang LM, Wu XP, Lin XC, Xie ZH. Pressurized capillary electrochromatography with indirect amperometric detection for analysis of organophosphorus pesticide residues. Analyst 2010;135(8):2150–6.

[79] Choudhury TK, Gerhardt KO, Mawhinney TP. Solid-phase microextraction of nitrogen and phosphorus-containing pesticides from water and gas chromatographic analysis. Environ Sci Technol 1996;30(11):3259–65.

[80] Farina Y, Bin Abdullah P, Bibi N. Extraction procedures in gas chromatographic determination of pesticides. J Anal Chem 2016;71(4):339–50.

[81] Sharma D, Nagpal A, Pakade YB, Katnoria JK. Analytical methods for estimation of organophosphorus pesticide residues in fruits and vegetables: a review. Talanta 2010;82(4):1077–89.

[82] Abdulra'uf LB, Tan GH. Review of SBSE technique for the analysis of pesticide residues in fruits and vegetables. Chromatogr 2014;77(1-2):15–24.

[83] Prieto A, Basauri O, Rodil R, Usobiaga A, Fernandez LA, Etxebarria N, Zuloaga O. Stir-bar sorptive extraction: a view on method optimisation, novel applications, limitations and potential solutions. J Chromatogr A 2010;1217(16):2642–66.

[84] David F, Sandra P. Stir bar sorptive extraction for trace analysis. J Chromatogr A 2007;1152(1-2):54–69.

[85] Guan W, Wang YJ, Xu F, Guan YF. Poly(phthalazine ether sulfone ketone) as novel stationary phase for stir bar sorptive extraction of organochlorine compounds and organophosphorus pesticides. J Chromatogr A 2008;1177(1):28–35.

[86] Liu WM, Hu Y, Zhao JH, Xu Y, Guan YF. Determination of organophosphorus pesticides in cucumber and potato by stir bar sorptive extraction. J Chromatogr A 2005;1095(1-2):1–7.

[87] Hu C, He M, Chen BB, Hu B. A sol-gel polydimethylsiloxane/polythiophene coated stir bar sorptive extraction combined with gas chromatography-flame photometric detection for the determination of organophosphorus pesticides in environmental water samples. J Chromatogr A 2013;1275:25–31.

[88] Simplicio AL, Boas LV. Validation of a solid-phase microextraction method for the determination of organophosphorus pesticides in fruits and fruit juice. J Chromatogr A 1999;833(1):35–42.

[89] Tette PA, Guidi LR, Gloria MB, Fernandes C. Pesticides in honey: a review on chromatographic analytical methods. Talanta 2016;149:124–41.

[90] Jin BH, Xie LQ, Guo YF, Pang GF. Multi-residue detection of pesticides in juice and fruit wine: a review of extraction and detection methods. Food Res Int 2012;46(1):399–409.

[91] Fuentes E, Baez ME, Labra R. Parameters affecting microwave-assisted extraction of organophosphorus pesticides from agricultural soil. J Chromatogr A 2007;1169(1-2):40–6.

[92] Yi LX, Fang R, Chen GH. Molecularly imprinted solid-phase extraction in the analysis of agrochemicals. J Chromatogr Sci 2013;51:608–13.

[93] Marx S, Zaltsman A, Turyan I, Mandler D. Parathion sensor based on molecularly imprinted sol-gel films. Anal Chem 2004;76(1):120–6.

[94] Zhu XL, Yang J, Su QD, Cai JB, Gao Y. Selective solid-phase extraction using molecularly imprinted polymer for the analysis of polar organophosphorus pesticides in water and soil samples. J Chromatogr A 2005;1092(2):161–9.

[95] de Llasera MP, Reyes-Reyes ML. A validated matrix solid-phase dispersion method for the extraction of organophosphorus pesticides from bovine samples. Food Chem 2009;114(4):1510–16.

[96] Yusa V, Millet M, Coscolla C, Pardo O, Roca M. Occurrence of biomarkers of pesticide exposure in non-invasive human specimens. Chemosphere 2015a;139:91–108.

[97] Yusa V, Millet M, Coscolla C, Roca M. Analytical methods for human biomonitoring of pesticides. A review. Anal Chim Acta 2015b;891:15–31.

[98] Kumar V, Upadhyay N, Kumar V, Sharma S. A review on sample preparation and chromatographic determination of acephate and methamidophos in different samples. Arabian J Chem 2015a;8(5):624–31.

[99] Krol S, Zabiegala B, Namiesnik J. Monitoring and analytics of semivolatile organic compounds (SVOCs) in indoor air. Anal Bioanal Chem 2011;400(6):1751–69.

[100] Ni Y, Kumagai K, Yanagisawa Y. Measuring emissions of organophosphate flame retardants using a passive flux sampler. Atmos Environ 2007;41(15):3235–40.

[101] Kemmlein S, Hahn O, Jann O. Emissions of organophosphate and brominated flame retardants from selected consumer products and building materials. Atmos Environ 2003;37(39-40):5485–93.

[102] Berijani S, Assadi Y, Anbia M, Hosseini MR, Aghaee E. Dispersive liquid-liquid microextraction combined with gas chromatography-flame photometric detection – very simple, rapid and sensitive method for the determination of organophosphorus pesticides in water. J Chromatogr A 2006;1123(1):1–9.

[103] Moinfar S, Hosseini MRM. Development of dispersive liquid-liquid microextraction method for the analysis of organophosphorus pesticides in tea. J Hazard Mater 2009;169(1–3):907–11.

[104] Tahboub YR, Zaater MF, Al-Talla ZA. Determination of the limits of identification and quantitation of selected organochlorine and organophosphorus pesticide residues in surface water by full-scan gas chromatography/mass spectrometry. J Chromatogr A 2005;1098(1-2):150–5.

[105] Goncalves C, Alpendurada MF. Solid-phase micro-extraction-gas chromatography-(tandem) mass spectrometry as a tool for pesticide residue analysis in water samples at high sensitivity and selectivity with confirmation capabilities. J Chromatogr A 2004;1026(1–2):239–50.

[106] Amendola L, Botre F, Carollo AS, Longo D, Zoccolillo L. Analysis of organophosphorus pesticides by gas chromatography-mass spectrometry with negative chemical ionization: a study on the ionization conditions. Anal Chim Acta 2002;461(1):97–108.

[107] Rossi S, Dalpero AP, Ghini S, Colombo R, Sabatini AG, Girotti S. Multiresidual method for the gas chromatographic analysis of pesticides in honeybees cleaned by gel permeation chromatography. J Chromatogr A 2001;905(1-2):223–32.

[108] Girotti S, Ghini S, Maiolini E, Bolelli L, Ferri EN. Trace analysis of pollutants by use of honeybees, immunoassays, and chemiluminescence detection. Anal Bioanal Chem 2013;405(2-3):555–71.

[109] Chafer-Pericas C, Herraez-Hernandez R, Campins-Falco P. In-tube solid-phase microextraction-capillary liquid chromatography as a solution for the screening analysis of organophosphorus pesticides in untreated environmental water samples. J Chromatogr A 2007;1141(1):10–21.

[110] Hayama T, Yoshida H, Todoroki K, Nohta H, Yamaguchi M. Determination of polar organophosphorus pesticides in water samples by hydrophilic interaction liquid chromatography with tandem mass spectrometry. Rapid Commun Mass Spectrom 2008;22(14):2203–10.

[111] Mol HG, van Dam RC, Steijger OM. Determination of polar organophosphorus pesticides in vegetables and fruits using liquid chromatography with tandem mass spectrometry: selection of extraction solvent. J Chromatogr A 2003;1015(1-2):119–27.

[112] Salm P, Taylor PJ, Roberts D, de Silva J. Liquid chromatography-tandem mass spectrometry method for the simultaneous quantitative determination of the organophosphorus pesticides dimethoate, fenthion, diazinon and chlorpyrifos in human blood. J Chromatogr B-Anal Technol Biomed Life Sci 2009;877(5-6):568–74.

[113] Sherma J. Review of advances in the thin layer chromatography of pesticides: 2008–2010. J Environ Sci Health Part B-Pesticides Food Contam Agric Wastes 2011;46(7):557–68.

[114] Sherma J. Review of advances in the thin layer chromatography of pesticides: 2010–2012. J Environ Sci Health Part B-Pesticides Food Contam Agric Wastes 2013;48:417–30.

[115] Srivastava M. High performance thin layer chromatography (HPTLC). Heidelberg: Springer Verlag, 2011:397.

[116] Jin N, Jin W. Review on the detection of organophosphorus pesticide residues. Zhongguo Huanjing GuanliGanbu Xueyuan Xuebao 2011;21(4):62–5.

[117] Mohammad A, Moheman A. A new spray reagent for selective detection and quantification of dichlorvos in bluish tinged maize grains by TLC-spectrophotometry. JPC J Plan Chromatogr-Mod TLC 2011;24(2):113–15.

[118] Liu D, Qian CF, Wang YH. Determination of organophosphorus pesticides in rice by TLC. Asian J Chem 2011;23(5):2011–13.

[119] Ganguru UM, Yalavarthy PD. Development of an enzymatic method for environmental monitoring of monocrotophos. Recent Res Sci Technol 2011;3:58–65.

[120] Akkad R, Schwack W. Effect of bromine oxidation on high-performance thin-layer chromatography multi-enzyme inhibition assay detection of organophosphates and carbamate insecticides. J Chromatogr A 2011;1218(19):2775–84.

[121] Kiguchi O, Oka K, Tamada M, Kobayashi T, Onodera J. Thin-layer chromatography/direct analysis in real time time-of-flight mass spectrometry and isotope dilution to analyze organophosphorus insecticides in fatty foods. J Chromatogr A 2014;1370:246–54.

[122] Iwamuro Y, Iio-Ishimaru R, Chinaka S, Takayama N, Kodama S, Hayakawa K. Analysis of phosphorus-containing amino acid-type herbicides by capillary electrophoresis/mass spectrometry using a chemically modified capillary having amino groups. J Health Sci 2010;56(5):606–12.

[123] Cao L, Liang S, Tan X, Meng J. Capillary electrophoresis analysis for glyphosate, glufosinate and aminomethylphosphonic acid with laser-induced fluorescence detection. Chin J Chromatogr 2012;30:1295–300.

[124] Sung IH, Lee YW, Chung DS. Liquid extraction surface analysis in-line coupled with capillary electrophoresis for direct analysis of a solid surface sample. Anal Chim Acta 2014;838:45–50.

[125] Dong YL, Guo DQ, Cui H, Li XJ, He YJ. Magnetic solid phase extraction of glyphosate and aminomethylphosphonic acid in river water using Ti^{4+}-immobilized Fe_3O_4 nanoparticles by capillary electrophoresis. Anal Methods 2015;7(14):5862–8.

[126] Liu SQ, Zheng ZZ, Li XY. Advances in pesticide biosensors: current status, challenges, and future perspectives. Anal Bioanal Chem 2013;405(1):63–90.

[127] Kumar P, Kim KH, Deep A. Recent advancements in sensing techniques based on functional materials for organophosphate pesticides. Biosensors Bioelectron 2015b;70:469–81.

[128] Van Dyk JS, Pletschke B. Review on the use of enzymes for the detection of organochlorine, organophosphate and carbamate pesticides in the environment. Chemosphere 2011;82(3):291–307.

[129] Yue GZ, Su S, Li N, Shuai MB, Lai XC, Astruc D, Zhao PX. Gold nanoparticles as sensors in the colorimetric and fluorescence detection of chemical warfare agents. Coord Chem Rev 2016;311:75–84.

[130] Wang XD, Yang YY, Dong J, Bei F, Ai SY. Lanthanum-functionalized gold nanoparticles for coordination-bonding recognition and colorimetric detection of methyl parathion with high sensitivity. Sens Actuat B-Chem 2014;204:119–24.

[131] D'Souza SL, Pati RK, Kailasa SK. Ascorbic acid functionalized gold nanoparticles as a probe for colorimetric and visual read-out determination of dichlorvos in environmental samples. Anal Methods 2014;6(22):9007–14.

[132] Paliwal S, Wales M, Good T, Grimsley J, Wild J, Simonian A. Fluorescence-based sensing of p-nitrophenol and p-nitrophenyl substituent organophosphates. Anal Chim Acta 2007;596(1):9–15.

[133] Van Houten KA, Heath DC, Pilato RS. Rapid luminescent detection of phosphate esters in solution and the gas phase using (dppe)Pt{S_2C_2(2-pyridyl)(CH$_2$CH$_2$OH)}. J Am Chem Soc 1998;120(47):12359–60.

[134] Zhang SW, Swager TM. Fluorescent detection of chemical warfare agents: functional group specific ratiometric chemosensors. J Am Chem Soc 2003;125(12):3420–1.

[135] Dale TJ, Rebek J. Fluorescent sensors for organophosphorus nerve agent mimics. J Am Chem Soc 2006;128(14):4500–1.

[136] Obare SO, De C, Guo W, Haywood TL, Samuels TA, Adams CP, Masika NO, Murray DH, Anderson GA, Campbell K, Fletcher K. Fluorescent chemosensors for toxic organophosphorus pesticides: a review. Sensors 2010;10(7):7018–43.

[137] Miao YQ, He NY, Zhu JJ. History and new developments of assays for cholinesterase activity and inhibition. Chem Rev 2010;110(9):5216–34.

[138] Songa EA, Okonkwo JO. Recent approaches to improving selectivity and sensitivity of enzyme-based biosensors for organophosphorus pesticides: a review. Talanta 2016;155:289–304.

[139] Yu GX, Wu WX, Zhao Q, Wei XY, Lu Q. Efficient immobilization of acetylcholinesterase onto amino functionalized carbon nanotubes for the fabrication of high sensitive organophosphorus pesticides biosensors. Biosensors Bioelectron 2015;68:288–94.

[140] Valdes-Ramirez G, Fournier D, Ramirez-Silva MT, Marty JL. Sensitive amperometric biosensor for dichlorovos quantification: application to detection of residues on apple skin. Talanta 2008;74(4):741–6.

[141] Crew A, Lonsdale D, Byrd N, Pittson R, Hart JP. A screen-printed, amperometric biosensor array incorporated into a novel automated system for the simultaneous determination of organophosphate pesticides. Biosensors Bioelectron 2011;26(6):2847–51.

[142] Alonso GA, Istamboulie G, Noguer T, Marty JL, Munoz R. Rapid determination of pesticide mixtures using disposable biosensors based on genetically modified enzymes and artificial neural networks. Sens Actuators B-Chem 2012;164(1):22–8.

[143] Campas M, Prieto-Simon B, Marty JL. A review of the use of genetically engineered enzymes in electrochemical biosensors. Semin Cell Dev Biol 2009;20(1):3–9.

[144] Sotiropoulou S, Fournier D, Chaniotakis NA. Genetically engineered acetylcholine sterase-based biosensor for attomolar detection of dichlorvos. Biosensors Bioelectron 2005;20(11):2347–52.

[145] Mishra RK, Dominguez RB, Bhand S, Munoz R, Marty JL. A novel automated flow-based biosensor for the determination of organophosphate pesticides in milk. Biosensors Bioelectron 2012;32(1):56–61.

[146] Pino F, Ivandini TA, Naikata K, Fujishima A, Merkoci A, Einaga Y. Magnetic enzymatic platform for organophosphate pesticide detection using boron-doped diamond electrodes. Anal Sci 2015;31(10):1061–8.

[147] Sun JF, Guo L, Bao Y, Xie JW. A simple, label-free AuNPs-based colorimetric ultra sensitive detection of nerve agents and highly toxic organophosphate pesticide. Biosensors Bioelectron 2011;28(1):152–7.

[148] Franko M, Liu MQ, Boskin A, Delneri A, Proskurnin MA. Fast screening techniques for neurotoxigenic substances and other toxicants and pollutants based on thermal lensing and microfluidic chips. Anal Sci 2016;32(1):23–30.

[149] Liron Z, Zifman A, Heleg-Shabtai V. Surface-enhanced Raman scattering detection of cholinesterase inhibitors. Anal Chim Acta 2011;703(2):234–8.

[150] Pundir CS, Chauhan N. Acetylcholinesterase inhibition-based biosensors for pesticide determination: a review. Anal Biochem 2012;429(1):19–31.

[151] Simonian AL, Grimsley JK, Flounders AW, Schoeniger JS, Cheng TC, DeFrank JJ, Wild JR. Enzyme-based biosensor for the direct detection of fluorine-containing organophosphates. Anal Chim Acta 2001;442(1):15–23.

[152] Istamboulie G, Fournier D, Marty JL, Noguer T. Phosphotriesterase: a complementary tool for the selective detection of two organophosphate insecticides: chlorpyrifos and chlorfenvinfos. Talanta 2009;77(5):1627–31.

[153] Zhang YY, Arugula MA, Wales M, Wild J, Simonian AL. A novel layer-by-layer assembled multi-enzyme/CNT biosensor for discriminative detection between organophosphorus and non-organophosphorus pesticides. Biosensors Bioelectron 2015;67:287–95.

[154] Rogers KR, Wang Y, Mulchandani A, Mulchandani P, Chen W. Organophosphorus hydrolase-based assay for organophosphate pesticides. Biotechnol Prog 1999;15(3):517–21.

[155] Mulchandani A, Chen W, Mulchandani P, Wang J, Rogers KR. Biosensors for direct determination of organophosphate pesticides. Biosensors Bioelectron 2001;16(4-5):225–30.

[156] Herzsprung P, Weil L, Niessner R. Measurement of bimolecular rate constants k_i of the cholinesterase inactivation reaction by 55 insecticides and of the influence of various pyridinium oximes on k_i. Int J Environ Anal Chem 1992;47(3):181–200.

[157] Zheng YH, Hua TC, Sun DW, Xiao HJ, Xu F, Wang FF. Detection of dichlorvos residue by flow injection calorimetric biosensor based on immobilized chicken liver esterase. J Food Eng 2006;74(1):24–9.

[158] Halamek J, Pribyl J, Makower A, Skladal P, Scheller FW. Sensitive detection of organophosphates in river water by means of a piezoelectric biosensor. Anal Bioanal Chem 2005;382(8):1904–11.

[159] Shao XS, Xia SS, Durkin KA, Casida JE. Insect nicotinic receptor interactions in vivo with neonicotinoid, organophosphorus, and methylcarbamate insecticides and a synergist. Proc Natl Acad Sci USA 2013;110(43):17273–7.

[160] Kolosova AY, Park JH, Eremin SA, Kang SJ, Chung DH. Fluorescence polarization immunoassay based on a monoclonal antibody for the detection of the organophosphorus pesticide parathion-methyl. J Agric Food Chem 2003;51(5):1107–14.

[161] Yan X, Li HX, Yan Y, Su XG. Developments in pesticide analysis by multi-analyte immunoassays: a review. Anal Methods 2014;6(11):3543–54.

[162] Yan X, Shi HY, Wang MH. Development of an enzyme-linked immunosorbent assay for the simultaneous determination of parathion and imidacloprid. Anal Methods 2012;4(12):4053–7.

[163] Skerritt JH, Hill AS, Beasley HL, Edward SL, McAdam DP. Enzyme-linked-immunosorbent-assay for quantitation of organophosphate pesticides – fenitrothion, chlorpyrifos-methyl, and pirimiphos-methyl in wheat grain and flour-milling fractions. J AOAC Int 1992;75(3):519–28.

[164] Garces-Garcia M, Brun EA, Puchades R, Maquieira A. Immunochemical determination of four organophosphorus insecticide residues in olive oil using a rapid extraction process. Anal Chim Acta 2006;556(2):347–54.

[165] Xu ZL, Dong JX, Yang JY, Wang H, Jiang YM, Lei HT, Shen YD, Sun YM. Development of a sensitive time-resolved fluoroimmunoassay for organophosphorus pesticides in environmental water samples. Anal Methods 2012;4(10):3484–90.

[166] Xu ZL, Wang Q, Lei HT, Eremin SA, Shen YD, Wang H, Beier RC, Yang JY, Maksimova KA, Sun YM. A simple, rapid and high-throughput fluorescence polarization immunoassay for simultaneous detection of organophosphorus pesticides in vegetable and environmental water samples. Anal Chim Acta 2011;708(1-2):123–9.

[167] Hashimoto JH, Ruvolo-Takasusuki MC, de Toledo VDA. Evaluation of the use of the inhibition esterases activity on Apis mellifera as bioindicators of insecticide thiamethoxam pesticide residues. Sociobiology 2003;42(3):693–9.

[168] Ivanov A, Evtugyn G, Budnikov H, Girotti S, Ghini S, Ferri E, Montoya A, Mercader JV. Amperometric immunoassay of azinphos-methyl in water and honeybees based on indirect competitive ELISA. Anal Lett 2008;41(3):392–405.

[169] Xu X, Ying YB. Microbial biosensors for environmental monitoring and food analysis. Food Rev Int 2011;27(3):300–29.

[170] Mulchandani A, Kaneva I, Chen W. Biosensor for direct determination of organophosphate nerve agents using recombinant Escherichia coli with surface-expressed organophosphorus hydrolase. 2. Fiber optic microbial biosensor. Anal Chem 1998;70(23):5042–6.

[171] Mulchandani P, Chen W, Mulchandani A, Wang J, Chen L. Amperometric microbial biosensor for direct determination of organophosphate pesticides using recombinant microorganism with surface expressed organophosphorus hydrolase. Biosensors Bioelectron 2001;16(7-8):433–7.

[172] Gaberlein S, Spener F, Zaborosch C. Microbial and cytoplasmic membrane-based potentiometric biosensors for direct determination of organophosphorus insecticides. Appl Microbiol Biotechnol 2000;54(5):652–8.

[173] Mulchandani AR. Microbial biosensors for organophosphate pesticides. Appl Biochem Biotechnol 2011;165(2):687–99.

[174] Pasco NF, Weld RJ, Hay JM, Gooneratne R. Development and applications of whole cell biosensors for ecotoxicity testing. Anal Bioanal Chem 2011;400(4):931–45.

[175] Lagarde F, Jaffrezic-Renault N. Cell-based electrochemical biosensors for water quality assessment. Anal Bioanal Chem 2011;400(4):947–64.

[176] Rotariu L, Lagarde F, Jaffrezic-Renault N, Bala C. Electrochemical biosensors for fast detection of food contaminants trends and perspective. TRAC Trends Anal Chem 2016;79:80–7.

[177] Stoytcheva M, Gochev V, Velkova Z. Electrochemical biosensors for direct determination of organophosphorus pesticides: a review. Curr Anal Chem 2016;12(1):37–42.

List of Abbreviations

AChE	Acetylcholinesterase
AF^4	Asymmetric field flow fractionation
AFID	Alkali FID
AhR	Aryl hydrocarbon receptor
APCI	Atmospheric pressure photoionization
APPI	Atmospheric pressure photoionization
ASE	Accelerated solvent extraction
BaP	Benzo(a)pyrene
BSA	Bovine Serum Albumin
BTEX	Benzene, Toluene, Ethylbenzene, o-Xylene
CALUX	Chemical Activated LUciferase gene eXpression
CE	Capillary Electrophoresis
CE-MS	CE-mass spectrometry
CGE	Capillary Gel Electrophoresis
CI	Chemical ionization
CIEF	Capillary Isoelectric Focusing
CITP	Capillary Isotachophoresis
CMC	Critical micelle concentration
CRMs	Certified reference materials
CZE	Capillary Zone Electrophoresis
DAD	Diode-array detection
DART	Direct analysis in real time
DBDI	Dielectric-barrier-discharge ionization
DESI	Desorption electrospray ionization
DGT	Diffusive gradient sampling in thin films
DLLME	Dispersive liquid-liquid microextraction
dl-PCBs	d(ioxin)l(ike) coplanar PCBs
ECD	Electron capture detector
EC_x	Effect concentration
EDA	Effect directed analysis
EGA	Evolved gas analysis
EI	Electron impact ionization
ELCD	Electrolytic conductivity detector
ELISA	Enzyme-linked immunosorbent assay
ELSD	Evaporative light scattering detector
EOF	Electroendosmotic flow

DOI 10.1515/9783110441154-012

ERA	Environmental risk assessment
ESI	Electrospray ionization
FID	Flame ionization detector
FLD	Fluorescence detection
FPD	Flame photometric detector
FRET	Fluorescence resonance energy transfer
FWHM	Full width half maximum
GC	Gas chromatography
GLP	Good laboratory practices
GPC	Gel permeation chromatography
HETP	Height equivalent of theoretical plate
HF-LLLME	Hollow fiber liquid–liquid–liquid microextraction
HILIC	Hydrophilic Interaction Liquid Chromatography
HPLAC	High performance liquid affinity chromatography
HPLC	High performance liquid chromatography
HQ	Hazard Quotient
IA	Immunoassay
Ig	Immunoglobulins
IMS	Ion mobility spectrometry
KLH	Keyhole Limpet Hemocyanin
LLE	Liquid-liquid extr
LOD	Limit of detection
LOQ	Limit of quantification
MAE	Microwave assisted extraction
MALDI	Matrix-assisted laser desorption ionization
ME	Micellar extraction
MECC	Micellar Electrokinetic Capillary Chromatography
MIP	Molecularly imprinted polymers
MISPE	Molecularly imprinted solid-phase extraction
MS	Mass spectrometry
MTP	Microtiter plates
NER	Non-extractable residues
NMR	Nuclear magnetic resonance spectroscopy
NPD	Nitrogen-Phosphorus Detector
OPs	Organophosphates
PAHs	Polyaromatic hydrocarbons
PCBs	Polychlorinated biphenyls
PCDDs	Polychlorinated dibenzodioxines

PCDFs	Polychlorinated dibenzofuranes
PEC	Predicted environmental concentration
PI	Photoionization
PID	Photoionization detector
PILS	Particle-into-liquid sampler
PLE	Pressurized liquid extraction
PM	Particulate matter
PNEC	Predicted no effect concentration
POM-SPE	Polyoxymethylene solid phase extraction
POPs	Persistent organic pollutants
ppb	parts per billion
ppm	parts per million
ppq	parts per quadrillion
ppt	parts per trillion
PSE	Pressurized solvent extraction
PTR	Proton-transfer-reaction
QC	Quality control
QuEChERS	A standardized extraction method
REMPI	Resonance-enhanced multi-photon ionization
RI	Refractive indexdetector
SBSE	Stir bar sorptive extraction
SDME	Single-drop microextraction
SEC	Size exclusion chromatography
SERS	Surface enhanced Raman scattering
SFE	Supercritical fluid extraction
SIM	Selected ion monitoring, Single ion monitoring
SOP	Standard Operation Procedures
SPE	Solid phase extraction
SPME	Solid phase microextraction
SPR	Surface plasmonic resonance
SRM	Selected reaction monitoring
TCD	Thermal conductivity detector
TEF	Toxicity equivalence factor
TEQs	Toxic Equivalents
TID	Thermoionic ionization detector
TLC	Thin layer chromatography
TOF	Time-of-flight detector in MS
TSP	Total suspended dust particle
TU	Toxic Units
UV	Ultra violet light
Vis	Visible light

Index

DOI 10.1515/9783110441154-013

www.ingramcontent.com/pod-product-compliance
Lightning Source LLC
Chambersburg PA
CBHW080715220326
41598CB00033B/5428